The Evolution of Industrial Districts

Contributions to Economics

www.springeronline.com/series/1262

Further volumes of this series can be found at our homepage.

Erik Lüth
Private Intergenerational Transfers and Population Aging
2001. ISBN 3-7908-1402-4

Nicole Pohl
Mobility in Space and Time
2001. ISBN 3-7908-1380-X

Mario A. Maggioni
Clustering Dynamics and the Locations of High-Tech-Firms
2002. ISBN 3-7908-1431-8

Ludwig Schätzl/Javier Revilla Diez (Eds.)
Technological Change and Regional Development in Europe
2002. ISBN 3-7908-1460-1

Alberto Quadrio Curzio/Marco Fortis (Eds.)
Complexity and Industrial Clusters
2002. ISBN 3-7908-1471-7

Friedel Bolle/Marco Lehmann-Waffenschmidt (Eds.)
Surveys in Experimental Economics
2002. ISBN 3-7908-1472-5

Pablo Coto-Millán
General Equilibrium and Welfare
2002. ISBN 7908-1491-1

Wojciech W. Charemza/Krystyna Strzala (Eds.)
East European Transition and EU Enlargement
2002. ISBN 3-7908-1501-1

Natalja von Westernhagen
Systemic Transformation, Trade and Economic Growth
2002. ISBN 3-7908-1521-7

Josef Falkinger
A Theory of Employment in Firms
2002. ISBN 3-7908-1520-9

Engelbert Plassmann
Econometric Modelling of European Money Demand
2003. ISBN 3-7908-1522-5

Reginald Loyen/Erik Buyst/Greta Devos (Eds.)
Struggeling for Leadership: Antwerp-Rotterdam Port Competition between 1870–2000
2003. ISBN 3-7908-1524-1

Pablo Coto-Millán
Utility and Production, 2nd Edition
2003. ISBN 3-7908-1423-7

Emilio Colombo/John Driffill (Eds.)
The Role of Financial Markets in the Transition Process
2003. ISBN 3-7908-0004-X

Guido S. Merzoni
Strategic Delegation in Firms and in the Trade Union
2003. ISBN 3-7908-1432-6

Jan B. Kuné
On Global Aging
2003. ISBN 3-7908-0030-9

Sugata Marjit, Rajat Acharyya
International Trade, Wage Inequality and the Developing Economy
2003. ISBN 3-7908-0031-7

Francesco C. Billari/Alexia Prskawetz (Eds.)
Agent-Based Computational Demography
2003. ISBN 3-7908-1550-0

Georg Bol/Gholamreza Nakhaeizadeh/Svetlozar T. Rachev/Thomas Ridder/Karl-Heinz Vollmer (Eds.)
Credit Risk
2003. ISBN 3-7908-0054-6

Christian Müller
Money Demand in Europe
2003. ISBN 3-7908-0064-3

Cristina Nardi Spiller
The Dynamics of the Price Structure and the Business Cycle
2003. ISBN 3-7908-0063-5

Michael Bräuninger
Public Debt and Endogenous Growth
2003. ISBN 3-7908-0056-1

Brigitte Preissl/Laura Solimene
The Dynamics of Clusters and Innovation
2003. ISBN 3-7908-0077-5

Markus Gangl
Unemployment Dynamics in the United States and West Germany
2003. ISBN 3-7908-1533-0

Pablo Coto-Millán (Ed.)
Essays on Microeconomics and Industrial Organisation, 2nd Edition
2004. ISBN 3-7908-0104-6

Wendelin Schnedler
The Value of Signals in Hidden Action Models
2004. ISBN 3-7908-0173-9

Giulio Cainelli
Roberto Zoboli
Editors

The Evolution of Industrial Districts

Changing Governance, Innovation and Internationalisation of Local Capitalism in Italy

With 26 Figures
and 83 Tables

Physica-Verlag
A Springer-Verlag Company

Series Editors
Werner A. Müller
Martina Bihn

Editors

Prof. Giulio Cainelli
Prof. Roberto Zoboli

CERIS-DSE
National Research Council
Via Bassini 15
20133 Milan
Italy

cainelli@idse.mi.cnr.it
zoboli@idse.mi.cnr.it

ISSN 1431-1933
ISBN 3-7908-0095-3 Physica-Verlag Heidelberg New York

Cataloging-in-Publication Data applied for
A catalog record for this book is available from the Library of Congress.
Bibliographic information published by Die Deutsche Bibliothek
Die Deutsche Bibliothek lists this publication in the Deutsche Nationalbibliografie; detailed bibliographic data is available in the Internet at <http://dnb.ddb.de>.

Physica-Verlag is a part of Springer Science+Business Media

springeronline.com

Softcover Design: Erich Kirchner, Heidelberg

SPIN 10941287   88/3130/DK-5 4 3 2 1 0 – Printed on acid-free and non-aging paper

Foreword

The essays collected in this book deal with the evolution of industrial districts and local systems of production in the Italian experience. The emergence of a production structure increasingly based on local agglomerations of small and medium-sized industrial firms has been one of the most important aspects of Italian economic development during the last few decades. Our introductory essay and the first part of the book illustrate in details why the current evolution of this production structure can be a source of research interest from a theoretical point of view and from an international perspective. The contributions presented in the book specifically address three aspects of the present change of Italian industrial districts, namely the changing firm ownership structure and the emergence of industrial groups of SMEs, the problems of formal and informal innovation, and the process of internationalisation. Besides being interconnected in a perspective of structural evolution, these issues are those we consider of critical importance in the present phase of the Italian economy. Most of the essays present the results of empirical analyses based on new extensive data sets at the very disaggregated level.

Various essays in this book originate from a line of research on local-level development in the Italian economy that has been carried out during the last few years by IDSE, the Institute of Research on the Dynamics of Economic Systems of the National Research Council of Italy. Some other essays are by colleagues with whom we are working in formal or informal research networks. Another book dealing with the structural change and competitiveness of Italian local systems of production, has been published in Italian in 1999. The two books together give account of a part of IDSE's research activity in the field of economic development. These evolving lines of research are now going on in the institutional framework represented by CERIS, the Research Institute on the Firm and Economic Development of the National Research Council, to which ex-IDSE belongs from 2003 as one of its research units. We are sure that CERIS, which has a wide set of capabilities and experiences in industrial economics and local development, innovation and firms, represents an environment in which our research on economic development will further evolve.

Giulio Cainelli and Roberto Zoboli

Milan, August 2003

Acknowledgements

We wish to thank Andrea Mario Pagani and Stefania Mantecchi for their patient, skilful, and effective work in the editing of this book. We also wish to thank Corinne Robinson for her excellent work in translating some of the essays and revising the English version of other essays. Ottavio Janni also contributed to the revision of the English version of some essays. The responsibility for all the remaining errors, including those of English language, must be attributed to authors and editors, who made the very final revision of the text as a whole.

Contents

Part III. Innovation 241

Introduction

The structural evolution of industrial districts and adaptive competitive advantages

Giulio Cainelli[1] and Roberto Zoboli[2]

[1] University of Bari and CERIS-DSE, National Research Council, Via Bassini 15, 20133 Milan
[2] CERIS-DSE, National Research Council, Via Bassini 15, 20133 Milan, and Catholic University, Milan, Italy

1 Industrial districts and local systems of production: An overview of the Italian experience and the theoretical debate

Italian industrial districts (IDs) and local systems of production (LSPs) are receiving increasing attention by economic research and economic policy even at the international level[1]. After the classical contribution by Becattini (1979 and 1989) and Brusco (1982), a great number of works have been published recently on this topic (see, for example, Bagella and Becchetti, 2000; Belussi and Gottardi, 2000; Guerrieri *et al.*, 2001; Quadrio Curzio and Fortis, 2002). ISTAT, the Italian Statistical Office, and the Bank of Italy have carried out extensive inquiries on various aspects of IDs and LSPs (see, among others, Signorini, 2000). At the same time, significant efforts have been made at the different levels of government, from national to local, to create specific instruments of industrial policy suitable for IDs and LSPs, up to arriving at the adoption of the Italian national Law 317 of 1991 on SMEs and industrial districts. Although the feasibility of defining IDs on legislative grounds based on statistical criteria has been widely debated – and the above-mentioned law institutionalising IDs did have little practical significance, this policy process suggests the great importance attached to IDs for the Italian economy.

This importance is based on a surprisingly good and stable growth performance that brought IDs and LSPs to become the backbone of the Italian industrial system. According to ISTAT classification, which combines indexes of manufacturing specialisation and share of SMEs in local economies, in the 199 'official' IDs 14 million people (around 26% of total Italian population) live and

[1] In this work, we will refer to industrial districts (IDs) and local systems of production (LSPs) in an almost interchangeable way. We will avoid too clear-cut a distinction between what could be classified as an 'industrial district' according to some criteria, and other forms of locally concentrated systems of (manufacturing) production – be they highly specialised or diversified – that are similar to districts or can enjoy similar mechanisms of development but would not be classified as 'industrial districts' following a 'purist' approach. The discussion in the text will further clarify our position on the need for keeping an open and flexible approach. See also the introduction to IDSE-CNR (1999) for a discussion on the concepts of ID and LSP.

there are 2.2 million of manufacturing employees (around 45% of total manufacturing employment)[2]. It is estimated that around 43% of total Italian exports originate from IDs – mainly specialised in textile and clothing (70 IDs), furniture and house-related products (37 IDs), leather and shoes (28 IDs), mechanical engineering (33 IDs), as well as other 'traditional' sectors, e.g. ceramic tiles – thus giving rise to the identification between IDs and export flows of high-quality products of 'made in Italy' (see also Quadrio Curzio and Fortis, 2000). The growth of Italian IDs and LSPs also stimulated a significant interest at the international level aimed at understanding the roots of their economic success and the possibility that a great number of SMEs and local economic systems in Europe may follow similar patterns of growth.

At the analytical and research level, the increasing interest for the concept of ID has been based on a sequence of motivations.

At the beginning, in particular during the 1970s, the interest was stimulated by the possibility to identify a different model of economic (self-) organisation reinforcing, together with new theories of firm organisation, the path away from the Fordist-Taylorist model. The attention of research, in particular by international scholars, focused on the ID as a 'model of capitalism' based on strong spatial interaction between the productive dimension, as represented by the local systems of SMEs or micro-firms, the social dimension, as represented by the 'local community' with its specific identity based on cultural, historical and social values, and the institutional dimension, as represented by the local authorities and a wide range of 'intermediate' private and public institutions (Best, 1990). IDs therefore represented a model of capitalism in which the role of dynamic knowledge spillovers and intangible assets, such as reciprocal trust, sense of belonging, jointly forming the so called 'social capital', was of central importance.

Later, the interest was motivated by the possibility, through reference to ID features, to account for the anomalies characterising the pattern of industrial development and international competitiveness of the Italian economy. The attempt to explain why the Italian industrial system was able to achieve significant growth performances despite the dominance of SMEs in sectors classified as 'traditional' with an allegedly weak propensity for innovation, found an interesting response in the peculiarities of IDs. The result was, not without exaggerations and undue idealisations, an emphasis on the merits of the small industrial dimension, the role of incremental innovations and tacit knowledge, as well as the role of historical and social specificities in the creation of the competitive advantage of IDs and LSPs.

More recently, the interest has been motivated by the similarities emerging between the ID model of self-organisation and the modes of development of high-tech industries that typically tend to spatially agglomerate in clusters

[2] See IPI (2002) for an account of the developments of the Italian law 317/1991 and a detailed description and mapping of the Italian IDs according to the different studies, for example Rolfo (1997). See also Brusco and Paba (1997) for the role of IDs in the post-war Italian economic development.

characterised by ID-like economic mechanisms, a point which we will return to below.

One of the results of this extensive analytical process, and the related debate at the academic and policy-making levels, has been a crystallisation of the 'district paradigm' around the picture of an organisational model having a significant component of abstraction (see Ferrucci, 2003), a picture which is actually realistic for a limited number of Italian IDs, in particular the extensively-studied case of Prato's textile ID (see Signorini, 2000). In the framework of this idealised conception of ID, research interest has been mainly directed towards the organisational structure of the ID and the nature of the relationships among its firms and socio-economic actors in a generally static framework.

However, this idealisation rapidly revealed its shortcomings upon accounting for the actual working and evolution of IDs, as in the case of the attempt to define the statistical criteria for classification of LSPs as IDs. In this regard, the difficulties encountered by the official criterion of classification, i.e. that by ISTAT mentioned above, in including among 'official' IDs some well-known historical IDs, for example agricultural machinery of Reggio Emilia and packaging machinery of Bologna (see Paniccia, 1998), stimulated many scholars to re-open the debate on the nature and the dynamics of IDs. Among the lines of research in favour of departing from a canonical view we may mention the works by Ferrucci and Varaldo (1993), Ferrucci (2003), Balloni and Iacobucci (1997 and 2001), Bramanti and Maggioni (1997), Bianchi *et al.* (2001) and Brioschi *et al.* (2002). The central idea of these works is the need to study, by using differentiated methodological approaches, the actual process of change of Italian IDs also in comparison, and continuous communication, with the 'canonical' view (see Ferrucci, 2003). In these lines of research, the focus is not on ID as a static organisational form but instead on its change and adaptation process from industrial, social and institutional perspectives, with a great role assigned to the study of firms and their behaviour. Although firms in IDs can enjoy 'systemic' positive influences from the ID itself ('Marshallian' externalities, knowledge spillovers, interaction with social and local cultural values, etc.) they assume a key role in adaptively responding to and guiding the process of change.

The other feature of these lines of research is their attention to the mechanism of growth and innovation deriving from economic concentration and local interaction processes. In this context, fruitful connections with urban and regional economics can emerge. More specifically, the works by Glaeser et al. (1992), Henderson et al. (1995) – who mention Italian IDs as a case in point – Cainelli and Leoncini (1999), Cainelli *et al.* (2001), and Maggioni (2002), address dynamic externalities in processes of co-localisation of firms in 'small' geographical areas. Two different forms of dynamic externalities emerge: those of localisation and those of urbanisation. Economies of localisation can favour intra-industrial spillovers, i.e. exchange of information flows among firms belonging to the same sector, whereas inter-industry spillovers associated to the variety of sectors in a local environment can give rise to processes of cross-fertilisation of ideas, information, competencies of firms belonging to different sectors. Italian IDs have been considered as a case of spillovers of knowledge arising from economies of localisation within production structures characterised by a high degree of competition, a kind of externality *à la* Porter according to Glaeser *et al.* (1992).

In conclusion, IDs can still be fertile ground yielding results for both economic analysis and economic policy if the paradigms of complexity and evolution are regarded as the most reliable guide to research (see also Quadrio Curzio and Fortis, 2002).

2 Italian industrial districts and European 'regional clusters'

In the framework of the 'Lisbon Strategy' launched in 2000, SMEs are increasingly seen as a strategic issue by the European Commission. The 'Observatory of European SMEs' covering 19 countries was established already in 1992 by the DG Enterprise and it is trying to create a shared base of knowledge on SMEs and 'regional clusters' in Europe (see European Commission 2002a, 2002b, 2002c).

There are around 20 million SMEs (defined as having less than 250 employees) in the 19 European countries considered, and they employed around 80 million people in 2000, or two-thirds of total employment. Micro-firms (less that 10 employees) represent half of total employment in European SMEs[3]. The analysis of indicators suggests that SMEs are relatively weak compared to large-sized enterprises (LSEs): they have lower productivity and efficiency, lower propensity to export and lower capabilities to innovate. These weaknesses not withstanding, SMEs are almost as profitable as LSEs while micro-firms have been the most dynamic group in terms of employment creation since 1995. At the EU level, between 1988 and 2001, the growth of real value added was 2.0% in SMEs and 2.5% in LSEs, but employment grew by 0.3% in SMEs whereas it dropped by -0.1% in LSEs, and profitability grew by 0.5% in SMEs compared to 0.3% in LSEs.

While being very useful for a global view of SMEs in Europe, standardised internationally 'comparable' data inevitably provide a mixed picture because of the limitations associated with considering firm size as the reference variable. Although there can be common structural features of SMEs, e.g. the key role of the single entrepreneur or the limited power in credit markets, there are also well-known great economic variations among SMEs of the same size. For example, to establish the common upper limit of 250 employees for defining SMEs can be acceptable to varying extents for different countries and industries; some SMEs simply represent, especially in some countries, an intermediate stage of the firm growth processes, and they are in a sense 'transitional' SMEs; SMEs are typically the area of most intense firms' birth and death processes, and this can be reflected

3 The industrial structure of enlargement countries of Eastern Europe is even more fragmented in terms of firm average size, but it is gradually converging towards the EU profile, whereas Japanese and US industry is based much more on large enterprises.

in their worse aggregate performance compared to LSEs, which are expected to be more structurally stable; many SMEs, in particular in recent years and in some countries, are the results of outsourcing strategies, based on either hierarchical or informal relationships, and their performance could be measured as a part of the enterprise system to which they belong.

Furthermore, aggregate data for European SMEs as a whole risk shadowing the huge variability and specificity of industrial structures at the national and regional level, which are well known. For example, according to data by the European Commission (2002c), over the period 1988-2001, Italian SMEs' value added grew by 1.3% per year (the same rate as LSEs), which is lower than SMEs in Germany and the UK (2.4% and 2.5% respectively), and employment in Italian SMEs decreased over the same period (-0.3% per year) while it increased in German, French, and the UK's SMEs. However, the profitability of Italian SMEs grew by an average 1.1% per year, which is the highest rate in the EU15 countries if Ireland (2.1%) and Portugal (2.4%) are excluded[4]. Both Ireland and Portugal are fast-growing small economies but the recent exceptional growth of Ireland is based on a unprecedented inflow of foreign direct investments representing 'international production' strategies and also including many innovative SMEs in high-tech sectors, whereas the Portuguese economy is among the less innovative in Europe according to most indicators – which sheds light on the different meaning of SMEs in different economies. In addition, the above-mentioned profitability results of Italian SMEs have been achieved despite the fact that Italy is systematically ranked among the worst countries in Europe – not far from Portugal – by international benchmarking exercises based on indicators of the conditions (e.g. legal framework, infrastructures, R&D investments, etc.) in which firms operate – which sheds light on the limits of indicators (see in particular European Commission, 2002d).

A step forward in investigating SMEs in the European economy is therefore to consider these firms in their own economic space, i.e. systems of local firms and/or firms rooted in a local economic systems, which is exactly the case with IDs, LSPs or 'regional clusters', as they are defined by the European Commission's 'Observatory'. This is also the perspective in which the evolution of Italian IDs can possibly be useful at the European level as a source of reference experiences.

The picture for European 'regional clusters', of which Italian IDs are a specific model (see European Commission, 2002b), suggests that, despite their very diverse configuration in different countries, they can be the suitable unit of reference for 'integrated' development policies, in particular for innovation policy. 'Regional clusters', be they traditional or high-tech, prove to be dynamic in terms of the growth of the firm and employment. In addition, they are able, in most cases, to overcome the 'structural' weaknesses associated to the small size of firms through economic mechanisms associated to their concentrated location,

---

[4] Profitability is defined as the difference between value added and labour cost, adjusted for the imputed wage of self-employed, as a percentage of value added (see European Commission, 2002c).

such as those discussed in Section 1. The 34 'regional clusters' studied by the European 'Observatory', half of which are classified as high-tech, generally perform better than the national average in some industries, and in many cases they are able to promote innovation and entrepreneurship, and attract foreign investments. Various policy issues, however, arise in terms of internationalisation Vs 'regionalism' (or 'localism'), the possibility of reproducing successful experiences in other locations, and the choice of the best instruments for injecting innovation capabilities in local systems.

Therefore, conceptual, interpretative, and policy issues raised by European 'regional clusters' are very similar to those already highlighted by the Italian experience in IDs and the theoretical debate, and similar to those addressed by this book.

3 The contribution of this book

In this book, 14 contributions have been collected that address four specific aspects of the evolution of Italian IDs and LSPs: (i) the possibility of dealing with IDs from differentiated theoretical perspectives also in the framework of development/growth economics and economic geography; (ii) firms' and entrepreneurial dynamics; (iii) the issues of formal and informal innovation; (iv) the issue of internationalisation.

Three main features characterise this book as a whole.

The first feature is that, although most of the contributions make extensive reference to a now internationally-known stream of economic literature on Italian IDs, the picture we present departs from 'canonical' views on IDs along the lines described in Section 1. We try to add new results from fresh empirical analysis, sometimes taking as a point of departure the extraordinary wealth of intuitions by the main economic thinkers on IDs but sometimes ending with non-conventional conclusions that emphasise the evolutionary properties of IDs and LSPs.

The second feature is a strong focus on empirical analysis. All the contributions, with the partial exception of those in the first part, are based on the elaboration of sometime massive datasets and the use of statistical or econometric techniques. This feature derives from exploiting the opportunities offered by newly-available datasets regarding firms and municipalities, i.e geo-economic disaggregation levels suitable for direct analysis of IDs and LSPs. This orientation can help to overcome the bias towards 'idealisation', or the relatively weak empirical basis of some theoretical constructions, often emerging in the debate on IDs.

The third feature is that, instead of celebrating the successes of Italian IDs, we highlight the factors possibly endangering the successful stability of the 'IDs model', the main issues open to debate, and the factors that can favour or hinder a positive process of adaptation to current changes. In this regard, we think that the current evolution of firms and ownership models, the weaknesses and specificity of innovation processes, and the ambiguous propensity towards different forms of internationalisation can constitute the key issues to be considered.

3.1 Three theoretical perspectives

In the first part, we present three different theoretical perspectives aimed at raising questions of interpretation (and possibly suggesting answers) about the current evolution of IDs and LSPs.

Antonelli places LSPs in the perspective of Italian economic development and the prospects for the deepening of the European integration. The form of economic organisation represented by IDs/LSPs seems to have represented an adaptive response to changes of macroeconomic conditions that reflected historical peculiarities of Italian local socio-economic systems and institutions, and it has been a cause-consequence of Italian specialisation in the international economy. The success of this adaptive response raises various interpretative and policy issues that are relevant for the many uncertainties surrounding future evolution. Firstly, the short-term visibility of these performances risk shadowing the fact that most important explaining factors are structural or idiosyncratic in nature, even though they are not necessarily 'genetic' as claimed by some scholars, and most lines of research on IDs/LSPs rightly consider structural interpretations of the past and present growth as well as structural responses to change. Secondly, Italian IDs and LSPs experiences are of significance in view of the increasing importance of SMEs and SMEs' clusters in both high-tech and traditional sectors of the European economy (see also Section 2), which is the result of technological innovation allowing 'lean' or 'international production', the rise of local competitive advantages in some emerging European regions, the sectoral and geographical dichotomization between large trans-national companies and systems of out-sourced or independent small firms. Thirdly, IDs and LSPs are intrinsically evolving through the interaction between localised informal and formalised knowledge, between the actions of institutions and economic agents, between internal conditions and external developments thus giving rise to a fundamental unpredictability of their 'next' adaptation as well as a limited possibility of rough transfer of the model to other localities. Fourthly, as has been the case in the past, economic policy does directly and indirectly matter for the possible evolution of IDs and LSPs in Italy and Europe. The single European currency reduces the freedom of 'national' economic policies and, at the same time, emphasises the role of competitiveness-oriented structural policies, which are often seen from sectoral perspective but must necessarily be 'localised'. In this framework, the three keywords of EU economic strategy, i.e. integration, convergence, and cohesion, suggest that the combination of a bottom-up 'open method of coordination' and 'international regional benchmarking', which assign a great role to regional policies, can be superior to top-down economic policies.

Nuti develops an in-depth critical re-assessment of key concepts, definitions, and interpretative hints produced by theories of IDs and LSPs. He claims that, although initially rooted in the empirical observation of reality, the flourishing of theories of IDs outpaced facts and moved towards idealisations that risk missing the links with evolving realities. Long-run history and recent history of Italian economic development can provide a great part of the basic necessary explanations about both take-off and the subsequent development of IDs and LSPs. In particular, the interpretations based on demand-led processes of development, which implicitly stress the role of 'external' factors, should be reconsidered as an important part of the story. Furthermore, a critical perspective could be taken towards the emphasis on cooperation, coordination, trust, and flexibility as the basic mechanisms of ID specificity, also by highlighting non-cooperation and the negative externalities at work. The key suggestion is to move away from any rigid model and to take an evolutionary perspective to understand change in local systems as a continuous process. The main basis for the author's sceptical position on the 'Neo-Marshallian' theoretical revolution, however, is based on the relatively insufficient development of convincing empirical analyses and evidence, a shortcoming that can lead to 'virtual modelisation'.

Maggioni proposes a different perspective on the dynamics of IDs and LSPs which is rooted in both the established tradition of regional economics and the more recent stream of '(new) economic geography'. This 'external' perspective is developed by a formal dynamic model of firm clustering based on economies of agglomeration, which combines the elements of neo-classical theorising on industry location with models of ecological dynamics. The same ecological models of population dynamics are then used to analyse the different outcomes of cooperation and competition between two different industries in the same ID and/or two different IDs specialised in the same industry. The aim of these models is to show that the evolution of an ID (thus its success or failure) is so deeply influenced by spatial and industrial interactions that, by concentrating on the mere sectoral specialisation of a single ID, one may well miss its most relevant causes, effects and laws of motion. Another model is employed for analysing the evolution of governance and hierarchical structures inside a district based on the intensity of interactions among firms and transaction opportunity costs. Without reference to a specific case history on the role of local key entrepreneurs, the model is able to show, through a series of simulations, that different forms of hierarchy spontaneously emerge thanks to the day-by-day structure of productive interactions. These models of long-run equilibrium of highly complex interactive environments can complement in different ways the 'traditional' approaches. In particular, they can pave the way to a more rigorous analysis of interaction processes, which, if good information were available, could enrich the empirical foundations of ID and LSP theories.

Implicitly or explicitly, the three contributions suggest the importance of keeping an open research attitude towards IDs and LSPs, avoiding specialisation towards 'IDs economics' (as mirrored, on industrial-policy grounds, by the above-mentioned Italian Law No. 317/91), considering the value of a plurality of hypotheses, methodologies, analytical, and policy frameworks, and assigning a significant role to empirical analysis (see also IDSE-CNR, 1999).

3.2 Firm re-organisation and growth

The second part of the book addresses 'institutional innovation' in Italian IDs and LSPs by looking, in particular, at the recently-produced empirical evidence on changing ownership and governance of industrial firms.

Varaldo and Ferrucci analyse the factors which make the change of firms in IDs slow and difficult. 'Normal' factors making non-marginal changes of firm organisation a rare event, i.e. costs and uncertainties associated to highly irreversible actions, are particularly significant in SMEs forming what is often named as 'local' and 'family' capitalism. In addition, even the successful performances by these firms in domestic and international markets can act as a stickiness factor by reinforcing self-confidence on the optimality of the present organisation as well as by feeding the aversion for radical changes if the latter are not a compelling necessity. However, most SMEs in IDs have to face at least three strategic problems requiring institutional adaptation: (i) the entrepreneur's age and the issue of succession; (ii) the need for process/product innovation; (iii) increasing international competition and/or the need to internationalise (the last two issues being considered in other parts of this book). The importance of these strategic concerns, and in particular entrepreneur's succession, is suggested by data on the Prato' textile ID where 40% of the entrepreneurs in the sample of Varaldo and Ferrucci expect to retire within two years (another 17.3% within 3-4 years). The possible ways of guaranteeing a firm's continuity are typically to hand over entrepreneurial responsibility to offspring and relatives, to involve selected employees already having a role in firm's organisation, to sell the firm to other entrepreneurs (mergers and acquisition). Although the preferred solution would invariably be to give offspring control of the firm, the realism of this solution can be questioned on different grounds. Other data presented in this book suggest the actual importance of mergers and acquisition as well as, in differentiated forms, former employees emerging as entrepreneurs. Arrangements leading to the development of a 'market for firms' in many IDs through tax incentives, specialised financial intermediaries and advisors are envisaged as a way to make adaptation easier.

Iacobucci addresses exactly what can be considered a specific aspect of 'market for firms' and, at the same time, a more general process of institutional change, i.e. the increasing importance of groups of SMEs. The development of financial or industrial groups including a number of formerly-independent or newly-created small and medium firms under the control of a single family/entrepreneur, is far from being a new phenomenon but only recently it has attracted attempts at the measurement of its importance, its qualification in terms of firms' reorganisation, and its implications for both firm strategies and public policy. The 'groupification' phenomenon is also important for the theoretical and empirical inquiry into IDs and SMEs because it further weakens the, largely idealised, picture of IDs as self-coordinated systems of independent small/micro firms with low concentration of ownership and weak hierarchical relationships. In other words, the role of the single firm might be weakened in favour of the 'group' of firms as the reference unit of economic analysis while individual-firm size might be abandoned as a criterion of industrial policy. This view, suggested by the contribution by F. Brioschi, M.S. Brioschi and Cainelli in this book, is discussed by Iacobucci from the broader perspective of firm organisation theory before presenting evidence on business groups in the Italian economy and IDs, together with an empirical analysis on the Marche region. Consistent sets of data on business groups in manufacturing industry are available only from mid-1990s and they show that groups of firms are widespread in higher size classes whereas they are relatively less significant in small size classes. However, there are not complete datasets on groups in IDs at the national level. Most information and analyses remain at the regional level and they suggest a significant and increasing presence of groups of SMEs corresponding to very different organisational models. The most important motivation behind the formation of business groups, often taking the form of 'entrepreneurial' and 'industrial' groups, is the growth process of a limited number of firms, which are pushed pursuing entrepreneurial scale as a result of internal learning processes and changing external conditions. The motivation for preferring the 'group model' for growth processes is that firms in IDs operate in highly segmented markets requiring great organisational differentiation even in presence of limited production diversification. It is therefore preferable to buy or create other (small) companies in segmented markets as separate legal entities. A third reason is the process of internationalisation of the largest companies that requires at least the creation of commercial companies abroad often followed by production units for direct market access and/or for the exploitation of localisation advantages in terms of costs.

F. Brioschi, M.S. Brioschi and Cainelli stress that, although the importance of firms groups has been recognised for at least three decades in economic analysis of IDs and LSPs, only new sets of regional data allow us to properly carry out empirical analysis of the phenomenon and its implications, i.e. firm Vs group as the most appropriate unit of analysis, economic concentration and its consequences for theory and policy. The presence of business groups in the Emilia-Romagna region is investigated by addressing different categories of groups and in particular the 'ID group', i.e. a group of firms under the control of a single family or entrepreneur headquartered in the ID and engaged in the ID's reference production sector. Out of 211 groups identified in the analysis, 112 are 'ID groups'. The second most important category is that of 'international groups' arising as a direct evolution of strong commercial presence in foreign markets (43 out of 211). The motivations behind group formation are not too different from those highlighted by Iacobucci, in particular in terms of growth, and are manifested mainly by the acquisition of companies in the same sector possibly having differentiated activities. In some cases, also vertical integration is important. The development of 'ID groups' directly affects the internal organisation and governance structure of the ID by increasing industrial concentration and/or hierarchisation. In some cases, hierarchisation tends to be the prevailing effect as the groups substitute for sub-contracting relationships inside the district. In other cases, instead, groups represent horizontal diversification processes that increase the degree of industrial concentration.

The importance of increasingly hierarchical structures in the systems of firms in Italian IDs, as represented by the increasing role of leading medium-large companies, is investigated by Balloni and Iacobucci based on the evidence for the Marche region. There are several factors explaining the 'rationality' of the permanence of small firms in the Marche's IDs. However, the past emphasis on small firms underestimated the evolution of local production organisation as well as the role of heterogeneity of firms within the IDs. Despite the high number of small firms, some IDs are dominated by one or few large firms for which the small ones are sub-contractors. The authors expect an increasing relative importance of large firms, and in particular a general organisational convergence towards the 'group form', as the importance of functions other than production increases during a firm's growth process. The evolution towards large firms or 'groups' will not eliminate the typical features of ID organisation and its advantages, as is already the case at present in most IDs. Instead, the evolution towards systems that couple few large firms with networked/controlled small firms may be the main way for preserving competitive advantages in a changing economic environment, especially given the competitive pressures from the EU single currency and the need for (vertical) internationalisation.

Omiccioli and Quintiliani give an account of the specific characteristics of IDs' labour markets and their relationships with entrepreneurial dynamics, based on direct surveys carried out by the Bank of Italy. Labour market 'flexibility', which is considered a non-secondary part of the competitive advantages of Italian IDs, must be also viewed in connection with evolving firm ownership/management structures. There is, in fact, a significant degree of permeability between these two areas so that 'social mobility' inside the IDs is an important mechanism for firm renewal and growth. In this regard, Omiccioli and Quintiliani also provide empirical evidence on one of the mechanisms of firm continuity and institutional change considered by Varaldo and Ferrucci. Comparing 700 firms in 15 IDs and 200 firms outside IDs, it emerges that, in ID firms a typical career starts with a payroll job in a different firm of the same ID after which the worker decides to start-up a new business. In outside-ID firms, the ownership of a firm is a less frequent point of arrival of individual careers, the decision to start-up entrepreneurial initiatives are less frequent, and careers with the same firm are more frequent. Evidence suggests on the whole that the ownership and management structure of ID firms are characterised by a stronger identification between ownership and management, and the flexibility in the use of the workforce in IDs is mainly the result of sub-contracting rather than flexible contractual arrangements inside the firm. Furthermore, in the recruitment of specialised workers IDs companies show a high degree of self-containment at the ID level, a higher internal mobility inside the firm as well as the use of more informal channels compared to non-IDs environments.

Nosvelli's contribution highlights a specific aspect of organisational evolution of IDs, i.e. the dynamics of a 'district-with-sub-district system', through the analysis of the textile district of Asse Sempione and the footwear sub-district of Parabiago. After 'demonstrating', based on the criteria established by Law 317/91 on IDs, the existence of the footwear sub-district of Parabiago (which also produces clothing as the district to which it belongs), the interrelated dynamics of the district and sub-district is examined by shift-share analysis. There was a diminished specialisation of the sub-district during the 1980s and 1990s and the latter has been attracted and integrated into the predominant activity of the district, i.e. clothing. The prominence of sub-contracting and slow entrepreneurial change in the sub-district creates the basis for increasingly negative results. At the same time, some difficulties in the textile and clothing sector create the basis for multiplicative negative effects on the two-sector production mix in Parabiago. The possible synergies of the coupled system 'district with sub-district' have not materialised mainly because of the lack of entrepreneurial vitality and coordinated growth strategies.

3.3 Patterns of formal and informal innovation

The third part of the book addresses technological innovation in Italian IDs and LSPs. As also emerging from other contributions in this book, a most debated feature of Italian IDs is the allegedly limited ability to produce innovations, a feature often directly associated the dominance of SMEs in sectors classified as 'traditional' – which is clearly an oversimplification. A number of contributions in recent years have contrasted this view by stressing the very extensive role of informal and difficult-to-measure processes of innovation intrinsically associated with 'Marshallian' and agglomeration externalities as well as learning dynamics, which can partly explain the otherwise non-understandable good performance of such local systems.

Here we present three contributions that, starting from sharing the view that non-conventional, non-formal processes of innovation are at work in most Italian IDs, try to highlight the role of formal innovation processes inside the Italian districts and suggest that, to some extent, formal innovation activity is greater then often acknowledged, it is growing, and it is positively interacting with informal innovation.

Cainelli and De Liso discuss on theoretical grounds the above-mentioned combination of 'intentional' and 'non-intentional' innovation processes by highlighting the role of the division of labour inside both firms and IDs as well as the possible significance of learning processes at the local level. They then develop an econometric analysis of the role of intentional, unintentional, and learning-related innovation for the performance of 1,218 companies operating in 'traditional sectors', both inside and outside IDs, by using data from three different archives: the Community Innovation Survey (CIS), the Italian Structural Business Statistics (SCI), and the Italian Business Register (ASIA). While suggesting that 'innovative' (i.e. formally innovating) ID firms have better performance indicators than 'non-innovative' firms both located inside and outside IDs, the results of econometric estimates confirm the existence of a 'district effect' but also highlight that formal product innovation in IDs' companies is more significant than expected. Furthermore, they suggest that formal and informal innovation are complementary, the two also interacting multiplicatively in improving firm performance.

Leoncini and Lotti investigate which structural and organisational features of the firms might affect the probability that they invest in R&D, register patents, and create innovative products. An econometric analysis is carried out on data from 'Technology Watch' of ASTER, the Emilia-Romagna's agency for innovation, using a sample of 1,026 firms belonging to 11 sectors and having from 50 to over-500 employees. The set of explanatory variables with regard to the likelihood of innovation includes firm size, sector, product type, specialisation, concentration, and international position together with variables indicating if the firm belongs to an ID or to a 'group' of firms. More competitive and more highly specialised sectors seem to lead their firms to higher investments in R&D, whereas to belong to an ID and to a group of firms seems to reduce the incentive to make formal R&D investments. This result can be explained, according to the authors, by the fact that individual firms in IDs and groups may need less formal innovation activity due to concentration of these activities in leading firms or the firm at the top of local hierarchical structure, e.g. the leading firm of a sub-contracting network. The results, which are similar to those mentioned above for patenting and product innovation, seem to confirm the importance of informal mechanisms of innovation in local systems of production but also highlight that competitive environment are more conducive to formal innovation investments. It is interesting that the econometric results suggest a significant role of public R&D funding, especially at the EU level, in supporting innovation, as well as an important role of cooperation with public research institutions, in particular the universities.

A different perspective to innovation in IDs is proposed by Montini and Zoboli, who address innovation in response to local environmental problems and policies. The possibility that policies against environmental 'negative externalities' of concentrated industrial activities can become a source of dynamic changes either hindering growth or stimulating innovative activity has been rarely investigated at the ID level. The authors look at the hypothesis that the 'positive' Marshallian and agglomeration externalities giving specificity to innovation in IDs may also be a source of specific approaches and advantages in innovatively responding to environmental policies. Starting from original empirical estimates of industrial air pollution in the Local Labour Systems and IDs of the Emilia-Romagna region, the authors examine the scattered evidence on organisational and technological innovations developed by Italian IDs companies in the environmental field. The results suggest that environmental innovation is more advanced in 'mature' IDs especially if they are specialised in highly polluting industries. There are interesting examples of 'cooperative innovations', i.e. investments and environmental infrastructures managed jointly by firms and/or local institutions, often motivated by economies of scale at the district level. However, the role of ID specificity for observed innovative responses to environmental policies is still weak, and IDs' 'cooperative advantages' are still under-exploited in the environmental field. Furthermore, the strategic consideration of environmental innovation as a possible source of competitive advantages in environmental (international) markets is weak. Cooperative responses to policies coupled with environmental leadership strategies remain an important possibility for the future because of the 'integrated' features of most 'new' environmental policies.

3.4 Diverging propensities towards internationalisation

The last part of the book addresses another issue often raised by the debate on advantages and weaknesses of IDs and LSPs, i.e. international competitiveness and internationalisation. The emphasis on the extraordinary international trade performances of Italian IDs is often paralleled by the concerns about the allegedly limited propensity of ID firms to become multinational through foreign direct investments (FDI) while internationalisation of firms is considered, also by other contributions in this book, as one of the most critical issues of IDs evolution. This same divergence is mirrored by the contributions of Menghinello on IDs' exports and Mariotti and Mutinelli on ID firms internationalisation.

Export performance is the most visible aspect of competitive advantages of Italian IDs and their firms, the latter covering very high shares of the world export market of some products, e.g. 28% for ceramic tiles by the Sassuolo' district alone, as well as very high shares of the total Italian export for the same products, e.g. over 90% for sport goods (see also Fortis, 1998; Quadrio Curzio and Fortis, 2000). Menghinello's analysis, based on an extensive original dataset developed at ISTAT, contributes new evidence for all the 784 Italian Local Labour Systems and 100 merchandise-groups disaggregation by elaborating on market shares, localisation quotients, export performances, and product quality indicators. The basic research hypothesis is that firms belonging to IDs enjoy competitiveness factors and their export performance 'reveals' the competitive advantages of such locally-specific economic organisation. Statistical analysis suggest that the 'ID effect' play a significant role not only in exports of low-tech products but also in medium-to-high technology exports, such as mechanical-engineering and electro-mechanical products. Furthermore, exported products from IDs have quality indexes systematically higher than the Italian national average.

Mariotti and Mutinelli analyse the issue of internationalisation through FDI by looking at evidence for mechanical engineering firms in 21 Italian IDs. By stressing the general weakness of Italian firms in FDI activity, the authors claim that the 'Italian model', based on IDs of successfully exporting SMEs may have been responsible for the negative divergence between Italy and other industrialised countries in the recent wave of FDI-based international integration. More specifically, the sectoral composition and firm size of the most successful

IDs is biased towards those sectors and sizes less favoured by the surge of FDI. Furthermore, the consolidation of successful exports from IDs reflects localised competitive advantages that cannot be easily reproduced in a geographically scattered production structure, while lock-in effects, also due to successful performance, pushed to look more inside than outside the ID for growth processes. These factors, which the authors define as 'intrinsic and genetically rooted weaknesses', are considered to be still powerful in preventing direct internationalisation by FDI. Empirical findings on mechanical engineering firms in IDs confirm this gloomy picture. There is a very limited propensity to FDI by these firms, with employment of foreign affiliates of IDs parent firms representing only 4.7% of their total employment. Inward FDI by foreign companies in the same IDs is more significant, the employment of foreign companies being 8.9% of total employment, but it is still limited. Very few IDs have high degrees of both inward and outward internationalisation and most of them are in the 'low inward – low outward' internationalisation area. Econometric analysis suggest that there is a relatively high correlation between outward internationalisation and industrial concentration and a high correlation also exists between inward and outward internationalisation, i.e. those IDs with the higher inward FDI are also those with the higher outward FDI. The hopes for the future are therefore based on concentration, i.e. the increasing role of medium-large firms and groups emerging from other contributions to this book, and openness to foreign investments as the authors consider them able to create an internationalised culture in otherwise 'inward looking' systems. We will shortly discuss below these clear-cut conclusions.

4 Evolving competitive advantages?

The variety of the specific results summarised above calls upon a discussion also to point out our different view on some issues.

The key factor attracting interest in Italian IDs and LSPs is the 'revealed' competitive advantage of such spatial models of socio-economic self-organisation. More precisely, a large part of economic research on IDs deals with a 'residual' of *observed* competitive advantages *not explained* by 'normal' or readily-measurable factors – an issue not too distant, from the analytical point of view, from that raised by empirical evidence on growth which led to the stream of endogenous ('new') growth models. In this regard, it is not surprising that a part of the interpretations of the unexplained 'residual' of competitive advantages of IDs and LSPs strongly focuses on the mechanisms of innovation in locally concentrated systems of production, or the 'endogenous' sources of locally sustained economic growth.

As described in Section 1, a better understanding of the nature of such unexplained 'residual' and its role for socio-economic success has been attained by looking at the 'silent' action of different forms of externalities, history and local civic culture, informal social interactions and intermediate institutions, and other 'systemic factors' able to give a valuable specificity to firms in their local environment, and leading to the disclosure of 'holistic advantages'.

This notwithstanding, interpretations about the extent and working of localised competitive advantages are still diverging and they centre around three very general issues:

- *measurability*, or the possibility to obtain empirical estimates of systemic (sources of) advantages or simply of a 'district effect' in observed performances;
- *sustainability*, or the capacity of stable performances through the adaptive evolution of competitive advantages in the presence of continuous change, and:
- *transferability*, or the possibility of replicating such models in other locations having either similar or very different socio-economic frameworks.

With this book, we have tried to contribute mainly to the issue of measurability and sustainability, whereas much less attention has been devoted to the issue of transferability, which is obviously extremely important from both analytical and, even more, policy perspectives. Transferability is actually of central interest for European and national policies, in particular on regional disparities and convergence, which are not, however, the main focus of this book.

4.1 Measurability and models

On measurability, we have tried to suggest that economic research, after having clarified and conceptualised the main factors behind the specific competitive advantages of IDs, should progress along the two routes of modelisation and empirical measurement. New possibilities of statistical and econometric measurement of the factors explaining IDs and LSPs performance are presented in the works by Cainelli and De Liso, and Leoncini and Lotti, while all other contributions investigate the main critical trends and issues by applying statistical techniques to large datasets. The evolutionary characteristics of systems such as IDs and LSPs make them suitable for analysis by other heterogeneous classes of models and methodologies. Models that can add formal understanding and measurement opportunities are those rooted in ecological dynamics, such as that presented here by Maggioni, as well as other models of complexity, e.g. those based on neural networks and other natural-science analogous, or adaptive dynamic models (see Lane, 2002; Quadrio Curzio and Fortis, 2002; Day, forthcoming). Although some of these models are very demanding in terms of data input, the increasing availability of information at a very disaggregated level gives new opportunities for useful implementation. In this way, also the risk of over-theorising or 'idealised virtual theories' can be reduced.

4.2 Sustainability of performances

Most of the essays in this book address the second key issue, i.e. 'sustainability' of performances or the ability of competitive advantages to evolve by adaptation to changing internal and external conditions. It is the issue which is closer to an evolutionary perspective of IDs as it deals with variation, learning, fitness, and selection mechanisms. More in general, it deals with the possibility of attributing a 'genotypic' or 'phenotypic' origin to observed behaviours, or the characterisation of change in such complex systems as 'Darwinian' and/or 'Lamarkian'[5].

In contrast with optimistic views, which range from enthusiasm for recent performance to the idea that, despite some difficulties, IDs still have a lot of resources for competing and growing, some different positions suggest that observed competitive advantages can turn into disadvantages if a deep structural adaptation does not rapidly take place in response to major internal and external changes. The latter are the 'maturity' of both reference markets and internal production structures, an increasingly innovation-intensive production regime at the international level, and the changing global geography of competition in manufacturing. A specific 'evolutionary' issue surrounds these positions: the role of positive performances of the recent past in either hindering, by adverse hysteresis or lock-in mechanisms making IDs 'unfit' to compete in the new framework, or instead supporting, through accumulated learning, capabilities, response capacity, a sustained process of further growth.

We will shortly discuss the three aspects of 'sustainability' and evolution addressed by this book: (a) changing firm ownership and governance; (b) relationships between formal and informal innovation in IDs; (c) diverging attitudes among firms towards export and FDI in internationalisation.

Firm dynamics

On the dynamics of firm ownership and local industrial governance, the results and positions emerging from the essays range from a pessimistic view (Varaldo and Ferrucci) based on the stickiness of firm organisation to the evidence of significant 'social mobility' in IDs (Omiccioli and Quintiliani), and from the evidence on the rise of entrepreneurial groups as the preferred way of growth

[5] See Hodgson (2001) for a general discussion on Lamarkian and Darwinian features of change in social systems. Economists are generally prudent in using close analogies with biology and ecology. Economics-specific evolutionary theories of change are pursued even in those areas of economics more influenced by evolutionary thinking as the economics of innovation (see in particular Nelson and Winter, 1982; Metcalfe, 1998; Saviotti and Metcalfe, 1991). This applies, with some exceptions (e.g. ecological dynamic models in economic geography, see Maggioni in this book), also to economics of local (industrial) development despite the well-known propensity of Alfred Marshall to look at biology and evolutionary doctrines as an inspiration for economics (see Limoges and Mènard, 1994; Groenewegen, 2001) as well as the close similarity between ID analysis and evolutionary concepts of (technological) change.

(Brioschi F., Brioschi M.S. and Cainelli) to the pivotal role of local medium and large-sized firms at the top of increasingly hierarchical economic structures (Balloni and Iacobucci).

The main suggestion emerging is that, despite the still central role of SMEs and micro-firms, entrepreneurial structure and firm-governance inside many IDs is more concentrated and hierarchical than expected by 'idealised' theoretical or narrative representations. Although the short time series of comparable data prevent us from claiming that this is a true novelty, there is the widespread conviction that the concentration of firms, predominantly in the 'group form', is increasing and it has already been a response to changing internal and external conditions.

If we accept the idea that increasing concentration and hierarchy is a positive adaptive factor in this phase, the concerns about SMEs having structural disadvantages for innovation and firm internationalisation can be alleviated. The key issue is, however, if a more concentrated firm governance and local economic structure can improve competitive advantages without impairing other 'traditional advantages' usually attributed to the flexible systems of independent SMEs or non-hierarchical networks of local firms' linked by contractual relationships and informal knowledge interactions.

In general, we believe that the present trend towards 'groupification' of firms can be an effective response to the need to achieve a larger firm's economic scale while maintaining the advantages of small technical and legal scale of single firms. This can be particularly true when the specific production processes do not have technical economies of scale, increasingly complex forms of functional organisation must be internalised, and the economic environment is changing. The 'group form' of firm governance may therefore be the right hybrid possibly bringing together different kinds of apparently non-compatible advantages, i.e. the strength associated with big scale and the flexibility of small scale.

Innovation and public policies

On the issue of innovation, a central one in the debate about both advantages and weaknesses of IDs and LSPs, the three essays presented here suggest that: (a) 'formal' innovation activity in IDs is probably higher than commonly acknowledged and its economic impact is *multiplied* by 'informal' innovation processes when it takes place inside IDs; (b) the chain of formal innovation (R&D, patenting, product innovation) is less significant and works in a different way in IDs as compared with non-IDs in the same geographical area; (c) an ID's specificity can emerge in developing innovative processes as a response to certain public policies, as represented by environmental policies and regulations.

As a whole, they confirm the idea that there is something specific in the innovation process in IDs or similar spatial models of economic self-organisation, even in the presence of formal innovative activity, and even amplifying, in a multiplicative way, the actual outcomes of the latter. This conclusion leads us to make some observations on the widespread concerns for the allegedly low innovation rate of Italian IDs.

Firstly, the increasing conceptual pluralism about the sources, the mechanisms, and the economic impacts of innovation tends to overcome the rigidity of measurable indicators and the often-biased picture they offer about what is innovative and what is not. More realistic and dynamic representations of innovation processes, generally at the crossroads between advances in microeconomics and evolutionary economics, are now well established as the most suitable approaches to systemic mechanisms of innovation and their consistent measurement. In this regard, it can be noted that the difficulties encountered in interpreting innovation in IDs and LSPs are experienced in studying innovativeness of the economic system as a whole, as in the case of innovation in the service sectors and their alleged role in slowing down the total factor productivity of advanced economies (see Antonelli and Zoboli, 2001).

Secondly, we believe that a prominent role of SMEs and IDs is not *in itself* a limit to innovation of the Italian economy, the main problem being the weaknesses of Italian large firms operating in high-tech and scale intensive sectors. There are various reasons for this.

SMEs are a typical form of innovative high-tech enterprises and the structural similarities between Italian 'traditional' IDs and clusters of innovative firms or high-tech districts are well recognised (see Section 2). Both follow paths of development rooted in dense networks of externalities, competition, imitation, interactive flows of knowledge, institutional connectivity, tacit localised knowledge, and the orientation towards international markets. The possibility that a 'district effect' amplifies the result of explicit and formal innovation efforts, as suggested by Cainelli and De Liso in this book, justify the joint interest for traditional and high-tech IDs but also makes clear that, in any case, we should expect higher rates of measured innovation in those industrial clusters where the dominating sectors are science-based or research-intensive, even irrespective of firm size.

However, even sectoral composition can be a source of misleading interpretations of the alleged low innovativeness of Italian IDs. To identify 'traditional' sectors with low–tech suffers not only from the issue of 'unmeasurable' innovation discussed above, but also from the issue of 'unmeasured' innovation. Some of the most traditional sectors of Italian IDs include differentiated high-tech segments, e.g. sportswear based on high-tech textiles and materials, while other traditional sectors innovate mainly in the form of industrial design and quality differentiation, which generally correspond to little-measured and little-protected intellectual property rights, such as copyrights and trademarks. The latter are the true backbone of international competitiveness for Italian IDs and the increasing international appropriation of these innovations, e.g. imitations in fashion-related products or food industry products, is now the centre of mounting issues on competition rules at the WTO.

A radical change of sectoral specialisation, with a rapid blossoming of high-tech districts paralleling or substituting for traditional IDs, cannot obviously be envisaged. This is not a consequence of 'traditional' ID dominance but a consequence of the structural weaknesses of large Italian enterprises in the rapidly growing sectors of high-tech, the extremely low private and public resources allocated to R&D, the shortcomings of Italian 'national innovation system' compared to international benchmarks, also as a result of some past (and present) public policies (see Quadrio Curzio, Fortis and Galli, 2002; Malerba and Lissoni, 2002). Without large and stable, private and public investments in R&D and diffusion of innovation, preferably in the core areas of the 'big science' and its technological off-shots, high-tech firms and districts will never come into existence.

However, efforts in R&D and appropriate innovation chains are also required if a higher rate of formalised innovation is to be achieved in 'traditional' IDs. This is an area in which public policy can have a significant role to play. In Italy during the 1990s, we were witness to a process of regionalisation, or even 'districtualisation', of innovation policies which were based on the full range of available instruments such as science parks, firm incubators, regional agencies for innovation, etc., also including high-tech sectors (see Rolfo, 2000; Calabrese *et al.*, 2002). The results are often mixed and the fruit borne is less than the seeds were sown, which is rather usual in innovation. However, the continuation of this wide range of micro initiatives is an important factor towards reinforcing the possibility that IDs become more formally innovative directing themselves towards more research-intensive product differentiation, specialisation, and market positioning.

Another open issue concerning the innovativeness of IDs is their poor infrastructural endowment – an area of contact with services and their role for innovation. Although a part of 'informal' innovation processes in IDs can be attributed to service sectors (e.g. intermediate institutions, local public agencies, and local banks), many Italian IDs suffer from a severe shortage, or conversely, high congestion of infrastructures and services, from transportation networks to local business services. Actually, the growth process of some LSPs is still underway in the presence of surprisingly adverse infrastructural situations (see Montini and Zoboli, 2003) so much so we cannot exclude that infrastructural difficulties might have favoured a geographically distributed economic development based on many small 'centres', i.e. IDs, instead of an economic process based on few big 'centres'. Some service sectors, and in particular network utilities, e.g. electricity, telecommunications, are areas of rapid innovation that can favour innovative processes at large, even by interacting with user sectors. The cooperation between the 'industrial pillars' in services/utilities and local systems of SMEs in IDs can therefore be envisaged as an important factor for boosting the formal and informal innovation capabilities of the latter, be they traditional or high-tech (see Quadrio Curzio, Fortis and Galli 2002).

Internationalisation and structural competitiveness

The last set of conclusions on 'sustainability' and adaptation of competitive advantages is that concerning internationalisation processes, which are obviously related to firm-ownership dynamics and innovation capabilities. The 'black and white' picture emerging from combining the analyses by Menghinello, who stresses the strong competitive advantages of IDs revealed by their export performance, and Mariotti and Mutinelli, who stress the low rate of FDI-based internationalisation by the firms in the same IDs, stimulates some comments in the light of the complex changes taking place in the international economy[6].

First of all, the conclusions about the unwillingness and/or inability of IDs firms to undertake FDI are mitigated by evidence emerging from other contributions to this book (Balloni, Balloni and Iacobucci, F. Brioschi, M.S. Brioschi and Cainelli, and, to some extent, by Mariotti and Mutinelli themselves). These contributions suggest that the creation of groups of SMEs and thc increasingly hierarchical governance structures are often, and increasingly, motivated by the advantages of direct internationalisation of those firm functions supporting foreign market penetration or exploiting localisation advantages through 'international production'.

The second observation is that there is a significant sectoral specificity of the propensity to FDI, with differences even between 'traditional' sectors, e.g. textile-clothing and furniture, that may be not related to location inside an ID, entrepreneurial culture, and firm size, and may simply derive from the different advantages firms in different sectors can enjoy from direct internationalisation. The most important firm internationalisation processes are taking place in specific sectors also reflecting the suggestions of the 'theory' of multinational enterprises[7]. According to UNCTAD (2001), the first 100 multinational enterprises structurally belong to specific broad sectors, i.e. resource intensive industries (e.g. energy, mining), medium-to-high technology mass-consumption goods (e.g. automobile, consumer electronics), research-intensive sectors (e.g. pharmaceuticals, ICT), and low-tradability services (e.g. banks, insurance). A similar sectoral pattern prevails for a large part of FDI flows and stocks with some significant exceptions, such as the 'international production systems' in textile and clothing. With the aforementioned exceptions, these sectors are not those of main specialisation of Italian IDs'. FDI respond to a combination of

---

[6] According to data on Italian multinational firms by the Polytechnic of Milan and CNEL (see CNEL, 2002), Italian parent companies are 1,017 with 2,573 foreign affiliates (both majority and minority interests). Affiliates of foreign parent firms located in Italy are 1,843 representing majority and minority interests of 1,046 parent firms. Out of 1,017 Italian parent firms, 830 (or 81,6%) have less than 500 employees, and 1,244 out of the total 2,573 foreign affiliates of Italian parent firms have less than 500 employees. An updated analysis of Italian FDI can be also found in ICE (2003). The figures for Italy suggest a rather low degree of direct internationalisation when compared with data for countries such as Germany (8,000 parent companies and 12,000 affiliates in the country), France (1,700 parent companies and 9,500 affiliates in the country) and the UK (1,100 parent companies and 2,700 affiliates in the country, but UK data are underestimated). This comparative picture for Italy is confirmed by almost all indicators on FDI activity (see UNCTAD, 2001).

[7] See Dunning and Narula (1996) and Markusen, (1995) for a description of the OLI paradigm, i.e. 'ownership', 'localisation', and 'internalisation' factors as necessary conditions for foreign direct investments. The 'internalisation' argument – the most intriguing in a theory of multinational enterprises (see Rugman, 1980) – is particularly interesting in an ID framework because it deals with the strategies for limiting the dissipation of firm-specific assets in the locations of the international affiliates.

rational and strategic decisions and, therefore, they are not always the best, or even the right, way of internationalising. Many firms in the dominating sectors of Italian IDs could simply lack one or more of the incentives to become FDI-oriented. Therefore, the true issue is the possibility that low FDI activity by Italian ID firms reflects non-rational behaviours or unexploited opportunities. Economic integration in Europe, firstly by the EU enlargement and then by Mediterranean integration, may actually point to the existence of unexploited as well as new opportunities, which can be further stimulated by the Asian competition.

The third observation is that analysis of ID internationalisation through FDI tends to disregard the process of 'internationalisation of ID labour markets'. Together with other significant changes of local labour markets (see Antonelli *et al.* 1999), many Italian IDs in Northern regions are experiencing conditions of almost-full employment and they are the destination of an increasing number of immigrants from Eastern Europe and LDCs. The acceleration of international labour inflow into IDs in combination with a low propensity to FDI by IDs' firms would seem to confirm that the latter suffer from a negative hysteresis and a problematic inability to go outside the ID itself thus remaining entrapped into export as the main form of internationalisation. However, convincing interpretations of relationships between export, FDI, and local labour market are not yet available, and the possibility that the advantages of keeping production in the local system are still very high whereas those of becoming multinational enterprises are (still) too low may also be envisaged. This is not the same as to claim that firm internationalisation would impair those critical advantages associated with concentrated localisation, a position sometimes emerging in the debate and which, in our opinion, reflects the rigidity of theoretical 'idealisations'.

In this regard, the new framework for export flows from IDs should also be considered. The combination of a strong Euro in the international monetary system and the rise of highly-competitive LDCs put the international competitiveness of many Italian IDs structurally under pressure[8].

---

[8] Internationalisation difficulties stimulate a fear that Italian economy may go towards a decline after what someone considers to be a 'New Renaissance' also on the basis of alleged historical continuities between ancient local civic culture and certain features of IDs (see, for example, Putnam, 1993). This fear seems to evoke what happened in XVIIth century when the development of Atlantic and Asian trade dominated by Northern European countries eroded the Italian role in the world economy. David Landes describes that decline in this way: "*Europe's shift in economic gravity northward obviously transcends the inglorious Spanish fiasco. The great old mercantile and industrial city-states of Italy – Venice, Florence, Genoa – also lost out. Italy had been at the forefront of the medieval commercial revolution and had led the way out of autarky into international trade and division of labour. As late as the sixteen century, Italy was a major player, splendid in its manufactures, preeminent in the commercial and banking services rendered to Spain and Northern Europe. Yet Italy never really sized the opportunities offered by the Great Opening: one does not find Italian ships in the Indian Ocean or crossing the Atlantic. Italy was centered in, caught in, the great Inland Sea. Caught also by old structures: guild control fettered industry, made it hard to adapt to changing tastes.*

The problem of extremely competitive exports from LDCs, in particular China, and its explanation mainly based on unfair competition is sometimes ill posed. Products imitation and violations of intellectual property rights (together with 'social dumping' and 'environmental dumping') are significant competition issues to be dealt with by completing and enforcing WTO rules and the action of other international organisations. However, the growth of LDCs is *unambiguously good* for the world economy, advanced countries must compete on advantages other than labour costs, and a significant part of competitive exports from LDCs, in particular from China, originates from the thousands of foreign affiliates of multinational enterprises investing in those countries, sometimes as a part of complex 'international production systems' dominated by Western companies[9]. There is increasing evidence that the surge of FDI in LDCs during the 1990s and the early-2000s is one of the causes of rapidly increasing trade flows from and to these countries (see WTO, 2003; ICE, 2003).

Competitiveness should therefore be increasingly different from 'cost competitiveness' and should be augmented by internationalisation of production if, as discussed above, FDI are justified, *inter alia*, by actual re-localisation advantages. The structural change associated with the single European currency pushes in the same strategic direction. Euro was born to pursue the advantages of a reserve currency, even at the cost of reducing the freedom of economic policies. However, a reserve currency must be sustained by a strong economy creating attractiveness for international saving flows and the trust in using that currency for transactions. In other words, it can be sustained only by 'structural competitiveness' mainly corresponding to efficiency and productivity in the presence of high labour costs, product quality and specificity 'enforceable' on the market, and continuous product and process innovation. If international specialisation of Italian industry cannot be easily moved towards sectors intrinsically having those characteristics, i.e. high-tech, then structural competitiveness must primarily be pursued in 'traditional' sectors and SMEs, which leads back to the arguments on innovativeness of local firms and environments as the key to evolving competitive advantages. As M. Porter says "*In fact, there is no such thing as a low-tech industry. There are only low-tech companies*" (Porter, 1998, p. 86)

Labour cost stayed high because manufacture was largely confined to urban, corporate workshops employing adult male craftsmen who had done their years of apprenticeships" (Landes, 1999, page 174). See also Braudel (1985) who highlights the role of trade strategies by Northern European cities as one of the causes of the declining role of Italian commercial power in the Mediterranean even before the 'great opening', and Cipolla (1965), who stresses the inability of Italian city-states to keep pace with other European countries in the innovation of military and sailing technologies which proved to be critical for the establishment of overseas colonial empires.

9 According to UNCTAD (2001), in China there are around 364,000 affiliates of foreign companies out of a world total of 821,800 foreign affiliates. China-based parent companies having affiliates abroad are 379 compared to 63,300 parent companies at the world level, 49,900 of which are based in OECD countries. It is estimated that around 60% of Chinese export originates from affiliates of multinational companies.

References

ANTONELLI G., *et al.* (a cura di) (1999), *Disoccupazione e basso livello di attività in Italia*, Il Mulino, Bologna.

ANTONELLI G., ZOBOLI R. (2001), "Ricerca e innovazione in una economia di servizi", in: GARONNA P., IAMMARINO S. (a cura di), *Economia della ricerca*, Società Italiana degli Economisti, Il Mulino, Bologna.

BAGELLA M., BECCHETTI L. (Eds.) (2000), *The Competitive Advantage of Industrial Districts. Theoretical and Empirical Analysis*, Physica-Verlag, Heidelberg.

BALLONI V., IACOBUCCI D. (1997), "Cambiamenti in atto nell'organizzazione dell'industria marchigiana", *Economia Marche*, vol. 16, n. 1.

BALLONI V., IACOBUCCI D. (2001), "Diffusione e caratteristiche dei gruppi di piccole e medie imprese nelle aree distrettuali", in: BRIOSCHI F., CAINELLI G. (a cura di), *Diffusione e caratteristiche dei gruppi di piccole e medie imprese nelle aree distrettuali*, Giuffrè Editore, Milano.

BECATTINI G. (1979), "Dal 'settore' industriale al 'distretto' industriale: alcune considerazioni sull'unità di indagine dell'economia industriale", *Rivista di Economia e Politica Industriale*, Vol. 5, n. 1.

BECATTINI G. (1989), "Sectors and/or districts: Some remarks on the conceptual foundation of industrial economics", in: GOODMAN E., BAMFORD J. (Eds), *Small Firms and Industrial Districts in Italy*, Routledge, London.

BELUSSI F., GOTTARDI G. (Eds.) (2000), *Evolutionary Patterns of Local Industrial Systems. Towards a Cognitive Approach to the Industrial District*, Aldershot, Ashgate.

BEST M. (1990), *The New Competition: Institutions of Industrial Restructuring*, Harvard University Press, Cambridge.

BIANCHI R., BRIOSCHI M.S. E CAINELLI G. (2001), "Gruppi di imprese e distretti industriali in Emilia Romagna", in: BALLONI V. (a cura di), *Piccole e grandi imprese nell'attuale contesto competitivo*, Giappichelli Editore, Torino.

BRAMANTI A., MAGGIONI M.A. (1997), *La dinamica dei sistemi produttivi territoriali: teorie, tecniche, politiche*, Franco Angeli, Milano.

BRAUDEL F. (1985), *La dinamique du capitalism*, ed. Arthoud, Paris.

BRIOSCHI F., BRIOSCHI M.S., CAINELLI G. (2002), "From the Industrial District to the District Group. An Insight Into the Evolution of Local Capitalism", *Regional Studies*, vol. 36, n.9.

BRUSCO S. (1982), "The Emilian model: Production decentralisation and social integration", *Cambridge Journal of Economics*, n. 2.

BRUSCO S., PABA S. (1997), "Per una storia dei distretti industriali italiani dal secondo dopoguerra agli anni Novanta", in: BARCA F. (a cura di), *Storia del capitalismo italiano dal dopoguerra ad oggi*, Donzelli, Roma.

CAINELLI G., LEONCINI R. (1999), "Externalities and Long-Term Local Industrial Development. Some Evidence From Italy", *Revue d'Economie Industrielle*, n. 4.

CAINELLI G., LEONCINI R., MONTINI A. (2001), "Struttura produttiva e sviluppo regionale di lungo periodo in Italia", *Moneta e Credito*, vol. LIV, n. 216.

CALABRESE G., COCCIA M., ROLFO S. (2002), "Analisi del processo innovativo nelle PMI italiane", *CERIS-CNR, Working Papers*, n. 11/2002, Torino.

CIPOLLA C.M. (1965), *Guns and Sails in the Early Phase of European Expansion 1400-1700*, Collins and Sons and Co Ltd, London.

CNEL (2002), *Italia Multinazionale. Inward and Outward Foreign Direct Invstments in the Italian Industry in 1998 and 1999*, CNEL, Roma, at www.cnel.it.

DAY R. (forthcoming), *The Divergent Dynamics of Economic Growth*, Cambridge University Press, Cambridge (chapter 3).

DUNNING J.H., NARULA R (eds.) (1996), *Foreign Direct Investments and Governments*, Routledge, London.

EUROPEAN COMMISSION (2002a), *High-tech SMEs in Europe*, DG Enterprise, Observatory of European SMEs, No. 6/2002, Brussels.

EUROPEAN COMMISSION (2002b), *Regional Clusters in Europe*, DG Enterprise, Observatory of European SMEs, No. 3/2002, Brussels.

EUROPEAN COMMISSION (2002c), *SMEs in Europe, Including a First Glance at EU Candidate Countries*, DG Enterprise, Observatory of European SMEs, No. 2/2002, Brussels.

EUROPEAN COMMISSION (2002d), *Benchmarking enterprise policy. Results from the 2002 scoreboard*, Commission Staff Working Document, SEC(2002)1213, DG Enterprise, November.

FERRUCCI L. (2003), "Il gruppo distrettuale: rafforzamento o destabilizzazione del paradigma distrettuale?. Un commento al contributo di Cainelli", in: PACE G. (a cura di), *Le disparità regionali nel Mediterraneo. Innovazione, sviluppo, apprendimento*, Franco Angeli, Milano.

FERRUCCI L., VARALDO R. (1993), "La natura e la dinamica dell'impresa distrettuale", *Economia e Politica Industriale*, n.80.

FORTIS M. (1998), *Il Made in Italy*, Il Mulino, Bologna.

GLAESER E., KALLAL H., SCHEINKAMAN J., SCHLEIFER A. (1992), "Growth in Cities", *Journal of Political Economy*, vol. 100.

GROENEWEGEN P. (2001), "The evolutionary economics of Alfred Marshall: An overview", in: LAURENT J., NIGHTINGALE J. (Eds.), *Darwinism and Evolutionary Economics*, Edward Elgar, Cheltenham.

GUERRIERI P., IAMMARINO S., PIETROBELLI C. (2001), *The Global Challange to Industrial Districts. Small and Medium-Sized Enterprises in Italy and Taiwan*, Edward Elgar, Cheltenham.

HENDERSON V., KUNKORO A., TURNER M. (1995), "Industrial Development in Cities", *Journal of Political Economy*, vol. 103.

HODGSON G.M. (2001), "Is Social Evolution Lamarkian or Darwinian?", in LAURENT J., NIGHTINGALE J. (eds.), *Darwinism and Evolutionary Economics*, Edward Elgar, Cheltenham.

ICE (2003), *L'Italia nell'economia internazionale, Rapporto 2002-2003*, Istituto Nazionale per il Commercio Estero, Roma, Giugno.

IDSE-CNR (1999), *Trasformazioni strutturali e competitività dei sistemi locali di produzione*, Franco Angeli, Milano.

IPI (2002), *L'esperienza italiana dei distretti industriali*, Istituto per la Promozione Industriale, Roma, www.ipi.it.

LANDES D.S. (1999), *The Wealth and Poverty of Nations. Why Some are so Rich and Some so Poor*, W.W. Norton and Company Inc., New York.

LANE D.A. (2002), "Complessità e interazioni locali. Verso una teoria dei distretti industriali", in: QUADRIO CURZIO A., FORTIS M. (a cura di), *Complessità e distretti industriali. Dinamiche, modelli e casi reali*, Fondazione Montedison 'Comunità e Innovazione', Il Mulino, Bologna.

LIMOGES C., MÈNARD C. (1994), "Organisation and the division of labour: biological metaphores at work in Alfred Marshall's Principles of Economics", in: MIROWSKI P. (Ed.), *Natural Images in Economic Thought*, Cambridge University Press, Cambridge.

MAGGIONI M.A. (2002), *Custering Dynamics and the Location of High-Tech Firms*, Springer, Heidelberg.

MALERBA F., LISSONI F. (2002), "I caratteri fondamentali del sistema di innovazione italiano", in: QUADRIO CURZIO A., FORTIS M., GALLI G. (a cura di), *La competitività dell'Italia. Volume I: Scienza, ricerca e innovazione*, Centro Studi Confindustria, Il Sole 24 Ore, Milano.

MARKUSEN J.R. (1995), "The Boundaries of Multinational Enterprise and the Theory of International Trade", *Journal of Economic Perspectives*, 9-2.

METCALFE J.S. (1998), *Evolutionary Economics and Creative Destruction*, Routledge, London.

MONTINI A., ZOBOLI R. (2003), Infrastrutture e sviluppo economico in Italia: un'analisi degli indicatori a livello comunale, in: PACE G. (a cura di), *Innovazione, sviluppo e apprendimento nelle regioni dell'Europa Mediterranea*, Franco Angeli, Milano.

NELSON R.R., WINTER S.G. (1982), *An Evolutionary Theory of Economic Change*, Harvard University Press, Cambridge Mass.

PANICCIA I. (1998), "One, a Hundred Thousands of Industrial Districts: Organisational Variety in Local Network of Small and Medium-Sized Enterprises", *Organisation Studies*, vol. 19, 4.

PORTER M.E. (1998), "Clusters and the new economics of competition", *Harvard Business Review*, November-December.

PUTNAM R. D. (1993), *Making Democracy Work*, Princeton University Press, Princeton.

QUADRIO CURZIO A., FORTIS M. (a cura di) (2000), *Il Made in Italy oltre il 2000. Innovazione e comunità locali*, Fondazione Montedison 'Comunità e Innovazione', Il Mulino, Bologna.

QUADRIO CURZIO A., FORTIS M. (a cura di) (2002), *Complessità e distretti industriali. Dinamiche, modelli e casi reali*, Fondazione Montedison 'Comunità e Innovazione', Il Mulino, Bologna.

QUADRIO CURZIO A., FORTIS M., G. GALLI (a cura di) (2002), *La competitività dell'Italia. Volume I: Scienza, ricerca e innovazione*, Centro Studi Confindustria, Il Sole 24 Ore, Milano.

ROLFO S. (a cura di) (1997), *Innovazione, piccole imprese e distretti industriali*, 3° Rapporto CNEL/CERIS-CNR, Rome.

ROLFO S. (a cura di) (2000), *Innovazione e piccole imprese in Piemonte*, Franco Angeli, Milano.

RUGMAN A.M. (1980), "A New Theory of the Multinational Enterprise: Internationalization versus Internalization", *Columbia Journal of World Business*, Spring, 23-29.

SAVIOTTI P.P., METCALFE J.S. (Eds.) (1991), *Evolutionary Theories of Economic and Technological Change*, Hardwood Academic Publishers, Chur.

SIGNORINI L.F. (2000), *Lo sviluppo locale. Un'indagine della Banca d'Italia sui distretti industriali*, Donzelli, Roma.

UNCTAD (2001), *World Investment Report 2001. Promoting Linkages*, UNCTAD, Geneva.

WTO (2003), *World Trade Report 2003*, WTO, Geneva.

Part I. Theoretical Perspectives

Regional governance and economic development: A European perspective[1]

Gilberto Antonelli

Department of Economics, University of Bologna, Strada Maggiore 45, 40125, Bologna, Italy

1 Introduction

In current debate the economic impact of local development can be traced back to two strains of perspectives and outcomes which are on the whole schizophrenic. On the one hand, when questions regarding disparity and inequality, or conversely cohesion and convergence, are being considered, what emerges are above all the potentially negative consequences, risks and uncertainties associated with it. On the other, when under investigation are the prerequisites of system competitiveness and therefore issues concerning the development of infrastructures, finance and good administrative practices, capital and technology, labour and human resources, natural and environmental resources or, in other words, the innovation in broad production factors, then what emerges are above all the potentially positive consequences, opportunities and trust associated with it.

The problem is that local and regional development assumes different meanings and dimensions: ranging from the urban or metropolitan level to that of the district area or province, and furthermore to the level of the great European regions. Administrative and functional territorial divisions are intertwined. The size of the local systems under consideration plays an important role in inducing its economic influence. Moreover, as I will argue in sections 2 and 3, local development is an essential determinant of structural dynamics which change over time the general framework of the economy, with effects that, in macroeconomic terms, are not easy to be foreseen.

As far as structural dynamics is concerned, local and regional development seems to support a *only partially endogenous* adaptation of institutions to the needs of the economic system. If this is true, structural dynamics currently in progress are far more complex than those retained in some recent studies on the evolution of the Italian economic system (Onofri, 2001). The complexity of results is due at least in part to the complexities of the processes of adaptation in progress. Following the internationalisation of national systems and the

---

[1] This contribution is part of a research project carried out in collaboration with A. Quadrio Curzio (Dean of the Faculty of Political Sciences at the Catholic University of Milan) and M. Fortis (Ufficio Studi Edison and Faculty of Political Sciences at the Catholic University of Milan). I am grateful to A. Pedone for his comments and suggestions. The usual disclaim applies.

globalisation of markets, basic components and internal functions of each economic system are being redefined. Likewise, if it can be true that some adaptation toward equilibrium is brought about by structural and institutional change, also disequilibrium and discordant elements are to be accepted as its by-products. This rises many doubts on the full capability of the market forces to lead to a sustainable configuration and on the assumed transitory nature of disequilibrium.

Karl Polanyi (1957) made an in-depth analysis of the historical evolution of economic systems that were profoundly different from each other in time and space and he highlighted the importance of the reciprocity principle[2] plays in the development of all these systems, side by side with the exchange[3] and redistribution[4] principles. Thus, he focussed on organisational models for economic activities as being complementary, substitutes or alternatives to the 'pure market' or to the 'pure state', taking into account in some way the potential failures in the coordinating mechanisms based on exchange and redistribution and the need for economic systems to achieve sustainability through different combinations of these three principles.

This principle cannot therefore be neglected when debate is focussed on the forms and modalities of macro-economic coordination of complex systems, such as the national and local ones, operating in contexts characterised by uncertainty and instability. Furthermore, the importance that the principle of reciprocity assumes in modern economies[5] results in those local and regional areas in which the production processes are actually carried out, playing essential roles in economy and social cohesion.

Furthermore, if the competitiveness of a system is based, on the one hand, on the commensurability and standardisation of the goods and the services it produces, on the other hand draws continually upon its ability to create new varieties of goods, services and processes. Therefore, if one of the principal effects of the current technological revolution is the development of an economy based on information and knowledge, then the easier it is to gain access to codified knowledge (global) the more important the role of implicit knowledge (local) becomes in order to improve the variety and heterogeneity of a firm's output. Three forms of dialectics are central to the dynamics of the learning process that influences competitiveness: between the uniformity and variety of goods and services; between codified knowledge and tacit knowledge; between local and global contexts. Therefore uniformity and variety, codified knowledge and tacit knowledge, local context and global context in the production systems are not antagonists or opposing pairs but rather they become two poles of a self sustaining process in which the local and global networks play a strategic role.

[2] This principle belongs to a civil society and can be managed directly by it.

[3] This principle, in its turn, is implemented by the market and firms governance.

[4] Finally, the management of this principle is in the hands of the different levels of government (supranational, national, local, sectoral).

[5] Mention can be made to the non profit or social economy, but also to the rise of industrial districts.

The local systems of production and local labour markets share this ambivalence even if with differing intensities according to the types of human resources under consideration.

This chapter is organised as follows. In section 2, the principal stages of development for firms and local systems of production in the Italian economy are outlined. Their innovation capacity is assessed and new policy needs explored. Section 3 examines the main problems connected with the diversified process of regional development in the short term (with an emphasis on divergence) and in the long term (with an emphasis on convergence). With reference to the success stories in European regional development a model for benchmarking is then proposed. Although this model is not easily transferable elsewhere, it is still useful as a stimulus for improvement. In section four two positions regarding the allocation of the tools of political economy and the role of the regional governance are briefly presented. Adopting a scheme which allows me to examine the impact of regional development on the following fields of reference, attention is focussed on the role of: (i) the single market; (ii) stabilisation policies; (iii) direct redistribution policies; (iv) public goods. Two views emerge on the process of re-allocation of economic policies which is worth-wile to compare: the first is based on the so called. 'economic theory of federalism' and the second on the so called 'open method of co-ordination'. The conclusion is that the second method opens up the opportunity for greater institutional flexibility and efficiency in the context of increased potential growth.

2 Firms and local systems of production in the Italian economy

It is convenient to begin by mentioning some aspects of the evolution of the Italian economy in order to have a better understanding of the particular categories of constraints/needs and the opportunities/degrees of freedom in the process of opening up and integration in progress at present in Europe.

The production framework of the Italian economy is characterised by the dominance of the service sector and some groups of products in the manufacturing sector and by the prevalence of micro and small to medium sized firms organised in local systems of production (LSP) mainly, but not exclusively, concentrated in the Northern regions. In spite of the considerable number of studies carried out, the mechanisms leading to the formation, consolidation, evolution and transferrabilty of such structures are still open to interpretation. Moreover, the implications of this uneven production structure on the features of the relevant macroeconomic model and policies have not yet been clarified. While this state of the art leaves ample material for scholars to investigate, it does not, however, help to define economic policy for Italy, at a time when there is a particularly profound and exacting transformation of production in the making caused by the process of integration in Europe and the globalisation of markets.

2.1 The phases of development in the post-war Italian economy

The period between the end of the sixties and the seventies may be considered a watershed for two important phases of development. Within the first phase (1945-1967), which has been called the 'era of calm', can be found 'the years of reconstruction' (1945-1950), a 'phase of internal growth' (1951-1957), 'the years of the economic miracle' or 'of expansion and of restructuring due to international economic integration' (1958-1963), 'the years of depression and of reconstruction' or 'strike of capital' (1964-1968/69). As is well known, at the end of this phase there was a series of shock events the impact of which was devastating for the world economy: in 1969 the possibility to convert dollars into gold was suspended; in 1971 the dollar was devalued and new parameters fixed; in 1972 the price of non energy-giving raw materials rose; in 1973 the system of fixed exchange rates was abandoned; at the turn of 1973/74 there was the first big energy crisis. In addition to these events, it is also necessary to remember the outbreak of union strife and social conflicts that took place internationally in the second half of the sixties giving rise to what has been called a 'labour strike' era. In this phase a nucleus of large firms was being consolidated and asserting itself in the North West which was, in part, the inheritance of a pre-war industrial structure and in part the reaction in the post-war period to the great expansion of national markets and to the first attempts to push the industrialisation of the country, especially in the South, through business companies participated by the State.

The combination of these different events paved the way for the transition to a new 'era of uncertainty' (1968-1991) that ended with the fall of centralised economies in the East. For the majority of industrialised economies this led to a clear structural break. As far as the Italian economy is concerned the effects were more ambiguous and diversified depending on the sector under consideration. The continuity and persistence of the manufacturing sector went side by side with profound changes in the tertiary sector. Furthermore, the dynamics of the tertiary sector contributed to the differentiation of the dominating features of economic development in the various regions in the country.

The era of uncertainty started with a 'phase of transition' (1968/69-1971) and divided then into a series of phases closely linked to oil crises: a 'phase of panic' (1972-1974), a 'phase of accommodation' (1975-1979), a 'phase of long recession' (1979-1982) set off by the second energy crisis in 1979. This crisis hit the Italian production system at the very moment, albeit with a delay in comparison to its European partners, it was beginning to try to move from a model based on decentralised production and increase of the intensity of capital to a model based on technological innovation and the re-organisation of work. With the beginning of the eighties the radical modification in industrial relations, the greater discipline in exchange rate policy due to the setting up of the EMS and the permissive management of the budget[6] launched a phase of 'long recovery' (1983-1988) further helped by the counter oil crisis of 1986.

At the end of the eighties and the beginning of the nineties a new phase of transition began that seems to have led a new era we might call 'regulated uncertainty' and dominated by the globalisation of markets and by the process of European integration.

We can observe three fundamental trends in this phase: (i) a global market of financial business is being formed up along with the progressive globalisation of consumer and service markets, including the labour markets; (ii) a process of economic integration in Europe which, at least in this phase, is led by financial markets and markets for goods to which human resources and the labour markets are expected to contribute by providing flexibility and essential real cohesion; (iii) an economy based on services[7], on knowledge and on information is asserting itself which is highly conditioned by continual technological and institutional changes and, more recently, by the business cycle.

In a first phase, the process of globalisation, led by financial markets, seemed to be self sustaining and in no need of any form of regulation or help by the institutions. However, first the recurrent crises of the European Monetary System between 1992 and 1993, Mexico at the end of 1994, Asian countries, Russia and Brazil between 1998 and 1999 and then the ensuing crisis brought about by the September 2001 terrorist attack (not to mention the emergence of inadequacies in structure of the so-called 'American model'[8]) have highlighted the importance of regulation, of efficient institutions and also policies towards the stabilising of macro-economies. In brief, also globalisation necessitates governance.

The era of uncertainty and that of regulated uncertainty have thus, on the one hand resulted in the consolidation of some of the peculiarities of the Italian productive system with respect to sector specialisation, the size of firms and organisation of local systems of production (LSP), and, on the other hand, in the beginning of profound endogenous transformations and innovation in the latter.

---

[6] With the approval of the Government many companies had recourse to the "Cassa Integrazione Guadagni" and early retirement. Heavy public transfers to families and firms, social security measures and wage policies in the public sector determined a sudden increase in the public deficit and the formation of an harmful shelter to structural adjustment.

[7] In Europe this sector represents nearly 70% of employment and in Italy over 60%.

[8] Here reference is made to the difficulties, however transitory they may be, of the 'new economy' which are further heightened by the crisis in the governance system and credibility of important financial organisations and business companies in the USA.

2.2 The competitiveness of industrial districts

In several of his studies[9] Fortis has pointed out how the Italian industrial system presents a very peculiar production specialisation in comparison to the other major industrialised countries. In fact, according to Fortis, Italian manufacturing system is principally based on the production of so-called 'made in Italy' goods (fashion, furniture, Mediterranean food products and its associated mechanical systems) in which the SMEs and the industrial districts (or LSP) constitute the supporting framework.

The phenomenon of productive specialisation's that are concentrated in specific geographical areas which are far from the big towns and cities and from those areas of traditional industrial importance has reached such dimensions as to draw the attention and admiration of observers from all over the world. The economic weight of the industrial districts in the various sectors has been, in fact, such as to consider them as important as large industrial groups of international reputation in terms of employment, productivity and their incision upon world markets.

We can count more than 200 local systems of production that specialise in typical 'made in Italy' sectors mainly located in Northern and Central Italy and to a lesser extent in the South. This number includes not only the main districts but also areas which specialise in one (rarely) or a few (more frequently) leading firms in a given industry and its connected businesses. The total employment in these local systems of production has been estimated to be, according to data from the 1991 census, 2.14 million employees, equal to about 58% of the total work force employed in businesses typifying the 'made in Italy' manufacturing sector which in its turn represents about 71% of the total of those employed in the Italian manufacturing industry (this being 5.26 million in the census year). In other words, more than 2/5 of the Italian work force is concentrated in these areas of specialisation or districts.(Quadro Curzio and Fortis, 2000; Fortis, 1998 and 2000).

In the opinion of Fortis 'made in Italy' represents the strongest force for employment that the country currently possesses. In the nineties while employment in the sectors such as the automobile industry, chemical industry, electronics, paper and cement industries fell sharply, in many sectors of 'made in Italy' employment remained stable or even increased such as happened in the shoe making industry, packaging machines, ceramic equipment industry, bathroom accessories, women's stockings, the tanning industry, glasses and many others. To give some example figures relating to those sectors principally based on district production (with one or more specialised districts) we can note that between 1993 and 1996 in the Italian sector of the packaging machine industry the total employment roses from 13,400 to 16,000; in the same period employment in the wood machine industry increased from 10,500 to 11,000, in the shoe making industry the number of those employed went up from 117,000 to

---

[9] Among which are those carried out by the Montedison-Cranec Observatory of the Catholic University of Milan and the Montedison Foundation for 'Community and Innovation'.

124,500; in the ceramic equipment industry employment increased from 2,600 to 3,200. In the Belluno district, a world leader in the glasses and frames sector employment rose from 7,700 to 11,700 between 1992 and 1996. Of no less significance is the faucet valve industry in Piedmont where the number of those employed (excluding those employed in associated businesses) rose from 6,000 in 1994 to around 6,900 in 1997 (Fortis, 2000).

Besides being an important factor for employment stability, 'made in Italy' plays a leading role in Italian exports: in 1997 the typical 'made in Italy' sectors recorded gross surplus of over 150 billion lire, enough to cover the deficit for energy, agricultural and industrial raw materials for the entire country, not to mention the deficit of other manufacturing sectors ensuring thus a trade surplus of more than 51,000 billion lire. This 'made in Italy' profit fell slightly in 1998 and to a greater extent again in 1999 as a result of a slowing-down in world trade but still remained a cardinal point in our balance of trade. The trends of the first 8 months in the year 2000, in confirmation of this fact and of the vitality of the small and medium industries and local districts, show a decided improvement in the surplus obtained by 'made in Italy' with an increase of no less than 8,500 billion lire compared to the same period of the previous year.

It has been estimated that about a third of all Italian exports comes from industrial districts or local production systems. If the proportion of some Italian industrial districts in world exports of some manufacturing products is analysed for the year 1996 (Fortis, 2000), it can be noted that the ceramics district of Sassuolo and the stocking district of Castel Goffredo in the province of Mantova manage to reach quotas of around 35-40% in their respective world markets. Prato and Biella, for woollen goods and Como for silk products reach quotas of almost 20% of the world market in their sectors. Many other districts, such as that of packaging machines in Bologna, goldsmiths in Arezzo and Vicenza, those of Novara and Lumezzane for fawcets and valves, Rimini-Forli-Pesaro for wood cutting machines, glasses in Cadore, Udine and Bari-Matera for chairs and sofas and Alto Livenza for furniture all represent quotas of between 10-15% in their respective business sectors. Besides these great sectors of traditional importance – and of particular fame – there are local systems of production that in spite of their small dimension occupy leading positions in world markets. These are districts specialising in special niche markets with turnovers of some hundred billion of Euros such as the button district of Bergamo, bicycle saddles in Vicenza and scissors in Premana near Lecce.

2.3 Questions of interpretations and the new needs of economic policy

Although it is relatively simple to trace the connections between post-war economic phases of development and changes in the structure of production, our understanding of the ways in which these have evolved is still limited, especially in terms of their territorial location and institutional background. The emphasis placed on the analyses of performance in the short term has led to an overshadowing of structural analysis which is central to any interpretation of performance as is suggested by the Italian economy's surprisingly frequent ability to survive and to adapt when faced by the difficult transitional phases over the last forty years. Therefore, we need to take into account not only the structure by economic sector but also the territorial and institutional features, placing an emphasis on characteristics which, if they have been consolidated over recent years, can lead us to an awareness of those factors which have in the long term constituted capabilities at a local level. This awareness is significant not only in order to comprehend some phenomena that are still not clearly explained, but also in order to be able to recommend a course of action for the main, unsolved structural problems for the Italian economy, including those conditioned by the dualism and regional differences to the process of European integration.

Following this line of argument, the production structure of the Italian economy may be considered only *in part* to be the outcome of responses by our economic system to the internal and international changes that came about in the years between the Seventies and more recent years. In other words, the particularity of our production system must *also* be interpreted as the expression of deeper and more stable mechanisms which operate in different ways in different stages of economic development throughout the entire century.

Several analyses agree that, however, the phase of spontaneous growth of the LSP is almost over and there are some signs of a potential crisis also and above all in some 'successful' areas, a crisis in the system is tending have an effect on the training of human resources, the reproduction of the entrepreneurial class and also regards the emergence of negative externalities for example the saturation of the infra-structural networks and the environmental capacities of local areas. Besides this, the new elements at an international level and the growing integration of micro firms, of SMEs and of Italian LSPs into the European economy bring about phenomenon that have, so far; hardly been analysed. Examples of this are outsourcing, decentralisation of production, delocalisation at an international level which have been effectuated by many small and medium sized companies in recent years. These generally result in the loss of some elements that characterise the production system when it is integrated locally and seem, in some cases to form 'internationalised districts'. In this case the coordination and services of these local companies, in the context of the network of firms subjected to the cycles that are typical of international systems, are accentuated as has been seen in the consequences of the recent international crises.

The limits in the interpretation of long term dynamics and the changes taking place appear even more significant at a moment in which the centrality of the SME and the LSP, normally associated with the Italian economy, seem to be extending on a European scale. The economic weight of the small and medium sized companies in their creation of value added and employment is tending to grow in the majority of advanced countries. Although there is a difference in the criteria adopted to realise their surveys, some OECD and EUROSTAT estimates show that micro firms and SME (less than 250 employees) represent about 99% of the number of firms in the EU, with employment shares ranging from 40% to

80% and over, with contributions to the GNP ranging from 30% to 66%. Such indicators are relatively higher for Italy where, besides representing 99% of the total firms, the SME represent 78.5% of total employment (compared to a EU of 15 average of 62.8%) with peaks of 95.2% of total employees in the commercial sector and with shares that are higher than the EU average in all sectors. Furthermore, those entities that are classified as self-employed could be taken as being micro companies, the importance of which is rising in many European countries, reaching a peak in Italy with about 18% of total employment at the beginning of the Nineties. To this can be added the fact that in recent years output from micro and small and medium firms has grown more rapidly than that of the large companies in most of the European countries.

These data show that a complex transformation of the European economy is in progress in which the relative expansion of small firms does not simply correspond to the rise of micro-firms and of SMEs as such, but also to intense technological changes that reduce the importance of technical scale economies and favour the 'widespread' company. The phenomenon contradicts the traditional view of the small size of a firm as being a pathological symptom of economic systems and leads one to think of micro-companies and SMEs not merely in terms of size, but as part of productive system in which small size plays a specific and rational role that goes far beyond that of saving on costs by decentralisation.[10]

Although these tendencies have brought about a growing international interest in the 'Italian model' as manifested in its local systems, they have, conversely, suggested considering the development of the LSP as a possible answer for Italy to the dualism between North and South which finds no equivalent in any other European country. This is happening in a phase distinguished by the contradictory co-existence of signs of economic re-birth in some Southern areas and the hypothesis of growing divergence between the Italian regions[11]. The LSP have therefore become, also due to the absence of an adequate understanding of the mechanisms which formed and evolved them, a centre of gravitation for economic policy.

Adhesion to the European single currency and the need to reduce the national debt place strong limitations on the discretional use of the tools of macroeconomic policy over the short term and increase, correspondingly, the importance of structural and infra-structural policies. The defining of structural policies in Italy still presents significant ambiguity. The 'structural' policies carried out in the past in the under developed areas or sectors may be interpreted as measures aimed at compensating the territorial differences in terms of employment, wealth and average production costs. The instruments used, ranging from support to supply (investments or incentives that might also be fiscal in nature) to backing for demand (increases in public spending or income transfer) may have lessened the inequalities but they have not eliminated the causes and

---

[10] See Harrison (1994) and Antonelli and Zoboli (1999).

[11] Of course, the extension of the underground economy contributes to rise doubts, in a way or another, on the available statistical sources.

have not exerted any influence over the production potential in the backward areas or sectors. A more effective, different approach and one that is more in line with the European Union's plan of action is to adopt those measures aimed at converging the different local areas and sectors the companies operate in. This objective aims at eliminating differences in efficiency, stimulating technological and organisational innovations. More specifically, it constitutes acting mainly as a lever on the social capacities and environmental conditions in which the companies operate, on the capabilities of single firms and individual workers. Such an approach presupposes, however, being able to draw a map of the economic vocation and relative constraints in fine detail.

At the same time recent years have seen the implementation of several policy actions aiming at the SMEs and LSPs which presuppose the same analytical capacity but which are often carried out without it, resulting therefore in that what appears to be a confused variety of interventions.[12]

The micro-firms, the SMEs and the LSPs have therefore become pivotal for a very wide range of policies, all of which goes to show that, apart from the overlapping and many inconsistencies, they are heterogeneous, systemic and open phenomena. Conversely, some of the important provisions previously mentioned seem to give rise to – given the central role assigned to the industry and the criteria used to identify industrial districts – a fossilisation of the LSPs as they have come to be arranged in last decades. This is taking place just as increasing evidence comes to light of how the LSP are intrinsically changeable and currently tend to take on a more complex set up internally, acquiring a richer range of relevant components (services, institutional infrastructures or social capital, local administrations), more integrated with the outside environment giving greater importance to interactions with other LSPs, through the mobility of human resources and a growing role played by to international variables. More specifically, with the change of political and administrative setting towards a

---

[12] In the so called 'White Book', the European Commission attributed SMEs a strategic role in the growth and competitiveness in European countries, while the guide lines towards employment which emerged from the 1997 Luxemburg summit include among its priorities the reduction of obstacles and encouragement of the setting up of SMEs. In many other fields concerning community policy and interventions, Structural Funding and EIB initiatives in particular, attribute central importance to SMEs, both as a possible starting point for the processes of development of backward regions and also as a model typically assumed by highly innovative companies. Besides national incentives, in the light of national economic policies all the regional authorities have laid out a large number of incentives for minor companies and many have given priority to SMI innovations. Many policies have been directed towards improving levels of employment and the increasing opportunities for young people and women in management. "Negotiated programming", such as "programme contracts" (law 64/86, law 488/92, approved CIPE 21/3/97), and also "territorial pacts" and "area contracts" (approved CIPE 21/3/97), are aimed at promoting local systems via co-operation between companies and institutions. Lastly, specific policies for industry and innovation have been elaborated for industrial districts. The 317/91 law, in particular, is beginning to take form, in spite of significant difficulties for regional authorities to identify the very districts and to implement interventions in areas which do not correspond to the traditional administrative ones.

more decentralised-federal model, local governments assume roles and duties that they had not previously held in the phase of the 'spontaneous' rise of districts or LSPs. Such a new institutional framework appears to be particularly relevant for those areas, especially in the South, where the political and administrative capacities are traditionally more limited but which at the same time manifest new productive vitality.

It is necessary to 'enlarge' the network of interactions on which the vitality of the SLP is based. This means promoting, also by means of the realisation of material and immaterial infrastructures, the amplification of and opening up of economic relations for production and trade beyond local area without losing the cohesion and the capacity for endogenous renewal. It is necessary to shorten the time needed for political procedures and strengthen functional economic policies for the extension of the network. All this leads me to maintain that it is regions that are destined to become 'adequate borders' for local governance and competitiveness.[13]

3 Regional economic development and convergence/divergence

Empirical studies (European Communities Commission, 2002) shows that there are significant economic differences between the current members of the EU, even if they have diminished considerably since 1998. Indicators of convergence principally concern those countries which receive benefits from the cohesion funds (Greece, Ireland, Portugal, Spain).

The disparities between regions of EU15 (Table 1) have diminished, even if to a lesser extent when compared to those between nations and, in some states, these disparities have widened. In any case, enlargement will bring about greater disparities and making up for lost ground will clearly be a long process.

The regions with the highest growth rates are mostly among those who are candidate members (Poland, Czech Republic, Slovenia). As far as the member countries are concerned there are areas of accelerated growth such as the regions surrounding the capitals Stockholm, Helsinki, Lisbon, Madrid, more widespread areas of Ireland, England and the Netherlands.

The disparities in the rate of employment figures have continued to diminish, even if in Italy, along with other Southern Mediterranean countries it is still under the threshold figure of 70% set as the target for 2010 by the European Council of Lisbon. The extent of these differences is due to an under exploitation of growth potential and persisting social inequalities.

Unemployment has decreased at a national level in all the member states in 2000, but the regional differences are still considerable and in Italy are great[14]. In

[13] See Maggioni and Bramanti (2001) for this concept.

[14] The difference between the rate of unemployment in Calabria (the highest) and Trentino–Alto Adige (the lowest) is of 25%.

addition to this long term unemployment among young people and women is particularly high in Italy.

Table 1. The gap between the Union's richer and the poorer regions, 1989-1999 GDP per head (SPP) as a percentage of the EU average

Regions	*EU15*	*EU15*	*EU25*	*EU27*
	1989	1999	1999	1999
10%+	153.3	157.4	70.3	176.5
10%-	55.6	61.7	38.3	30.7
Ratio	2.8	2.6	4.5	5.8
25%+	132.9	138.1	147.2	152.6
25%-	66.5	68.7	54.2	45.6
Ratio	2.0	2.0	2.7	3.3

10%+ and 25%+: the regions with greater GDP per head which represent respectively the 10% and the 25% of the total population;

10%- and 25%-: the regions with lower GDP per head which represent respectively the 10% and the 25% of the total population.

Source: European Commission 2002 on Eurostat data and DGREGIO processing.

Factors relating to regions' competitiveness are reflected by the sectoral structure of employment. The service sectors have acquired a key role in the knowledge based economy, while much of agriculture and traditional industry is in need of a significant overhaul. The lack of young people and the low level of education and training reveal problems regarding the supply of a crucial resource: human capital. Research and innovation are held to be of such importance that the Commission has adopted a Plan of action for regional research projects in Europe.[15]

The key fields that regional authorities are to aim at are the following: (i) administration and technical skills for the management of the economic development strategy; (ii) finance at a particularly difficult time for public budget; (iii) the economy, because financial investments by the EU must stimulate national or local, public and private investments and not substitute them.

Table 2 shows some figures for some selected regions. They are successful regions such as Bavaria in Germany, Wales in Great Britain, Emilia Romagna and Veneto in Italy. In brief, they represent success stories based on the capacity, in various stages of development, to extend local networks and cut the time taken for political procedures.

In order to simplify the analysis and to focus one's attention it is useful to have a point of reference as an example, in other words a benchmark. The case of Bavaria lends itself to this easily (Romano, 2002).

[15] October 3rd 2001.

Table 2. Main indicators for the selected regions

Region	*GDP per head (in SPP, average 1997-98-99, UE15=100*	*Percentage of employment in services, 2000*	*Applications for European patents (per million of inhabitants, average 1998-99-2000)*	*Unemployment rate among young people, 2000*	*Employment rate (15-64 years old), 2000*	*Percentage of population <15 years old, 1999*	*Percentage of population (25-59 years old) with a low level of education, 2000*
Emilia-Romagna	131.3	58.5	146.6	13.0	66.0	11.1	49.0
Bavaria	122.6	60.7	439.5	5.1	71.3	16.5	20.0
Catalogna	98.6	60,7	51.3	18.6	63.0	14.1	56.0
Wales	82.2	69.3	63.0	14.8	66.7	19.1	22.0
Veneto	121.2	54.0	98.0	9.1	62.3	13.2	54.0

Source: European Commission 2002 Eurostat data and DGREGIO processing

This is a region that has understood that human resources are the real main resources, the raw material in today's economy. In Bavaria [16]the region's GDP at a regional level has increased at a higher rate than the GDP at a national level over the last thirty years. In 2000 the investment share on the public spending was only a little less than 20%, more than 6% of the employees were in the high tech sector, the rate of regional unemployment (5.3%) was almost 50% less than the national average (9.6%).

From being a mainly agricultural region it has rapidly become 'high tech' with a highly differentiated production system. It is a world leader in the fields of biotechnology, mass media finance and information technology.

At the beginning of the 90's an intense privatisation programme was launched, the proceeds of which were re-invested into the educational system and in projects aimed at the development of the biotechnology and information technology industries. These investments have given rise to an innovative network of around 30 thousand high tech companies and 120 institutes of science.

Its educational system is one of the best in Germany. It is the headquarters of the European Union Patent Office and has a wonderful Science Museum.

The local government is a stronghold of economic policy. Its managerial class did not hesitate to take protectionist measures or to attract foreign investments. Twelve offices have been set up to deal with international relations with the aim of attracting foreign capital. In fact around 1,030 firms have their headquarters in Bavaria[17].

Bavaria provides advanced material and non-material infrastructures and offers generous fiscal schemes. Furthermore, it is the only German region after the Second World War to become a *Land* donor. Its agricultural origins have enabled it to reduce the expense involved in a costly industrial conversion that regions such as the Ruhr and Baden-Württemberg have had to uphold.

This is not to say that there have not been some rather controversial[18] rescue operations of some businesses and the model would be difficult to apply at a national level in Germany or in other European regions. This would in fact seem to be the general rule as far as local systems of production are concerned.

However, even if the model is not transferable to other areas it does still remain useful as a stimulus towards improvement.

If the appropriate benchmark is this, then, notwithstanding the many merits that that regions such as Emilia Romagna can boast of, they cannot rest on their laurels, above all if we are to look at not so much at the levels of the variables but at their rate of variation, which expresses the dynamic concerned.

The situation that was shown in the report published by the Federation of Industry in Emilia Romagna (1995) is still valid today albeit approximate.
According to this report the GDP per inhabitant is very high and rising, but at a slower rate compared to the more industrialised regions in Europe and Italy. The

[16] The biggest and oldest German 'Länder'.

[17] 600 from the USA and 90 from Japan.

[18] Grundig and Schneider in the electronics sector, for example.

rate of business is very high but on the decrease. The material and non material infrastructures are of a good level and well equipped and significant investment has gone into their upkeep also in recent years.[19] There are, however, signs of obsolescence and inadequacy when compared to competitor regions. Exports are particularly high but there appear to be signs of slowness in penetrating new markets. The training system, while being effective and diffused, is not always capable of anticipating the requirements in terms of human capital in time.

The fact is that Emilia Romagna shares some characteristics with the rest of the nation that make it slow to adapt. In fact, in spite of the great structural changes that have seen important sectors for example in the textile, clothing and food industry lose ground to mechanical engineering, the 'productive specialisation' in Italian and regional manufacturing system has turned out to be very persistent over time (IDSE-CNR, 1999).

Of equal persistency appears to be the high degree of provincial and regional polarisation bearing witness to a trend of unbalanced growth. In 1911 five regions (Lombardy, Piedmont, Veneto, Emilia-Romagna and Tuscany) provided work for more than 63% of the total number of employees and in 1991 the same regions provided work for over 67% of the total. Only during the years of the economic boom was there a 'relative convergence' recorded. With the 1970s the process of divergence resumed.

The regional impulse towards development can be identified as being located within the 'industrial triangle' as far as the first half of the century is concerned up until the 1960s. Afterwards the centre of gravity changed in a definite move towards the North-East. The latter has the highest average rate of variation of employees per decade. As far as single territorial divisions are concerned, a first analysis shows that there are quite diversified models of development in terms of sectors. In fact, there is a high degree of specialisation both in the North East and Central Italy, but accompanied by large firms in the first case and small to medium firms in the second. High diversification of production over the territory characterises the North East model while the model applicable to the South apart from being less efficient in terms of growth appears to be characterised by 'production de-specialisation'. There is therefore the co-existence of heterogeneity in production accompanied by scant specialisation. Besides the significant impact made by variety, more generally the differentiation of systems of production on performance[20] of local districts have been highlighted.

From the empirical analysis into tertiary sectors (IDSE-CNR, 1999) it emerges that on the whole it is the services to businesses and people on the one hand and the public administration on the other hand the divisions that have sustained the major impetus to the rise in employment in the tertiary sector in Italy. However, utilising empirical evidence to get a whole picture of one field of reference can mislead and result in distortion. What is read as an aggregate is in fact the result of very different dynamics in highly heterogeneous local systems.

---

[19] For example, the rapid development of the university system throughout the region.

[20] At least that measured in terms of the rise in employment.

In the context of a significant rise in the number of firms and the employees, the structure of the service sector underwent great changes in the period between 1951-91. In general there was a rise in the average size of firms, with the exception of the transport and insurance sector and at the same time there was a dramatic growth in 'other services', including some of the more innovative businesses such as services to businesses, while the sectors linked to commerce remained stable. Such changes are differentiated in the two great phases of development of the Italian economy which saw 1971 as a watershed. While two thirds of the increase in the number of firms is concentrated in the period 1951-71, only a fraction under two fifths of the increase in the number of employees is of the same period. An increase of similar dimensions is recorded for the decade 1971-1981. There seems, therefore, to have been a compensatory absorption by the private tertiary sector in a phase of a difficult adaptation to modifications brought about by the repeated shocks events that have taken place since the end of the Sixties.

Examination of the changes in the geographical distribution of the structure of services reveals two main phenomena. Some divisions show increases both at a national level and also in the different circumscriptions: wholesale trading, some other services for commerce, credit and other services. Transport shows decreases in all areas except the South. The different circumscriptions are clearly differentiated as far as trends in the other divisions with the same sectors which are decreasing in some regions and increasing in others.

These trends may be re-interpreted by using a classification based on the infra-structural function of services. In some divisions of the tertiary, services of a prevalently tradeable nature are provided. In a part of these the rise in productivity, as regards supply, is high as so far as companies are in search of higher market quotas rather than an increase in the size of the whole market. It can be added that this subset of services tends to generate 'long network' operations. In other divisions the production of services which are of the prevalently tradeable type have only modest increases in productivity made possible by introducing information and telecommunication technology. In this case too they are services which tend to generate long network interactions. Lastly, there are services which are prevalently non tradeable which remain at a low level of productivity also due to their imperviousness to new technology. These latter tend to generate 'short network' operations.

An analysis of the specialisations in the various geographical areas based on these categories provides some interesting information. From 1981 to 1991 the North-East tended to specialise in and consolidate itself in that range of sectors defined as tradeable with high productivity rates, or 'long network', and with systemic impact. The North-East seems to have specialised and consolidated in that range of sub sectors defined as tradeable but with a low increases in the productivity rate, or 'long network' and with systemic impact. Furthermore, the potential for infra-structural services seems to have grown even if probably not in proportion to the increase in demand. The Centre seems to collocate itself half way between the range of sub sectors of a tradeable type with only a modest growth in productivity rates , or a 'long network' and the range of sub sectors of the tradeable type with low productivity rates or 'short network'. In these areas too, the impact of services seems to be mainly systemic. Furthermore, in concomitance with the North-East, the Centre hosts influential employers' associations and unions although this is of greater importance in the North-East. In the South a strong connection is shown indirectly between the services offered by private and public sectors and a greater presence of traditional short network services but with some differentiations internally. The South-East, in particular, appears to be collocated at a half way point too between the range of services of a tradeable type with modest growth in productivity rates, or 'long network' and the range of services of a non tradeable type with low productivity and 'short network'. The impact of services seems to be mainly systemic. The South–West, on the other hand, maintains specialisation in those sub sectors of a non tradeable type with low productivity, or 'short network' and with internal effects.

As has been first noted and then further investigated in this section, in current debate regional development is connected to two types of perspective and economic effects which are contrasting and upon first appraisal schizophrenic.

When problems of disparity and inequality or conversely cohesion and convergence are taken into consideration then potential negative consequences, risks and uncertainties are revealed.

When, on the other hand, the roots of the competitiveness of the system are taken into consideration and so therefore issues relating to the development of infrastructures, finance and good administration practices, capital and technology, labour and human resources, natural and environmental resources then positive consequences, opportunities and certainties come to the fore. In particular, in the knowledge based economy, innovation in broad production factors revolves around knowledge.

This dichotomy highlights phenomena with foundations in both cases on empirical evidence and realistic overviews. But how can the two be reconciled?

To answer this question it is helpful perhaps to note that: (i) the two types of issues differ in terms of time span – while disparity and inequality relate to the short term, the sources of competitiveness of a system refer to the long term; (ii) the components of the economic system the problems refer to are themselves different – while disparity and inequality deal with the effective exploitation of the productive potential, the sources of competitiveness of a system regard above all the future development of that potential.

This means that we are not faced with evaluations that can be added together and from which the final sum can be mathematically deduced. It is advisable to proceed by examining the two types of phenomena side by side and to avoid misleading simplifications.

4 The allocation of the tools of economic policy and the role of the regions

The terms *cohesion, convergence and integration* have become the 'three-pronged' mission for European construction (Maggioni and Bramanti, 2001).

As often happens in very complex processes there are two main approaches that may be adopted when evaluating the most appropriate institutional framework for the sustainability of the process of convergence. In our case, in both approaches a crucial (implicit or explicit) role is played by human resources and by the labour market.

For both approaches a single labour market in Europe is difficult to conceive, at least if we consider the time required for the political decisions made and procedures to be carried out. Local labour markets are generally recognised to be of relevance. Indeed, it is often the local labour markets range that enables us to define the borders of local systems of production. Furthermore, labour markets are in constant evolution, linked as they are to demographic, economic and institutional dynamics. Moreover, the local dimension of the labour markets is even more relevant when the quality of its human resources is crucial; and this can wield significant influence on the competitiveness of local systems.

In a top-down approach the focus is placed on to the best possible allocation of political competences in terms of national and European competitiveness.

Local borders are to some extent residual. The advantages of centralisation are constituted mainly by the absorption of spillovers and of the exploitation of economy of scale. While the advantages of decentralisation are constituted by the capacity to make heterogeneous preferences and to exploit local information.

Table 3 represents a possible pattern of such an approach (Tabellini, 2001). The regional and local governments have little autonomy and play only a residual role in the process of integration. The labour markets act as a 'sponge' to what happens in the rest of the economy. In some way, the labour markets are destined to be dominated by conservatory organisations which defend particular privileges and interests.[23] While waiting for the beneficial competitive effects of a single goods market to be felt on the degree of efficiency of the labour markets, it is better to limit oneself to interventions aimed at increasing of the mobility of workers between nations.[24] Unemployment is mainly a political problem and not an economic one and it is the flexibility and mobility of the labour force that

---

[21] This is, for example, the case of 'local systems of labour' delineated as being so by their auto-sufficiency. This term is used to indicate (by ISTAT and IRPET) the capacity of an area, including several localities, to concentrate a certain number of jobs and to provide work (or to change jobs) for the majority of the local residents without their having to move home.

[22] Reference is made here to the variety of competences in different areas and the so called 'tacit knowledge' or 'local know-how'.

[23] As happens in agricultural markets and CAP.

[24] Accepting also possible segmentations in the labour market in the phase of transition.

represents the primary tool for economic convergence. Therefore, interventions above the national level must be limited to partial corrections to health insurance systems and other components of the social state in so far as they hinder mobility.[25] With this aim private health schemes may be arranged at a European level and made transferable between nations. Benchmarking may be useful. But the redistribution of resources by the EU should be limited to transfers between national governments.

In a bottom-up approach territorial governance is called upon to give tangibility to the principle of vertical and horizontal subsidiarity. Regional and local governments are autonomous and play an influential role in the construction of economic policies (Table 4).

The labour markets are very specific and they distinguish the entire territory they operate in clearly. This has even greater validity in the knowledge based economy. Inequality and divergence can be contained by organisational solutions and the contractual agreements between companies and workers. Participation can be a fundamental tool to stimulate performance and increase competitiveness for companies. 'Intermediate' organisations of entrepreneurs and workers can play an important role towards cohesion and economic development. They can have a more or less innovative role but are embedded in the mechanisms of markets. The local dimension then strengthens their role.

Table 3. Prescriptions of the economic theory of federalism[a]

Columns = Level of government appropriate for the different tools of economic policy
Rows = Degree of decentralisation of structures and behaviour

	European Union	Nation	Regional government
Low	Policies aimed at reinforcing the single market[b] and international trade Regulation policies Monetary policy Stability pact	Cohesion funds Redistribution policies (lump sum compensation transfer) Levels of national social protection[c]	Repeal of structural funds (in EU 15) Gradual elimination of the community agricultural policy Allocative functions
High	Policies for the supply of pure public goods (defence, foreign policy, law and order, environment) Policies for the development of private pension schemes (to improve mobility of labour) and immigration Benchmarking	Fiscal policy (slackening of the Stability Pact) Policies for the supply of impure public goods (education, Vocational training, culture, R&D) Policies for infrastructures	Health policies Transport policies

[a] These prescriptions take into account distortions due to policy failure and governments' policy expedience; [b] With special reference to public utility services (telecommunications, postal services, energy) and financial services; [c] Reference is made here to the rejection of the so-called social charter.

[25] Also immigration needs to be regulated and managed from the centre.

A common minimum basis of rights (of protection and assistance) at a European level gives a greater sense of belonging and motivates participation.

Unemployment is also an economic problem, given that it leads to waste and destruction of present and future human resources.[26] Mobility is made possible by adequate infrastructures as well as by territorial differentials in wages. Efficient and sustainable systems and contracts must know how to make the best of the human resources available both in order to create greater variety of products and processes and also so as to reduce the monitoring costs incurred by single firms.

Table 4. Prescriptions deduced from the open method of coordination

Columns = Level of government appropriate for the different tools of economic policy
Rows = Degree of decentralisation of structures and behaviour (horizontal and vertical internal subsidiarity)

	European Union	Nation	Regional government
Low	Single market Regulation Monetary policy	Redistribution policies (lump sum compensation transfer) Cohesion funds	Community agricultural policy
High	Stability Pact Policies for the supply of pure public goods (defence, foreign policy, environment) Establishment of a minimum level for social protection (harmonisation of labour markets legislation and welfare state)[a]	Fiscal policy (slackening of the Stability Pact, reduced crowding out effect) Policies for infrastructures Structural funds (also in the EU 15)	Work policies (benchmarking) Employment policies Immigration policies Policies for the supply of impure public goods (education, vocational training, culture, R&D) Health policy Transport policy Policy to combat poverty

[a] Reference is made here to the acceptance of the so called 'social charter'.

A transparent process of benchmarking along the lines of the so-called 'open method of coordination' is an important lever not to force undesirable choices but rather to determine the best solution offered by the various territorial systems and also those outside national confines.

As upheld by Maggioni and Bramanti (2001), in the second approach the regions represent: (i) the natural place for the generation, evaluation and

[26] Via hysteresis.

mediation of the interests of economic actors; (ii) the tool by which competences can be acquired to translate local interests into concrete community projects.

However, to avoid any miscomprehension, the role of the regional governments must be translated into a clear and precise expression of intent and formulated as a detailed coordination project. A clear expression of intent and a detailed coordination project have gradually been elaborated since the member states' measures to sustain employment initiated at the Amsterdam summit meeting in June 1997 and the extraordinary summit meeting on employment in Luxemburg in November 1997 which were based on the formula of the *national action plan* and their *benchmarking.*[27] The intent and the coordination project were then extended to other fields of social policy and rendered more credible with the so-called 'Lisbon strategy' proposed at the summit meeting there in 2000 and which gave more tangibility to the 'open method of coordination'.

References

ANTONELLI G., GUIDETTI G. (1999), "Il monitoraggio delle politiche del lavoro in Emilia Romagna: primi elementi di analisi", in: *Istruzione, formazione, lavoro*, Rapporto 1999, Regione Emilia Romagna, Bologna.

ANTONELLI G., CAINELLI G., GIGANTE P., GREGORI C., GUIDETTI G., MARZANO G., MONTINI A. (1999), *Monitoraggio delle azioni per l'istruzione, la formazione, il lavoro. Anno 1998*, Regione Emilia Romagna, Bologna.

ANTONELLI G., ZOBOLI R. (1999), Introduzione, in: IDSE-CNR, *Trasformazioni strutturali e competitività dei sistemi locali di produzione*, Franco Angeli, Milano.

EUROPEAN COMMISSION (2002), *Prima relazione intermedia sulla coesione economica e sociale*, Bruxelles.

FEDERAZIONE DELL'INDUSTRIA DELL'EMILIA ROMAGNA (1995), *Il differenziale competitivo: l'Emilia Romagna a confronto con le Regioni d'Europa*, Confindustria, Bologna.

FORTIS M. (1998), *Il made in Italy*, Il Mulino, Bologna.

FORTIS M. (2000), *Il made in Italy nell'economia italiana e mondiale: il rilievo delle piccole e medie imprese e dei distretti*, in: QUADRIO CURZIO A., FORTIS M. (a cura di), *Il made in Italy oltre il 2000. Innovazione e comunità locali*, Il Mulino, Bologna.

FORTIS M. (2000), *Innovazione tecnologica e sistemi territoriali: specializzazioni produttive e interdipendenze*, CRANEC, Milano.

IDSE-CNR (1999), *Trasformazioni strutturali e competitività dei sistemi locali di produzione*, Franco Angeli, Milano.

---

[27] In fact upon proposals made by the commission the Council adopted the guidelines for the formulation of "national action plans" which each member state has presented since 1998. For an example of this type of monitoring of labour policy and training at a regional level see: Antonelli, Cainelli, Gigante, Gregori, Marzano and Montini (1999)

MAGGIONI M., BRAMANTI A. (2001), "Le macroregioni europee e le politiche per lo sviluppo locale", in: QUADRIO CURZIO A. (a cura di) *Profili della costituzione economia europea*, Il Mulino, Bologna.

ONOFRI P. (2001), *Un'economia sbloccata*, Il Mulino, Bologna.

POLANYI K. (1957), "The economy as insituted process", in: POLANYI K., ARENBERG C. M., PEARSON H. W. (Eds.), *Trade and markets in arcaic societies*, Einaudi, Torino.

QUADRIO CURZIO A., FORTIS M. (a cura di) (2000), *Il made in Italy oltre il 2000. Innovazione e comunità locali*, Il Mulino, Bologna.

ROMANO B. (2002), "La ricetta Stoiber per vincere a Berlino", *Il Sole-24 Ore*, 31 January.

TABELLINI G. (2001), "The Assignement of Tasks in an Evolving European Union", *Europeos*, n.9.

Italian industrial districts: Facts and theories

Fabio Nuti

Department of Economics, University of Bologna, Strada Maggiore 45, 40125, Bologna, Italy

1 Introduction

It is not difficult to understand why Italian industrial districts have been one of the most popular and debated issues both in economics and other fields over the last two decades. The scenario, reminiscent of the Industrial Revolution, astonished everyone in the '70s, when the phenomenon began, or, rather, became conspicuous. Interest was aroused not only in Italy, due to the close proximity of the the phenomenon but also in other places, because it challenged prevailing economic thought and opened up new horizons for economic policy. It is only natural that the topic of industrial districts has generated a flood of literature since then, but *libelli habent sua fata*. Scientific speculation sometimes poses no lesser problems than reality itself. With the advantage of the hindsight, one is led to suspect that speculation may have superseded observation, and may eventually have created a virtual *ersatz* of a reality which was (and is) in itself fascinating, but riddled with problems. In this work, I will try to separate those, which, in my view, are the original facts from the theoretical elaborations which followed, and to offer some comments about the latter.

Over the years, many different justifications have been offered in literature for the continuing interest in Italy's industrial districts. Many authors have expressed hopes that the district could provide an altogether new model for economic policy and, perhaps, even a new paradigm in the theory of the firm. The majority of this literature, however, while initially upheld by a certain amount of descriptive works, has then adopted a very speculative tone, thereby encouraging the overgrowth of theory – rather than the appraisal of facts.

This state of affairs may be partially due to the difficulty in submitting mainly qualitative observations on Marshallian industrial districts to quantitative tests, but, at the same time, the difficulty can be regarded as the consequence of an unclear definition of reality. The concept of Marshallian industrial district, as codified by Becattini over the years, does not lend itself too easily to empirical analysis. This is not so much because a number of important features, such as industrial atmosphere, cannot be translated into quantitative expressions (in fact, they have), but because these features tend to be postulated, rather than proven or at least described.

Although almost all commentators agree that 'structural' variables *per se* (e.g., size, specialization, etc.) cannot help us to distinguish a district from a mere cluster of firms, nevertheless, many of them derive conclusions from studies where districts are only defined in these terms.[1] One could reply[2] that districts can nonetheless be defined in terms of the values of structural variables, without ambiguity, because (i) the prevalence of a given specialization within the same area is proof enough of the district-like nature of the area, and (ii) firms lying outside the strictly defined district network also benefit from the general surrounding environment. This may, however, be mixing the notion of Marshallian industrial district with that of local specialization. The latter has advantages in itself, and can be used to explain the success of certain areas, but it does not imply the *same* definition of advantages.

Unfortunately, given the current state of the literature, there is no obvious solution to this problem. The progressive abandoning of descriptive studies is regrettable, because – as some people feel – we still have only scant real knowledge of the stylized facts connoting the Marshallian industrial districts phenomenon, if such a phenomenon exists indeed as a homogeneous whole. The evidence on which we can base our considerations covers a comparatively limited span of time (roughly from the early 1970s to the late 1980s at best, with very few exceptions beyond that date), and covers many different areas and sectors, and the more recent contributions do not allow systematic comparisons. This makes of Marshallian industrial districts a rather unusual matter of investigation. The resumption of research in the field might certainly be welcome, but the problem is deeper, and it lies in the fact that local studies (which, by definition, refer to often extremely heterogeneous realities) are very demanding and costly. Methodological issues, then, are often dodged on the assumption that the object ('the' Marshallian industrial district model) is unambiguously defined - and understood by everybody. But, from a perusal of the recent trends in Marshallian industrial districts analysis, one may get an uncomfortable feeling: the model seems to contain much more in terms of idealizations than in terms of generalizations from actual facts. Moreover, it seems overburdened with an assumption of 'optimality' which is even less easy to maintain, since (i) it implies a comparison with alternatives which may exist only in theory[3], and (ii), it sistematically overlooks the potentially negative aspects of reality, only to emphasize its alleged virtues.

---

[1] See, typically, Iuzzolino (2000).

[2] For example Fabiani *et al.* (2000)

[3] Obviously it is not always the case – historically and economically speaking – that both the manufacturing district and the big plant were available at the same time in a given condition. But, on top of that, the very idea of optimality seems to me to have given a typically academic twist to the literature, away from empiricism and toward ideology, thereby weakening the original vitality of the notion.

2 Why should we study industrial districts at all?

In general, we are interested in new ideas because we believe that they can lead us to more effective actions. The policies envisaged are inspired by and depend heavily on the assumptions embodied in the theory behind Marshallian industrial districts.

Many recent contributions incline toward a certain amount of *laissez-faire*. This may not be a simple matter of coincidence, if the mainstream theory is itself imbued with stylised tales about perfect competition, the spur of competitiveness, etc and there exists a current bias in literature against intervention (and perhaps economic policy in general). It is worthy of note that Marshallian industrial districts were initially studied as a possible alternative to traditional regional policies, and are now used to uphold the idea that the market should take care of itself. Hopefully, some critical considerations on this point will emerge from the following notes. But surely the main reason for studying Marshallian industrial districts is still their enormous capacity to produce a continuing theoretical spinoff. Therefore, even if we firmly believe that the policy implications have been and are important, our considerations will be generally of a theoretical nature.

2.1 What are Marshallian industrial districts?

Contrary to what is sometimes assumed, there is little *de facto* agreement on the basic definition of Marshallian industrial districts. It is perhaps safe to say that the focus of the investigation has shifted from the initial issue of small firms *per se*, to embrace clusters of *manufacturing* firms located in specialized areas, and, thereafter, clusters of (often technologically advanced) firms, many of them small, concentrated in the same economic region. Many features of the current definition of Marshallian industrial districts seem to have been grafted onto a body of knowledge made up essentially of findings collected during the early phase of the debate.[4]

At the present stage, Marshallian industrial districts can no longer be seen to be 'mainly' involved in manufacturing, 'fashion', or 'light' industries, etc. Nevertheless, in the prevailing view, hi-tech firms are bracketed together with low-tech ones, leading firms with subcontractors, capital-intensive with labour-intensive. In my opinion, interesting facts can be missed in so doing, and the chances of developing alternative views of the phenomenon are substantially reduced.

---

[4] Certain American authors, for instance Storper and Harrison (1994), have pushed this concept so far as to include things as differing among themselves as *all what connotes a 'system', or a network* – or is just supposed to.

2.2 A single or many concepts?

Many authors acknowledge that what constitutes a Marshallian industrial districts is far from homogeneous (see, Signorini, 2000). The cases which have been studied as examples of industrial districts since the mid-'70s presented indeed a great variety of attributes. Small size was an obvious fixed point of reference, but large firms were not unknown, although their role might differ from place to place. Certain districts consisted essentially of subcontractors (with the subcontracting enterprises located elsewhere, perhaps even abroad), while others were self-contained systems. Some industrial districts were essentially made up of exporting firms, while others produced for the domestic market. Capital intensity differed greatly, not only within the areas, but above all between areas, to such an extent that cases could be seen where highly capital-intensive firms and very limited division of labour prevailed, *vis-à-vis* others, where unassisted workforce was the main input and co-operation among firms was evident to everybody. Of perhaps more importance is the fact that *historical features* differed substantially from area to area. A very rough distinction could be made between 'old' districts, and more recent ones. It was the latter which, in general, attracted the greatest attention because they seemed to promise a completely new recipe for industrial development.

Applied analysis has come to rely mainly on *size* and *degree of specialization* (per area) as the required conditions needed to define Marshallian industrial districts. *Qualitative characteristics*, which in many cases were the result of casual observations, and which may have been found in a limited number of instances, tended to be taken as given, provided size and specialization conditions could be fulfilled.

2.3 History of an idea

The long debate started in the late 1960s. It began with a re-assessment of the relative roles of big and small firms, and flourished in the following decades with ever increasing intensity and scope (including ideological overtones). In part, this was due to the uncanny ability of the debate itself to produce and attract intriguing notions, among which 'flexibility' is perhaps the most ubiquitous. Reference to small firms, industrial districts, 'local' systems of production and so on, has become, with the passing of time, a sort of banner[5].

[5]The development of the idea of industrial districts – and of related concepts – has been illustrated in various places. Here, I offer my personal reconstruction of how the debate evolved, for the benefit of the newcomers. However, it may be missed out without any loss of continuity. *The awakening of interest in the small and medium enterprises (SME).* Throughout the 1960s, studies in industrial organization centered on large enterprises. During the late 1960s, however, big firms were at the centre of fierce struggles (the 'Hot Autumn' of 1968). Observers noticed that a number of them (notably, FIAT) were subcontracting parts of their production to smaller firms. This observation shaped

The approach that prevailed in the first stages of this evolution was strictly neoclassical,[6] in the sense that virtually all structural changes were seen as reactions to the historical rise of wages. Even if, over the time, this has diminished, the search for a satisfactory model of the firm (or the network) has remained the prominent concern for scholars. A model is not necessarily a bad thing *per se*, provided it can accommodate the relevant features of reality. As it happens, however, the model which has emerged — in a number of variants — in recent literature is the model of an organizational form, and as such it concentrates on features relative to supply. An important alternative view, which has been overlooked through the years, would take the role of demand into consideration, as we shall see in due course.

I maintain that, in the literature concerning Italian industrial districts,[7] two main potential problems seem to emerge: (i) statements which are valid for *some* realities are extended (by analogy) to *other* realities; (ii) statements which are unproven as such are applied to actual cases.

The two are connected. In the examination of the main tenets of the theory, which we present here, it will not always be possible to distinguish which one is under discussion.

---

subsequent interpretations of the relationships between large and small enterprises. *The negative interpretation of the nature and role of SME.* Empirical studies on SME started to appear. These, however, were still concerned with the complementary roles of large firms and small subcontracting units, mainly in the most industrialized areas of Italy. *The emergence of the territorial dimension.* Scholars noticed that groups of SMEs, located outside the traditionally industrialized regions, achieved very good results in terms of production and employment. The 'negative' interpretation, though, still prevailed and this was seen as a carefully crafted response by part of the leading companies to offset declining profits. *The 'discovery' of industrial districts.* A number of studies highlight the unusual distribution of certain concentrations of SME, as well as their sociological and political peculiarities. *The codification.* Becattini underlined the analogies between those concentrations and Marshall's 'industrial manufacturing districts' (IMDs), and dictated the list of the characteristics by which a genuine IMDs was denoted. The 'negative' vision was abandoned altogether, and IMDs from that moment on were regarded as *positive* example of a 'third way' to the industrialization of backward areas (following large firms and state entrepreneurship). The socio-economic literature became flooded with empirical - and largely qualitative - studies, which are mostly monographic. *Studies of 'virtual' IMDs.* Since the end of the '80s, empirical investigations diminished. At the same time, the IMD started to be considered as a model of industrial organization, and sometimes as a sort of meta-model of a firm, which could be studied in itself. Lately, it provided the foundations for a theory which was likened to a brand of the endogenous growth theory (see Camagni and Capello, 2002)

[6] Becattini would not consider 'neoclassical' a derogatory adjective. Nevertheless, one suspects that the neoclassical way of thinking, which perhaps lingers under the idea of a *model* of district, is in itself somewhat inconsistent with an evolutionary view of the *development* of districts from the generally small clusters of firms of the late 1960s into the much bigger entities which have attracted scholars' attention since the end of the 1970s.

[7] I shall confine my observations to cases which can be labeled as *manufacturing* districts.

Drawing from continued observations of a number of different areas, I would suggest that at least two main distinctions are in order: (i) historical and recent districts, and (ii) clusters of technologically advanced firms, and clusters of '*traditional*' firms.

The first distinction is relevant for the analysis of the causes of the rise of industrial districts and for the identification of the main inflluences on their development. The second one deals essientially with ideas such as 'industrial atmosphere'and 'trust'.

2.4 Historical and recent districts

The distinction between 'older' districts – those which have been operating ever since the period of Italy's first industrialization – and 'newer' ones, namely, those which have reached a critical size only around the 1970s, can hint to the working of different historical and economic causes. The districts of the former type can perhaps be explained quite satisfactorily in terms of classical location theory. Generally, these districts developed through a long series of transformations into real centres for their industries, thus adding new sources of advantages to the initial ones. Como, for example, which can be equated with the silk industry of Italy, or Sassuolo, which accounts for a major part of the tile industry, are just two examples among others of places that grew from their initial nucleus to become internationally recognized leaders in their sectors. Such districts appear to have taken a long time to gain control of their relative markets, by shaping production factors to their requirements, and by establishing a steady network of firms whose dimensions were dictated mainly by the size of the corresponding business. Recent literature insists on the comparative advantages of the district *gestalt* (= of the organizational form *per se*), but this may be an example of the neo-classical way of reasoning to which we referred above.[8] To acknowledge the importance of history is to tread firmly in Becattini's footsteps: culture and tradition are not built overnight.

What we need in order to explain the birth and rise of districts of the former kind is actually little more than an update of the basic theory.

On the hand, districts of a more recent origin seem to have developed within a well defined time span (which may vary, from case to case, from the late Sixties and troughout the following decade), and to have grown under the pressure of particular conditions which lasted long enough to produce a structure.

Certainly there have been surprising phenomena, which challenge the scholars' capacity for fitting a number of facts into the classical framework. Not all of these facts can be explained in terms of conditions relating to supply only. Many of

---

[8] Krugman (1991, p. 61) has noticed that in most cases, localization resulted from a combination of an initial accident in a particular location, after which "cumulative processes took over." In the subsequent stages of regional growth, different arrangements can develop, as a consequence of endogenous and exogenous factors.

them can be better explained by reference to factors lying on the demand side: for instance, the evolution of consumers' tastes and habits, the new trend away from customized consumer goods toward standardized goods, the pull exerted by external buyers, etc. Thus, we can say that a new model of organization has emerged, but this model should perhaps make some allowance for the sensitivity of the emerging systems to external factors.

Although the observation that most industrial districts of the *new* vintage have been made up essentially of subcontractors is true only to a certain extent, in the current analysis it could constitute a useful pointer. Finding a niche in the prevailing division of labour is a strategy that is typical of firms operating in prevalently rich environments, and/or with limited access to finance, and therefore unable to invest large sums in long-term commitments. This had a number of consequences, which I shall try to summarize: small size was typical, as it reflected limited access to the market, low capacity to collect capital and individualistic appraisal of the risks of trade; growth could take place in terms of either additions of units of a similar type, or of complementary plants, but almost never in terms of internal growth; actual and potential markets remained generally unknown; firms consisted essentially of productive units, with no strategic capabilities or attitudes.

At the beginning, (in the 1960s and 1970s) the problem of coping with barriers to entry, typical of SMEs, was generally dealt with through effective fundraising in the form of inter-personal loans. This reinforced ties with the local environment, but also limited sectoral and geographical mobility, thus generating a reputation for short-sightedness and immobility. Mobility could be improved in the horizontal sense by abandoning declining enterprises in favour of new ones. In summary, it appears that the typical Marshallian industrial district of the time suffered a number of serious limitations, while at the same time producing undeniably good effects in terms of income and local employment.

In many cases, Marshallian industrial districts did not appear out of the blue, 'new' industrial districts emerged from a small pre-existing nucleus, and reached a remarkable size in a comparatively short time, thereby attracting much attention. One of the best-known examples is the leather and skin district of S. Croce, Valdarno Inferiore not far from Prato, which has been and is still the dominant concern of Becattini's research. Although it has existed almost since the Middle Ages, however, it was not until the 1970s that it reached the size and the economic importance for which it is known. The factors which are central to Becattini's considerations, culture, labour, ethics etc. certainly belong to the past history of the place, but its shape as a district has evolved under the constraints of historical circumstances.

At times, larger firms *have* been the catalyst for a number of recent districts. This is not the same as saying that a general explanation for the district phenomenon (if such an explanation can indeed exist) can be found there. However, in determining the 'explosion' of the new districts, growing *demand,* either domestic or external (and sometimes both), has certainly played a role, whether channelled by public or private buyers. For instance, bulk purchases of shoes or clothing for the Army, or massive orders of furniture by certain foreign governments, have often injected the initial impetus that unleashed the incredible expansion of the original trade. On the other hand the effects of the continuing devaluation of the national currency, in those decades, should not be overlooked. The fact that districts bought input from weak currency areas (developing countries) sheltered them from the worst effects of devaluation, while, by selling their products in strong-currency areas, they could effectively compete with most foreign competitors (In this respect, it must be noted that import-export agencies representing entrepreneurial initiatives have played a crucial role in the development of certain local industries).

Still from the demand perspective, a general explanation of the explosion of the phenomenon, could be found in the persisting growing demand for certain manufacturing goods which characterized Italy during the 1960s-1970s. For example, the evolution of the clothing industry from bespoke tailoring into industrial large-scale activities was an unexpected feature of Italy's post-war economic landscape. This could not take place without enormous consequences on the structure of the industry itself. Under the impetus of increasing demand, often at a breathtaking pace, the industry had to relocate outside its traditional urban areas where, within a limited area, the workforce had become scarce.[9] In other cases (e.g., shoes and leather, furniture, etc.) the emphasis was more on the need for more mechanized or more modern techniques, so that relocation was less frequent. In all cases, however, we saw that the new firms were born with an overwhelming commitment to 'capacity-widening' with a correspondingly lesser concern for the inner organisation of the firms themselves.[10] All these factors were peculiar to a particular moment in the historical development of Italy.

After the initial phase, districts have met with different fates, in certain cases acquiring a stable configuration whereas others have been more fluctuating. Scholars at the University of Pisa, such as Varaldo[11] and Mariti, among others, underlined the need for an explanation of these facts in terms of a dynamic

---

[9] There can be an explanation for the proliferation of SMEs in the Centre of Italy, in areas with little manufacturing specialization, and as a result of subcontracting from large firms of the North. In many cases, the SMEs of Central Italy were directly connected with larger firms of the North, and worked as subcontractors for the latter. In such cases, the distance between the two points could not stretch beyond one traveling day (for the round trip). In other, more relevant cases, however, small nuclei of manufacturing firms began to grow under the influence of swelling orders. The fact that many areas had no previous *specialization* is not inconsistent with the presence of large incubator plants, which on the contrary were a recurring feature of those places.

[10] When Brusco (1991, p. 26) acknowledeges that an important difference between the North and the South of Italy in the post-war period until the mid-Seventies was the higher degree of 'openness' of the big firms of the North vis-à-vis the small firms of the 'Mezzogiorno', he may be implicitly emphasizing the role of mass consumption - and large scale demand - in that period. But that period witnessed the rise of the North's small firms.

[11] From the vast production of these early scholars of the district phenomenon, we would signal as particularly relevant Varaldo, 1988, Varaldo and Ferrucci, 1997, and Mariti, 1980.

theory. The Adam Smith-George Stigler line of thought explained the tight division of labour that was shaping those areas as the result of increasing demand. A *demand-led* model tends to shift the emphasis from *qualitative* adjustment, which underline innovative activities as focus — and the main problem — of the district, to *quantitative* adjustment, i.e., the attitude of the system to adjust supply to demand by having its various parts respond consistently to external impulses. This view, while for a certain time received a deserved consideration, subsequently ceased to do so. As a consequence, we are unable to empirically test the hypothesis of a demand-led growth *at least as a general model of Marshallian industrial district growth.* However, the attitude to respond in terms of quantity seems to be a relevant property of an economic organization, as it is responsible for at least a part of its efficiency. A system affected by serious exposure to redundancy, and therefore to unemployment, of productive factors, sluggishness in matching market demand, etc., can still be called efficient – in the sense of being capable to adapt to changing conditions – but it might not commend itself as a model of local devolpment.

2.5 The cyclical interpretation

One could speculate as to what might happen in response to a slowdown of demand, or a slump. A whole range of consequences of a subsequently flattened demand, from changes in inter-firm relationships [12] to deeper structural changes[13] have been documented, although they have generally taken place only after many districts had already undergone significant transformations (more about this later). Becattini (2000, p. 77) agrees that districts can decay and even die.

Finally, from the point of view of demand it must be remembered that in the years when the new districts were born other European countries were abandoning the less capital-intensive manufacturing industries thus providing welcome opportunities for Italian producers. Here typical Marshallian industrial districts started to prosper.[14] In the late 1960s Italy might have been expected to follow in the footsteps of its European counterparts and to move toward the high-tech sections of modern industry. To the great surprise of most observers *mature* sectors became Italy's most dynamic ones. [15]

---

[12] Some evidence is contained in a former work of mine (see Nuti, 1992), where the emergence of hyerarchized *groups* of firms (as suggested also by Lorenzoni, 1983, 1987) is described.

[13] There are examples (Castelfidardo) of areas where small firms have been merged anew into larger units.

[14] The difficulty to keep at pace with progress in hi-tech sectors seems to be a peculiarity of the late economic history of Italy - a peculiarity which still persists, and which, as certain observercs argue, is likely to raise serious problem in the near future.

[15] This required a massive mobilization of labour and capital, which took place in different areas with an ingenuity and a truly capitalistic aggressiveness which deserve much more than just cursory interest.

In summary, there *is* a historical dimension to the phenomenon and ignoring it can amount to denying the same assumptions which underlie Becattini's research. In such literature, very few contributions treat history as anything but a prologue. Many of Becattini's closest followers write in a vein that stresses the model rather than elements relating to evolution, giving birth to a rather hypostatic categorization.[16] Different approaches, evolutionary ones, path-dependency, etc. might have been pursued. But, in my opinion, even if confined ourselves to introduce demand into the model, the problem of adapting the Marshallian industrial district-system to changes in quantity would be of some relevance, at least because it would imply some allowance for a sort of potential failures which has been overlooked (as we shall see in due course).

A possible relationship between the economic cycle and the development of districts was suggested years ago (Nuti, 1992), but never submitted to empirical tests. However, with the advantage of the hindsight, today we can rely on some empirical accounts which seem to confirm that Marshallian industrial districts are far from impervious to external factors. Brioschi and Cainelli (2001), for example, have pointed out that (in the shoe-making district of S. Mauro Pascoli) declining demand has swept away many small subcontractors "but has also produced the beneficial effect of clearing the district of all of those weak units which had been born on the wave of enthusiasm and easy profits" (p. 146, our translation). Similar evidence can be cited in other cases of districts which not long ago were mentioned as typical representatives of Marshallian industrial districts, Castelfidardo (musical instruments) for example, as well as other, extremely successful ones, like Castel Goffredo (stockings) and others.[17]

The number of examples would perhaps rise, if more empirical fieldwork were available. As a whole, it seems that the basic (structural) features of the Marshallian industrial district model, like the generally small dimension of

---

[16] Sforzi (2000, p. 12) recently wrote: ".. talking of 'the district of Bolognese leather-working' hints that that particular category of commodities ... must be understood as integrated within the local social context, Bologna, so that we can be able to understand its evolutionary characteristics, be it the ability to innovate or the difficulties it faces in the market; [it] tries to express its locally peculiar manner of producing those commodities, which belong to the same class as those which are produced in Florence, but are not exactly the same, as they are rendered different by the economic, productive, social and cultural constraints created by the territorial context".

[17] Becattini (2000, p. 26) insists that location factors and economies of integration cannot *per se* explain the whole phenomenon of districts, since these have showed a greater resilience and endurance than the above interpretation could ever account for. It is difficult to disagree upon that. This, however, seems to imply that contingent phenomena affecting the structure of an industry cannot explain districts completely. Becattini himself, in the definition which appears for the first time in his seminal essay of (1979), equates Marshallian industrial districts to a "localized thickening ... of inter-industral relationships showing a *reasonable stability in time*" (our italics). From this, it seems to follow that, wherever such stability is not found, we are not dealing with real districts. But this can perhaps be better applied to older districts, which have *reached* such stability with the passing of time, than to the ones which were born more recently.

producing units, the low capital/labour ratio, the tight co-operative network, etc., have undergone substantial changes undern certain circumstances, like the declining importance of external buyers, and other changes in the level and/or composition of market demand – all facts which have severely affected most districts since the mid-Eighties. All this only adds to the already stinging perplexity which one is likely to get when inquiring into the actual properties of the model.

2.6 Traditional and technologically advanced districts

A second source of confusion may lie in grouping together many different kinds of clusters, assuming that all a cluster needs in order to become an actual district, is specialization and small dimensions of firms (or units). As a matter of fact, the cases which have attracted universal attention since the late '70s, because of the incredible rate of growth they exhibited (which seemed to suggest a possible 'new way' to regional development) were concerned with the manufacturing industry and, to a great extent with the 'fashion' industry. As such, they employed traditional technologies,[18] had their competitive edge in the handicraft skills of local workers, and at most produced 'incremental' innovations, in the sense of small refinements capable of injecting additional efficiency into an otherwise well established body of knowledge, and responded essentially to changes in external demand.

Admittedly, the emphasis on industrial districts of the traditional kind may be ill-posited, now that we have gained a greater awareness of the far-reaching implications of the idea of local system. Still, even the most recent accounts of actual districts seem to tell a familiar story, not very remote from the one that would be regarded as typical of the original idea of district.[19] At any rate, the primitive idea of district was subsequently generalized to all clusters composed of firms working in homogeneous sectors. In this way, the Marshallian industrial district idea was projected onto areas characterized by their homogeneity and small scale firms, but not necessarily by division of labour. In many cases, these SMFs do belong to hi-tech activities. Small size, however, is not unusual there, while division of labour, not to mention co-operation, is. Marshallian industrial districts would not attract such curiosity, if it were not for the look of 'scattered factory' (*fabbrica diffusa*) that districts of the former kind exhibit.

[18] This is not at odds with the fact that significant innovations were being introduced into those sectors at the end of the decade. Such innovations, however, were drawn from the stock of existing technologies and were essentially intended to enlarge the sectors' productive capacity.

[19] See, among others, the empirical material contained in Camagni and Capello (2002).

2.7 The concept of Marshallian industrial districts revisited

To sum up, are there features which can be said to be truly 'typical' of Marshallian industrial districts?

The cases which have been studied are certainly characterized by certain similar aspects, but they seem to be limited to: (i) *the nature of the industry*: all the success stories that have been collected and analyzed refer to manufacturing industries, generally with high standards in terms of craftsmanship; many areas operated in the fashion industry; almost all produced commodities for fast developing sectors, like home furniture, clothing, etc.; success cases which can be found in other sectors generally lack some of the following properties; (ii) *parallel development paths*, including a phase of accelerated – sometimes incredibly so – growth between the end of the 1970s and the beginning of the 1980s, with remarkable transformations (from both the qualitative and the quantitative point of view).

We might be tempeted to add *structural features*: small dimensions are certainly a salient characteristic of industrial districts (although not much can be inferred from this about the relationships existing within industrial districts); but, on the other hand, we have cases of districts made of capital-intensive, insulate firms, co-operating among them to a very limited extent, perhaps only in the sense of catering from the same 'human pool'.

Thus, common causes seem to have operated in roughly the same manner on similar realities, for a certain time. It must be noted that such shared characteristics are not enough to completely describe the Marshallian industrial districts phenomenon. This latter, in its codified form, includes a number of features which could not be found in all the areas which nevertheless have been subsumed under the Marshallian industrial district label. Many features have been assumed to be shared, and pinned to the general description of the modern Marshallian district.

2.8 The 'wishful residual' : industrial atmosphere and innovativeness

A district is typically made up of a collection of units – several among them real enterprises – whose dimensions are on the whole well below the average of the corresponding sector, as measured in other industrialized countries.[20] But small size apparently excludes economies of scale. Moreover, as these firms seem to co-operate towards common ends, there should be high transaction costs in dovetailing the decisions of so many different entities. In considering the nature of industrial districts, Becattini has been continually suspicious of explanations based on integration and location forces alone. In his view, there had to be something which could account for the *limited* integration of firms which were

---

[20] Of course, this generalization conceals the fact that there have always been ties among the various units of a typical Italian district.

located in such close proximity. Marshall makes his fundamental point about why companies in the same industry (e.g., high fashion garment) congregate. According to Marshall, industrial districts *enjoy the same economies of scale that only very big firms can obtain.* This happens because specialized suppliers get to know where to go to minimize their efforts and maximize their business. At the same time, skilled workers have a well defined place to come to play their trade. Furthermore, everybody can benefit from the spillovers of specialized knowledge. As Marshall put it, "The mysteries of the trade become no mysteries, but are as it were in the air". Marshall's "something in the air" was the way he described the added virtue of the prototypical industrial district of the 19th century Manchester.[21]

Most modern Marshallian industrial district interpretations emphasize the importance of an intangible factor in the development of districts, namely, something not accounted for by the standard theories of the firm, but which should help make up for the lack of scale advantages and for the (presumably) high level of transaction costs. This factor is usually identified with concepts like 'co-operation', 'industrial atmosphere', etc.

Yet, the existence of this factor has not been subjected to any deep scrutiny. In empirical research, it has been treated as a mere residual, hence the term. There is scant and anecdotal evidence in qualitative reports of single areas (at any rate, field researchers would have a hard time in confirming – or disproving – the existence of this factor). However, it is a fact that casual observation could provide as many (and, perhaps, more) examples of lack of co-operation among district firms: producers refusing to share their information with their neighbours, or refusing to spread news; firms duplicating their investment in order to gain advantage over their competitors, and so on, are not entirely unkown in these areas. *Individualism* and selfishness are often quoted by insiders as the greatest hindrance to greater development. The same concept of individual 'amorality' which, along the line of Banfield, has been noticed by some authors (Signorini, 2000) as a *strength* of the model, could indeed be used to stress the pervasiveness of non-co-operative behaviour in those areas.

'Industrial atmosphere' certainly exists in some sense (at worst, as a differential quality in the local people's attitude toward enterprise), and one – certainly not the least – merit of Becattini's work is having rescued it from the vaults of

[21] It is not always clear whether the emphasis on industrial atmosphere serves the purpose of explaining the structure of Marshallian industrial districts (firms do not coalesce into large units because they all like freedom and personal enrichment and, even more, because they do not really need to merge), or is used just to underline a source of advantage *vis-à-vis* the 'fordist' firm. Probably both. Becattini recently wrote (2000, p.31) : "Motives which push a host of former workers, hired or independent, urban or rural, toward small business, are of varied nature, ranging from typical desire for personal enrichment, to hate for subordination, to personal fulfilment in the face of their peers ...". The social form that such motivations are conducive to, is a 'thick network' of relationships, where the community feeling features at a premium vis-à-vis individual values. All this is likely to lead to innovativeness.

Unfashionable Economics. This is not to say that the *use* made of this concept by many subsequent commentators is wholly satisfactory. What, then, is the exact content of the idea of 'industrial atmosphere'? A significant refinement of the initial intuition was made when ideas like 'externalities' and 'trickling down' of innovative activity crept into the debate.

2.9 Externalities and trickling down

Marshall's description quoted above portrays a situation where ideas circulate and information is a public good. The literature on Marshallian industrial districts has initially labeled these as externalities.

Judging from the sense of many contributions, externalities may have been used from the start as an *ersatz* for integration, co-ordination and like concepts. Actually, this concept has been superseded by a vision which concentrated on interrelatedness and, lately, by the advantages of proximity.

There is no denying that all studies of the firms per se, or of the industry as defined according to some conventional criterion, are doomed to be unsatisfactory (Becattini, 1979). Externalities have been considered as part of the bundle of factors which can effectively and uniquely explain a given devlopment process. However, (i) externalities must be explained; (ii) concentrating on positive externalities can be unfair, as long as there can also be negative externalities (Nuti, 1992 and Varaldo, 1997). Strictly speaking, externalities are by-products of a 'principal' good. Thus, firm A (the principal) might confer an externality on firm B, and in turn firm B could confer an externality on firm A, and so on, but, unless we explain where is the advantage that each firm is enjoying in its own right, the whole thing may look like a conceptual sleight of hand. The main danger here is circularity: externalities must exist, because otherwise comparative advantages enjoyed by Marshallian industrial districts over other forms of industrial organization would be unwarranted. Empirical research should perhaps clarify the point, indicating where such advantages lie. Since, in the latest literature, the ultimate source of externalities has been identified in the innovative activity, one is led to assume that much of the competitive advantage of Marshallian industrial districts should come from a stream of innovations taking place in a high number of the component units. Then, one notices with a certain disappointment that - for instance in the activitities filed under 'made in Italy', which account for a large part of the district phenomenon - the role of Marshallian industrial district firms is still largely relegated to manufacturing tasks, due to the fact that most of the creative process takes place outside them. It seems then that reality is again a bit reluctant to be bracketed in too narrow frames.

Even taking for granted the externality concept, one is led to believe that the existence of *negative* externalities may have been overlooked in the literature. Pollution, congestion, the denial of a role for trade unions (with the lack of safety in the workplace which follows) are all features of most manufacturing industrial districts. The fact that local communities seem to be fit in with the set of values which inspire this somewhat chaotic model of development is certainly beneficial from the point of view of economic success. But this is not to say that negative externalities are without importance. Even leaving aside all value judgements, a model *cum* negative externalities appears to deviate from a real optimum, both in economic and social terms.

Moreover, we could mention the issue of *redundancy*, which in a sense can be labeled as a different sort of negative externality. Certain expenses may be replicated by many different firms far beyond what the optimal level would dictate. Scale or integration economies could not be pursued, even when available. And we could go on. Evidence of such negative externalities are themselves provided through casual observations, but this should have been an incentive for further research.

2.10 What kind of innovative activity?

In this respect there are at least two 'grey zones', which warrant further research. On the one hand, we might ask whether the 'spillover' hypothesis is fitting whenever information or a technical solution to a problem could give considerable advantage to an entrepreneur. Upon closer scrutiny, in fact, it would seem that the 'spillover-public good hypothesis' can only hold for *certain* situations, perhaps those intended to enable one to catch up with the average level of knowledege, and only in particular (and perhaps, early) stages of the industry cycle.

On the other hand, solutions which grant significant advantages to single entrepreneurs might not be so important in industrial districts. In fact, what sort of innovation is relevant for the sectors which are typical of Marshallian industrial districts? Product innovation, for instance, is not always left to acts of creation by manufacturing enterprises located at the 'end of the pipe'. Rather, they are channelled through the pipe following formal patterns, where proximity is not crucial. This is especially true in the fashion industries, which constitute a large part of the reality of industrial districts (especially at the time of their boom of the '70's and part of the '80s), and where the firms' innovative activity is actually restricted to the task of translating into practical terms the prescriptions of the strategists. Here, innovations – even if we dare call them so – simply cannot be circulated, as they are the product of a learning-by-doing (or perhaps a trial-and-error) approach which does not easily lend itself to codification: e.g., how to obtain a certain 'novelty' appearance of the skin in the tanning industry. But, at the same time, it must be conceded that in these cases fragmentation may result in a *lesser* responsiveness, and that, if the opposite happens, it cannot always be taken for granted.

With process innovations, things might be different, and in fact we have examples of innovations made possible, or at least easier, because of spatial proximity (e.g., the adaptation of agricultural machinery to industrial processes in the tanning industry). But, even here, it seems that information and ideas which are being circulated in the typical industrial district environment are those which can help 'follower' firms to keep in line with the general effort of the area, which, more often than not, is originated outside the district - rather than those capable of putting someone in the lead. This sort of circulation can well have had a relevance in a context where the ability of the area to respond to buyers' expectations constituted a common – and perceived – advantage. Possibly it is in this respect that we have an example of the confusion between proper Marshallian industrial districts and clusters of small firms, which in several cases, in the region of North and Central Italy, can indeed be technologically advanced. Such clusters, made up of firms which may be small for a host of reasons that do not pose any real challenge to the analysis, form an important feature of Italy's industrial landscape, but they are conspicuously lacking in the characteristics which are – and have always been – regarded as typical of Marshallian industrial districts, such as interrelatedness and co-ordination.

On the other hand, these clusters do present peculiarities which are worth investigating: the rather unstable configuration of firms help ideas and intellectual resources circulate throughout the area, with significant 'breeding and spawning' effects. All this, however, has little to do with the industries where the most significant cases of Marshallian industrial district were identified initially. The fact that different situations have been embraced in the same vision seems to be more the result of a generalizing approach (with an emphasis on the 'structure') than the consequence of observations.

Rullani (2002) went even further than this, in order to explain why the clustering of many independent producers should help the district as a whole to overcome problems in creating and circulating knowledge and ideas. If I may dare to resume his explanation in just a few words: 'tacit' knowledge is more efficient (less time-consuming, less cumbersome) to transmit than codified, explicit, one, thus creating the advantages of 'complex adaptive systems', like the district.[22]

A certain – perhaps only a small – degree of inefficiency could perhaps be accomodated for in the newest literature without too much harm. The frequent references made to the *cognitive* aspects of the process of economic growth should suggest to leave some room for the limits that rationality may be subjected to,[23] and, personally, I feel that every scholar should be inclined to look rather suspiciously at this usage of cognitivist references aiming to imply that what belongs to the realm of cognitive learning is by definition rationale (and *therefore* optimal)!

---

[22] Calafati (2002), for instance, assumes that "A local system is a human system whose components ... are by definition rational (in a procedural sense). .. It follows that not only the decisions, but also the innovations that such units introduce are rational" (p. 92).

[23] Kahneman has been awarded the Nobel Prize in 2002 for showing exactly this.

2.11 Co-ordination

Literature is full of concepts which are relevant to those which we have considered above. The postulated *co-ordination among firms* in industrial districts deserves specific treatment. In the prevailing literature, such co-ordination took place between equal partners, and helped preserve their small size. To a certain extent, this concept is unconvincing with regard to all the early stages of the industrial districts expansion (thc '70s and the early '80s), when co-ordination was not really an issue, given that the relationships existing at that time among firms took the form of *subcontracting*. In the following period, however, different paths have been followed: while some districts have indeed produced hierarchized structures, others have not. Among the latter, a return to integration within the same firms has not been unusual. Among the former, actual co-ordination (if it ever existed) gave way to division of labour between unequal partners.

According to a different view, which we would like to remember here, since it seems to be a good starting point for futher empirical research, during the initial phase, clustering might have been the product of a 'spawning''process, made up of a myriad of individual decisions, driven not by strategic consciousness of the 'need' of the districts as a whole, but rather by a sort of 'portfolio-like' strategy. As reported in the empirical works of the time, new entrepreneurs who entered the market did not weigh the pros and cons of small size vis-à-vis large dimensions, but rather tried to get a return from investment which seemed profitable, quick and comparatively safe. Building a new business would be demanding enough, in conditions when external finance is scarce,[24] and this certainly is proof of the enterprising nature of men (and women) who entered industry for practically the first time.[25] What they produced, however, although a remarkable feat in its own rights, looks different from the conscious building of a new strategy, and can at best be regarded as a very lucky stroke of serendipity. For a while, this proved successful, and the increasing number of firms operating in the same area was itself a condition for further growth.[26] But at the same time

[24] Banks remained conspicuously absent from this early phase of industrial district growth. However, in most areas they were being effectively substituted by individuals (often businessmen, wholesalers and the like) who were willing to loan money to producers. In many cases, the former were involved in the latter's trade, and helped form almost invisible groups of firms. But such groups responded essentially to a 'capital-widening' strategy, and could not be equated with real networks.

[25] As a matter of fact, many of them did come from other industrial activities, although it must be remarked that a high number came from trade, or had no previous experience at all. The primitive theory, according to which most of those new entrepreneurs were former industry workers or sharecroppers, has not been proved in general terms.

[26] Dimension can attract customers (a sort of *signalling effect*, as it has been called by Camagni, (2000). Buyers knew that they could count on the area having capacity and enough skills to satisfy virtually any requirement posed by them. (This was particularly appreciated in the fashion industry). The attainment of a certain size by the lot of

this hypothesis, if confirmed, *could prove* that 'industrial atmosphere' of a sort does (or did) exist in IDs.

Our doubts concerning the optimum fragmentation are compounded by the empirical findings provided by writers such as Barbetta (1994) or Buratti *et al.* (1996) which underline the existence of *costs* of fragmentation, often well above the corresponding benefits. It may well be, as suggested by some authors, (Lorenzoni, 1990) that such networks were, after all, nothing more that a straightforward manner of replicating the more familiar small firm on a higher scale, but, again, it is not clear whether this is to be regarded as a form of entrepreneurial myopia or a superior strategy.

If the creation of new firms from the point of view of the incumbent entrepreneur could be regarded as a sort of 'portfolio investment', we might expect it to behave procyclically. This would increase the weakness of the districts in a slump (on this point, more below under *flexibility*).[27]

But perhaps the aspect which is likely to cast the strongest doubts on the idea of the flexibility of industrial districts in a quantitative sense, is the recourse, in later times, to *international subcontracting*, which has been acknowledged as an ubiquitous policy of today's districts, and for commodities in which price competition is important. Of course, one is not necessarily led, by such considerations, to conclude that the Marshallian industrial district model is on the whole less appealing. After all, the same evidence could be cited as a proof of adaptability. But to label as 'optimal' a model even when production is liable to be curtailed, and employment to shrink dramatically, probably amounts to using the term in a rather unusual fashion.

2.12 Co-operation and trust

Another controversial aspect of the Marshallian industrial district model is the assumption of *co-operation* among firms. This in itself could be regarded as a *de facto* statement, but, in the mainstream literature, it has taken on deeper overtones. Co-operation has been at times described as a sort of functional division of labour, and at times as a stronger (if more elusive) kind of trust.[28]

---

producers can, in other terms, produce a 'scale' effect which at times can overweigh the 'efficiency' aspect.

[27] The issue of 'groups' (see Brioschi and Cainelli, 2001 and the literature referred there) could be explored in this perspective: a widening of scale and scope can be attained not by 'inner' growth, but by the creation of new separate firms, connected among themselves and with a dominant (usually larger) enterprise. The point which is still to be settled as far as groups are concerned, is whether such strategy can lead to better results than the alternative, or must be considered just a 'satisfying' one, as the network which is created remains intrinsically weak.

[28] The late Harrison (1994), held that the difference between Marshall's model and its Italian counterpart was to be seen in trust. According to him, regional growth in Italy

Trust, (in Harrison's words, 'an outgrowth of experience'), has been the subject of many speculations over the 'systemic' nature of Marshallian industrial districts. The idea behind the use of this concept is that firms can establish relationships among themselves whenever needed, and do not have to internalize different tasks within the same unit, *because they trust each other*, so that transaction costs can be kept at a minimum even with small dimensions.

As it happens, trust is a very ambiguous concept.[29] In all economic theory, relations which are held together by trust can prove unstable, and defection is always a possibility. Thus, explanations which assume trust, must also assume defection. But there are even greater ambiguities. Do we get along without steady relations with others because we trust them, or because we do *not* trust them enough ? As a matter of fact, the answer which seems to prevail in many writings is that in Italy's Marshallian industrial districts, trust is what (in a rather *Italianate* fashion) [30] drives entrepreneurs close enough to one another as to establish informal relations – but no more than that.

According to the prevailing view, even when firms need subcontractors for the routine provision of parts,[31] entrepreneurs refrain from signing binding agreements with others, and continue with a reduced outfit, because this does not prevent them from reaping the benefits of large scale production (on the contrary, they gain from this).

On the other hand, in clusters or networks of firms, steady relations can also serve useful purposes. There may be advantages to be gained for the former by maintaining a regime of *loose coupling* with the latter, but these must rely on regular demand in order to survive. Turning too frequently to new providers may prove too costly in many respects. Furthermore, from the point of view of the leading firm, it might not be efficient to 'choke' its providers because of short-term considerations. If this became the generalized response of the majority of the leaders to a temporary crisis, it could lead to to the destruction of much needed capacity, and surviving the crisis can become more difficult.

Thus, we might propose an alternative explanation: in industrial districts, firms *sometimes* avoid regular relationships not because they want to be flexible, but because they are short-sighted. Stable relationships have been documented in many cases, and it does not seem far-fetched to assume that this is what many entrepreneurs look for. Of course, a network configuration allows for a certain degree of malleability – especially in the sense of *quantitative* adjustments – and this is an interesting feature of industrial districts, but the alleged benefits of networking can be at odds with the pursuit of more articulate strategies.

---

proceeded "from proximity to experience to trust to collaboration to enhanced regional growth."

[29] Interesting contributions on this theme have been collected by Gambetta (1988), which includes an essay by Lorenz on trust in firm networks.

[30] Bergman and Feser (1997).

[31] As some authors observe, trust is relatively easy among firms and their suppliers. A much rarer phenomenon – with the apparent exception of Italy – is trust among competitors.

2.13 Flexibility and flexible specialization

In *The Second Industrial Divide,* Piore and Sabel (1984) characterized 'flexible specialization' as the developmental stage succeeding Fordism, stressing the superior qualities of the Italy of the districts, as featured by its clusters specializing in high-fashion products. They saw virtuous networks emerging among rival firms, which managed to cooperate in the performance of activities of mutual benefit such as training, marketing, and research.

The term 'flexibility' has come up again and again in literature since the Eighties. The main point seems to be that SMEs can enjoy scale economies without necessarily evolving into big, integrated firms. As one of the main disadvantages of large scale is the difficulty to adapt to changes in (level and content of) demand, it should follow that SMF systems can effectuate mass production (as if they were large firms) even in a condition of changing demand. This idea presents similarities with others, which have emerged from the recent debates, such as 'flexible automation'. However, flexible specialization (FS) is not the same as flexible technology. As it will be obvious from the previous references, there is something more: use of technologies – which occasionally can be of the 'flexible' kind – and organizational adaptiveness in order to obtain great quantities of varied products.

In real terms, we have witnesses to the ability of SMFs networks to accommodate a *growing* demand. This was obtained, in the age of the maximum development of districts, through the entry of many new producers into the expanding sectors, generally as subcontractors. At the same time, we would have difficulty in maintaining that those systems are – or have been – so peculiar as to allow the producers take up and leave their jobs according to the level of demand. This could happen in very primitive conditions, which can be inconsistent with the requirements of today's industry (especially fashion).

Certainly, SMFs may have a greater capacity than large firms to adapt to changing demand in quantitative terms, that is, they can indeed adapt, if the neighbouring economy is wealthy and offers a surplus of vacancies in other sectors. But, while an *increase* in demand calls for greater exploitation of capital and men, or for an entry of new firms into the market, a *decreasing* demand cannot be met other than by falling profits, or by closure. This amounts to saying that flexibility implies a certain steady level of scale.

In general, then, periods of boom do not raise the same problems as slumps, in so far as expanding the capacity, with or without diversification, surely calls for less stringent measures and conditions than curtailing it.[32] Again, very little has been proven, in empirical terms, in order to confirm or deny these points.

[32] There are certainly sociological implications in what has been said, and sometimes one gets a feeling that economics alone may be unable to grasp them all. A preference for free lancing could be an important feature of a whole class of individuals, without them necessarily becoming entrepreneurs in the strict sense. Such preference has been noted by many scholars in different economic sectors and social contexts (e.g., the attitude of the younger generation to skip factory work in favour of less paid, occasional jobs).

As an alternative hypothesis to the prevailing one, flexibility in the *upward* direction can indeed be shown to have existed as a generalized peculiarity of many specialized areas - and it was provided by ease of entry. Flexibility in the *downward* direction, however, may be counterproductive for the system: a closed firm is a dead firm. SMFs had to look for ways to survive bad times.

This interpretation is compounded by the fact that SMF systems grew at a time when demand was channelled toward the Italian producers by (and through) foreign buyers. When this stopped, the search for new markets became the crucial variable in the districts' life. Otherwise, firms had to devise completely new strategies, e.g., merging into groups under one leader, and with different assignments.

From another point of view, it could be suggested that the whole issue of flexibility has been slightly overworked, at least as an exhaustive explanation of the initial thrust shown in industrial districts. In full analogy with what happens with flexible automation, we know that mass production is not completely inconsistent with product differentiation. A variety of final products can be obtained from the same production line, with only limited (and inexpensive) adaptations to margin. Production lines become longer, rather than shorter, and exploit scale economies to a great extent. At the same time, however , smaller changes are introduced at the bottom of each line, thus giving the production the appearance of variety. Therefore, at least in certain industries, standardization and product diversification need not be so inconsistent with each other as a certain reading of the idea of Fordism would suggest[33]

On the other hand, and in closer proximity to the current condition of many industries, we know that, especially in the fashion industry – where vagaries of demand are certainly significant – the whole of each year's variable investment (which can account for a great part of total investment) is targeted for the same year's production. Therefore, changes which take place from one year to another are dealt with through the adoption of the 'right' machinery. Investment is risky because it may be either a success or failure, not because it may 'freeze' too much capital within the firm for too long.

3 Conclusions

These are the points to which, in my opinion, it might be useful to devote some additional research: (i) in spite of a debate which has gone on for little less than 25 years, evidence about the actual meaning of the Marshallian industrial district theoretical 'revolution' is still unconvincing; however, the need for empirical

---

[33] In many cases, we know that firms were *not* exposed to such a rapidly changing demand. As one manufacturer once said in the course of an interview in the mid-Fifties, "shirts are 80 % white, even in the top price rank…'Fancy' shirts account only for a minor slice of the total, and can be regarded as marginal after all".

material has been replaced lately by a rather self-indulging propensity for models of virtual realities; (ii) these models tend to assume an idea of optimality as their starting point; leaving aside the results obtained so far by this literature - a point which I have not commented here - the assumption is likely to bias the research in many respects; (iii) there are ideological implications of the model, which one may or may not like; we dare say that they verge upon a certain *laissez-faire* in politics and on a re-evaluation of 'weak thinking' in general; (iv) it is difficult to say whether the renaissance of a certain municipal egotism, in Italy, has been completely unrelated with the revision of academic common sense about large and small firms; at any rate, such implications go beyond their most apparent scholarly meaning, and affect the very way we look at facts in economic history; by denying relevance to the dynamic approach, we constrain ourselves to take the most rigidly neoclassical standpoint; (v) categories are perhaps to be re-defined; the prevailing *models* sit rather badly with what we know of districts in 'traditional' sectors which, after all, were the ones which generated the interst for the Marshallian industrial district model, because of their astonishing performances in terms of income and employment; at the very least, districts operating in the fashion industry (which, however, account for a substantial part of the phenomenon) should be set aside; (vi) quantitative research is badly needed; what we have at our disposal presently is either flawed by a certain amount of conceptual confusion, or inappropriate for meaningful comparisons; on the other hand, we lack shared criteria to judge the reliability of such research work; things being as they are, qualitative results are still a central issue; a satisfactory combination of the qualitative approach and of the intellectual *finesse* that is typical of so much of the recent literature, with more systematic hypothesis-testing could produce very interesting results.

References

BECATTINI G., (2000), *Il distretto industriale*, Rosenberg & Sellier, Torino.

BERGMAN E. M., FESER E.J., (1997), "Industrial, Regional or Spatial Clustering?" in: *OECD Workshop Position Paper on Cluster Analyses and Cluster-based Policies* , OECD Industrial Cluster Focus Group, Paris.

BRIOSCHI F., G. CAINELLI, (a cura di), (2001), *Diffusione e caratteristiche dei gruppi di piccole e medie imprese nelle aree industriali*, Giuffré Editore, Milano.

BRUSCO S., (1989), *Piccole imprese e distretti industriali*, Rosenberg & Sellier, Torino.

BRUSCO S., (1991), *La genesi dell'dea di distretto industriale*, in PYKE F., BECATTINI G., SENGENBERGER W. (a cura di) (1991), *Distretti industriali e cooperazione fra imprese in Italia*, Rosenberg & Sellier, Firenze.

CALAFATI A.G., (2002), *Apprendimento collettivo e sviluppo locale*, in: CAMAGNI R., CAPELLO R. (a cura di), (2002), *Apprendimento collettivo e competitività territoriale*, Franco Angeli, Milano.

CAMAGNI R., CAPELLO R. (a cura di), (2002), *Apprendimento collettivo e competitività territoriale*, Franco Angeli, Milano.

CAMAGNI R. (2002) *Competitività territoriale, milieux locali e apprendimento collettivo: una contro-riflessione critica*, in: CAMAGNI R., CAPELLO R. (a cura di), (2002), *Apprendimento collettivo e competitività territoriale*, Franco Angeli, Milano.

CAPECCHI V., (1991), "Una storia della specializzazione flessibile e dei distretti industriali in Emilia-Romagna", in: F. PYKE, BECATTINI G., SENGENBERGER W. (a cura di), *Distretti industriali e cooperazione fra imprese in Italia*, Rosenberg & Sellier, Firenze.

FABIANI S., PELLEGRINI G., ROMAGNANO E., SIGNORINI L.F. (2000), "L'efficienza delle imprese nei distretti industriali italiani", in: SIGNORINI L.F. (a cura di), (2000), *Lo sviluppo locale. Un'indagine della banca d'Italia sui distretti industriali*, Donzelli, Roma.

GAMBETTA D., (1988), *Trust. Making and Breaking Cooperative Relations*, Basil Blackwell, Oxford.

HARRISON B., (1994), *Lean and Mean*, Basic Books.

IUZZOLINO G. (2000), "I distretti industriali nel censimento intermedio del 1996: dimensioni e caratteristiche strutturali", in: SIGNORINI L.F. (a cura di), (2000), *Lo sviluppo locale. Un'indagine della banca d'Italia sui distretti industriali*, Donzelli, Roma.

LANZARA R. (1988), "I rapporti strategici con le imprese conciarie", in: VARALDO R. (a cura di), *Il sistema delle imprese calzaturiere. Struttura e strategie competitive*, Giappichelli, Torino.

LORENZONI G., (1983), "La costellazione di imprese: una base di indagine sui processi di sviluppo", *Economia e Politica Industriale*, vol. 38.

LORENZONI G. (1987) "Costellazione di imprese e processi di sviluppo", *Sviluppo e organizzazione*, vol. 102.

MARITI P., (1980), *Sui rapporti tra imprese in un'economia industriale moderna*, Franco Angeli, Milano.

NUTI F., (1992), *I distretti dell'industria manifatturiera in Italia*, Franco Angeli, Milano.

PIORE M. SABEL C. (1984), *The Second Industrial Divide*, Basic Books, New York.

PYKE F., BECATTINI G., SENGENBERGER W. (a cura di) (1991), *Distretti industriali e cooperazione fra imprese in Italia*, Rosenberg & Sellier, Firenze.

RULLANI E., (2002), "Il distretto industriale come sistema adattivo complesso", in: QUADRIO CURZIO A., FORTIS M. (a cura di), *Complessità e distretti industriali. Dinamiche, modelli, casi reali*, Il Mulino, Bologna.

SIGNORINI L.F. (a cura di), (2000), *Lo sviluppo locale. Un'indagine della banca d'Italia sui distretti industriali*, Donzelli, Roma.

VARALDO R. (a cura di), (1988), *Il sistema delle imprese calzaturiere. Struttura e strategie competitive*, Giappichelli, Torino.

VARALDO R., L. FERRUCCI, (a cura di) (1997), *Il distretto industriale tra logiche di impresa e logiche di sistema*, Franco Angeli, Milano.

Modelling the structure and evolution of industrial districts

Mario Agostino Maggioni

Faculty of Political Sciences, Catholic University, Largo Gemelli 1, 20131, Milan, Italy

"Should an entomologist observing a moving line of ants try to understand what their leader sees in a certain new obstacle in its path and how it describes that obstacle in 'antish' (sic) to those following behind? Should the entomologist do no more than study the behaviour of the ants on the basis of an 'objective' description of their movements, of the terrain, of the obstacles in their path, etc, as they appear to human eyes?"
Becattini (2002), "Industrial Sectors and Industrial Districts: Tools for Industrial Analysis", *European Planning Studies*, vol. 10, n.4.

"Regional or 'spatial' economics can be summed up in the question: "What is where and why –and so what?" (...)?" Where refers to location in relation to other economic activity; it involves questions of proximity, concentration, dispersion, and similarity or disparity of spatial patterns (...). Until fairly recently, traditional economists ignored the where question altogether, finding plenty of problems to occupy them without giving any spatial dimensions to their analysis. Traditional geographers, though directly concerned with what is where, lacked any real technique of explanation in terms of human behaviour and institutions to supply the why and resorted to mere description and mapping".
Hoover (1971), *An Introduction to Regional Economics*, A. Knopf, New York.

1 Introduction

The two initial quotations compose an ideal 'dialogue' between the Italian tradition of analysis of 'industrial districts' and the American tradition of regional economics on a very crucial question. How should be local production systems studied in order to get the most interesting (for academia) and useful (for both firms and public authorities) insights? Should they be studied from an 'objective and external' perspective or from an 'subjective and internal' one?

In this dialogue, violating the chronological order, the two quotations have been arranged in such a way that Becattini poses a question and Hoover gives, what I believe to be, a convincing answer.

In my opinion, this is not the time in which a regional economist should devote himself thoroughly to the use of the internal perspective[1] (i.e. to the study of 'ants

[1] Although a working knowledge of the insider perception of rules, institutions, habits and even dialect spoken within industrial districts may be of some help when dealing with applied analysis of specific cases.

language', to use Becattini's image), but the time in which he/she should take the external perspective and try to build on the plethora of interesting case studies and analyses them with the economic theory's toolbox, leaving aside other interesting academic traditions and approaches (such as the sociological and the ethnological one).

Aim of this chapter is therefore to follow the Hooverian tradition (explaining what is where and why – and so what) by focussing on three crucial characteristics and phenomena which originate and shape the evolution of industrial districts, namely: agglomeration economies, inter-industry relations, and governance structures from a theoretical perspective.

2 How do industrial districts come about? Agglomeration economies and firms' clustering

In this section we theoretically deal with the macro-economic aspects of the location process which originates an industrial district, concentrating on the patterns (and causes) of local industrial clustering and development, on the relationship between the development of different industries in the same clusters and on the intertwined development of clusters and industries.

If one starts the analysis from this macro-economic perspective, the first issues to be addressed concern the existence and the extent of spatial clustering processes, the main questions to be answered being the following: why do firms cluster in an industrial district? and why do districts have a finite size?

2.1 The effect of previous locations on clustering dynamics

In the following sections, firms' decisions to locate into a district (and consequently the entry rate and the development path of the industrial district) are explained in terms (as a function) of the 'economic mass' of the cluster, measured through the number of firms already located there. From a mathematical viewpoint, the core of this situation can be represented by a differential equation (or by a system of differential equations, when multi-regional a/o multi-sectoral interactions are considered) which describe the changes over time of a variable (birth minus deaths, inflows minus outflows) as function of: the level of the same variable (the 'economic size' of the district) at each moment in time; a ceiling level (which takes into account the limit imposed by the available amount of resources, both inputs and output); and the level of other variables (which represent the interactions between different industries and different districts).

However, these models do not merely postulate the existence of a stock-flow mechanism between the number of incumbent firms in the district and the entry rate, even if one may reasonably assume the number of spin-off firms, which are generated in each period within the district, to be a proportion of the number of firms already established there. Neither do they assume that the number of firms is the only relevant variable considered by potential entrants when taking locations decisions. Instead the number of already located firms reduces potential entrants' search costs, signals the existence of geographic and agglomeration benefits, diffuses information, reduces uncertainty, and causes informational cascades.

What follows is thus a family of models aimed at explaining location decisions and the development path of an industrial cluster when firms interactions (both effective and perceived) are relevant. For the sake of exposition, we first present a model describing the effects of the location decisions of a firms within an isolated district (i.e. when there is only one potential site for the location of firms belonging to a given industry). In the other sections, the scope of the analysis is enlarged to encompass a number of interactions between two (or more) clusters and/or industries.

2.2 Locational benefits and costs and the development of an industrial district

Firms decide to settle in a district on the basis of the expected profitability of being located there. This profitability depends on net locational benefits – obtained as the difference between gross locational benefits and costs – which, in turn, are based on both observable and unobservable elements.

For simplicity it can be assumed that, in an uncertain world – with limited information regarding local costs and revenues available to the outsiders – profitability expectations for any particular location will be based solely on the number of firms already located there (the number of previous locations being the only observable variable).

Let us assume, as in Arthur (1988 and 1990), that locational gross benefits B_{fq} for firm f locating in district q are composed of geographical and agglomeration benefits[2].

Geographical benefits G_{fq} depend on the intrinsic features of the site (such as the quality of local factors of production: capital k_q and labour l_q; the efficiency

[2] For analytical convenience we split locational benefits in two classes: geographical and agglomeration benefits. The first class refers to those components which are unaffected by the number of incumbents; while the second refers to those components which depend on the number of incumbents. By adopting this formulation, however, we do not intend to state that agglomeration benefits refer only to spillovers of scientific and technological knowledge and know-how. On the contrary we are convinced that relevant agglomeration benefits derive also from external economies of scale in the use of local resources. The same variable (i.e. labour productivity) has a fixed geographical component, which depends on the quality of local workers, and a variable agglomerative component which depends on the number of firms already located in the district.

of the local network of specialised suppliers and business service firms s_q; and the quality of urban and industrial infrastructures u_q). Agglomeration benefits $A_{fq}(n_q)$ are a concave non monotonic function of the number of incumbents (i.e. firms already established in district *q*) n_q. Thus:

$$B_{fq} = G_{fq}(k_q, l_q, s_q, u_q) + A_{fq}(n_q) \qquad [1]$$

The assumption of concavity and non monotonicity in A_q implies that, as the number of firms located in district *q* increases, gross benefits firstly increase because of agglomeration economies (due to productive specialisation; scientific, technical and commercial spillovers; reduction in both transport and transaction costs, increases in the quality of the local pool of skilled labour force and in the efficiency of the local credit market); then decrease when congestion more than compensates for agglomeration economies.

Locational costs c_{fq}, symmetrically, include two components: geographical costs g_{fq} (reflecting the cost structure of the cluster in terms of locally prevailing wage w_q and interest rate r_q; average price of business services d_q; and level of land rent and taxation t_q), and agglomeration costs a_q, which are assumed to be a convex non monotonic function of the number of incumbents in the district n_q.

$$c_{fq} = g_{fq}(w_q, r_q, d_q, t_q) + a_{fq}(n_q) \qquad [2]$$

The assumption of convexity and non monotonicity in a_q implies that, as the number of firms in district *q* increases, locational costs initially decrease until some 'optimal' number of users for a given set of urban, industrial and environmental infrastructures and resources is reached. Then they increase due to the competition, between a larger number of firms, for a limited pool of local inputs (i.e. capital, labour, business services, land and public infrastructures) which raises their prices[3].

---

[3] An alternative explanation for the convexity of the locational costs function for firm *f* runs as follows: the locational costs function is compose by a 'fixed' and a 'variable' component. The fixed part of the costs (geographic costs) decreases as the number of entrants increase; while the variable part increases (because of competition) as the number of entrants increase. The combination of these two effects produce an U-shaped (convex) cost

Net locational benefits can now be calculated as the difference between equations [4.1] and [4.2].

$$N_{fq} = B_{fq} - c_{fq} = H_{fq}\left(w_q, r_q, d_q, t_q, k_q, l_q, s_q, u_q\right) + h_{fq}\left(n_q\right) \qquad [3a]$$

Assuming that the geographical benefits and costs do not change overtime, if we focus the analysis of the location process on the dynamics of the interactions between the level of available locational benefits, what becomes relevant for describing firms' location decisions is just the net benefit function N_{fq} in the incumbents' space. We can therefore summarise the geographic components H_{fq} with a parameter α_q, which vertically shifts the locational net benefits function, and write the following expression:

$$N_{fq} = B_{fq} - c_{fq} = \alpha_q + h_{fq}\left(n_q\right) \qquad [3b]$$

It easy to see that the locational net benefits function [4.3b] is always concave, since N_{fq} is equal to the difference between a concave function $B_{fq}\left(n_q\right)$ and a convex one $c_{fq}\left(n_q\right)$. In other words, each marginal firm, which enters the district, increases the average profitability of locating in the district only up to a threshold. After that point, any new entrant lowers the average net benefits available to each resident firm and new entrant[4].

This formulation recalls some general results, obtained in the industrial location and urban/regional economics literature (Weber, 1929; Isard, 1956; Richardson, 1978; Papageorgiou, 1979; Tauchen and Witte, 1983; Miyao and Kanemoto 1987), which show the existence of an optimal dimension of a given spatial agglomeration of firms and/or households because of the concavity of the various benefits functions.

---

curve as the interaction between fixed and variable costs of production in standard microeconomics textbooks. A symmetric reasoning may also explain the inverted U-shaped benefits function. This interpretation is surely more realistic than the one used in this paper, however it is not as theoretically efficient as the other one since both components become dependent on the number of incumbents.

[4] However, as it is made graphically evident in Figure 2, because of the inverse *U* shape of the marginal benefits function, there is a range, within the number of incumbents, where marginal net benefits are already decreasing, but still higher than average ones, and average net benefits are still increasing.

We can therefore state that – if the number of potential entrants is sufficiently large and there are no relevant entry barriers – as net locational benefits initially increase, the number of incumbent firms increases; then they decrease and are finally driven to zero.

If we assume that the number of potential entrants (i.e. the number of firms belonging to a given industry) is not constant but it is changing overtime, than the entry pattern will be determined by the rate of birth of potential entrants, since a proportion of such firms will locate in the district. If we allow the number of actual entrants to be proportional to the average locational benefits available in the district – as a first approximation (or, alternatively, as a macroeconomic hypothesis without explicit micro-foundations[5]) – and we assume that the entry rate of firms into the district is proportional to the current level of locational net benefits[6], then the district growth will initially be fast, then will slow down and finally will stop[7].

If this is the case[8], one would expect the growth of the industrial district to follow an S-shaped path with a slow start (when locational benefits are still low), an 'explosive' central period of rapid increase (when the average net benefits in the region are highest) and a final part when the district gradually reaches and then settles to its 'equilibrium'[9] size as in figure 1.

---

[5] Another part of this story refers to the role played by spin-off firms. If we believe that average net locational benefits give incumbent firms some extra profits, workers (or, better, managers) of incumbent firms, which see these extra profits, will decide to start a new firm to appropriate them. Therefore the number of spin-off firms, which are generated by parents firms, is positively related to the level of locational benefits in the cluster, generated by the incumbent firms.

[6] It is evident that, for the moment, we are implicitly assuming firms to behave in a non-strategic manner (i.e. although their entries modify radically the locational benefits available in the district, they do not take into account such a modification) and to be totally myopic (i.e. their entries are only dependent on the current level of locational benefits and not on some discounted value of the flow of locational benefits they enjoy from the moment of entry onward).

[7] This is not to say that, from a certain moment onward, entries will not take place. On the contrary, once the equilibrium level of the district is reached, new entries are still possible, but these happen at the expenses of some incumbents which are driven out of the district (or, more probably, out of business). This continuous series of entries and exits may determine a continuous oscillation of the district's industrial mass around the equilibrium level.

[8] And we collected some empirical evidence supporting this story (see Maggioni, 1993; Maggioni and Porro, 1994; Gambarotto and Maggioni, 1998).

[9] It should be borne in mind, however, that this notion of equilibrium value can be changed in the long run by radical innovations, exogenous demand shocks and appropriate regional policies.

Figure 1. The development path of an industrial district

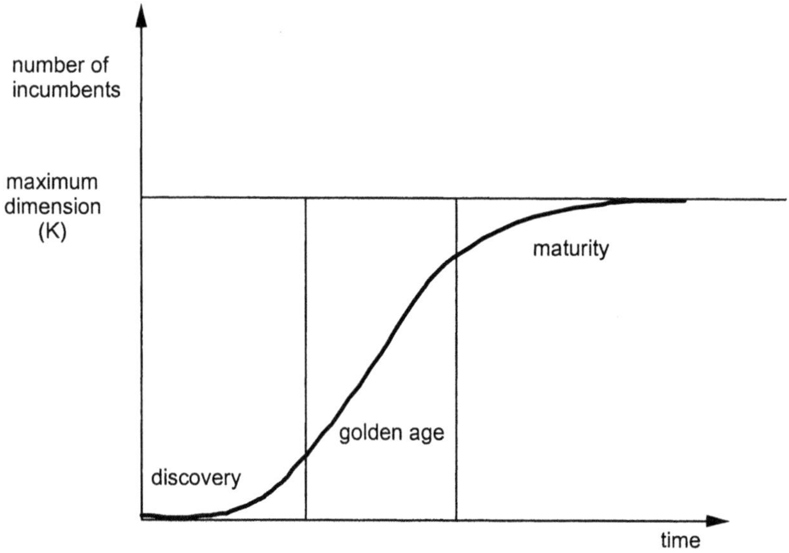

What is needed now is a formal model of the location process and the development of the industrial district, which may allow a rigorous empirical testing of all these assumptions.

2.3 A logistic model

The simplest growth model for an industrial district *q* – which stresses the relevance of firms spatial interactions – can be expressed in the following format: "the rate of growth of the industrial mass equals the product of the individual firm's contribution[10] to the district population's growth and the number of firms already in the district" (Maggioni, 1993).

If only agglomeration economies and positive spillovers are taken into account (and these are assumed to be constant), then each individual firm's contribution to the level of average locational benefits and, consequently, to the growth of the cluster would be equal to a constant r_q. In this case the district industrial growth would follow an 'explosive' exponential path[11], formally:

[10] In terms of changes in the average locational net benefits, due to the interactions of agglomeration economies and diseconomies.

[11] The higher is r_q, the faster is the growth process.

$$\frac{dn_q}{dt} = r_q \, n_q(t) \qquad [4a]$$

Alternatively one can solve the equation for $n_q(t)$ as a function of the exogenous initial industrial mass of the district $n_q(0)$:

$$n_q(t) = e^{r_q t} \, n_q(0) \qquad [4b]$$

On the other hand, if congestion and competition effects are included, then some modifications to this simple model are required to allow for some 'density dependent' factors to progressively depress the level of locational benefits and to slow down the process of industrial growth of the district. A simple dynamic model, which takes into account these features is the logistic equation[12], which can be written as [4.5a]

$$\frac{dn_q}{dt} = r_q \, n_q(t) \left(1 - \frac{n_q(t)}{K_q}\right) \qquad [5a]$$

where r_q is the incipient (or maximum) rate of increase and $K_q = \lim_{t \to \infty} n_q(t)$, is called the district 'equilibrium' level[13]. Integrating equation [5a] and solving for $n_q(t)$ one obtains [5b]:

$$n_q(t) = \frac{K_q n_q(0) e^{r_q t}}{K_q + n_q(0)\left(e^{r_q t} - 1\right)} \qquad [5b]$$

[12] The logistic equation – firstly developed by Verhulst (1845) and Pearl and Reed (1920) for demographic studies, then adopted by the ecological literature since Lotka (1925) – "is the simplest model containing negative density dependence interaction. Further, it is the first two terms in a power series expansion of a more general growth model where the growth is a function of the actual size of the population" (Dendrinos and Mullally 1985, p. 38).

[13] *K* in the original ecological jargon is called *carrying capacity*: "a measure of the amount of renewable resources in the environment in units of the number of organisms these resources can support" (Roughgarden 1979, p. 305).

Here the individual firm's contribution to regional growth decreases as a linear function of the district population size and is equal to $r_q - \frac{r_q}{K_q} n_q(t)$ (Roughgarden, 1979).

Plotting $n_q(t)$ against time yields an S-shaped curve due to the counteracting roles played by r_q and K_q. When the district is small ($n_q(t)$ is near to zero) the term in brackets in equation [4.5a] is close to one (hence the logistic equation approximately describes an exponential growth path); but as $n_q(t)$ approaches K_q, the term in brackets tends to zero, driving the growth rate to zero and terminating the entry process. Both K_q and r_q play a major role in shaping a logistic growth path: the greater is r_q the steeper is the S shaped curve, the larger K_q the higher the ceiling level of the function (and the equilibrium size of the cluster).

r_q is the incipient rate of growth or the intrinsic rate of industrial growth of the district which is calculated as the difference between total entries (or start-ups) and exits (or bankruptcies) in the time period considered. The same value of r_q can therefore correspond to two very different situations: a steady growing district where few new firms enter and no one exits, and a perturbed cluster where a high 'birth' rate is almost compensated for by a high 'death' rate. Hence r_q is a composite index that describes the cluster growth 'potential' and the probability that firms, once entered, survive in the cluster.

K_q defines the district carrying capacity: the maximum number of profitable firms the region can sustain in isolation (i.e. when inter-regional interaction are not considered). K_q will depend upon: (i) the finite quantity of geographical benefits (which is related to the limited availability of local 'resources' such as: labour, capital, land, and infrastructures); (ii) the decreasing part of the agglomeration benefits function (which depends on the strategic interactions between firms: competition, congestion and lobbying of incumbents).

K_q is therefore determined by the relationship between the amount of resources (inputs) available in the district and the (technical and organisational) efficiency of incumbents in the use of these resources. Therefore in the long run K_q may change as result of the inflow of additional skilled workers, the provision of new advanced public infrastructure, the diffusion of (technical, organisational, etc.) innovations.

For a given district q and a given population of M_q outsider firms[14], therefore, we assume that there is an equilibrium level $K_q \leq M_q$ acting as an upper limit to

[14] Composed of two main categories: firms already established outside the district and potential entrepreneurs inside the district looking for the right moment to start their own business.

the cluster's growth. In each period t, the number of entries therefore depends both on the actual number of potential entrants $K_q - n_q(t)$ (i.e. the number of outsider firms which can enter the district in time t and still make profits) and on the number of firms already located there $n_q(t)$. K_q and $n_q(t)$ in fact determine the level of average locational net benefits available to incumbent firms in each period of time.

The logistic equation thus establishes a parametric relationship between the level of locational benefits and the entry rate through the introduction of the term K_q. But what exactly is K_q and how is it determined? Is it an endogenous or exogenous parameter of the model? For it to be endogenous, some assumptions about the locational cost structure of a region are needed.

In section 2.2 we described the functional forms of locational benefits and costs. In this section, with the help of Figure 2 (which shows both marginal and average locational costs and locational benefits schedules), we want firstly to highlight the existence of several 'optimal' sizes[15] of the district, and secondly to show how K_q is endogenously determined by the structure of locational benefits function.

Let us consider the costs and benefits derived from entering a district; for the sake of simplicity let firms outside the region experience zero locational benefits[16] and assume that geographical benefits are set to zero. Firms are assumed, for the moment, to be locationally identical (i.e. the agglomeration economies and dis-economies, locational benefits and costs are the same for every firm). Therefore we can study the behaviour of a representative firm *f* and analyse its average net benefits function[17].

---

[15] Throughout the chapter the 'industrial size' or 'economic mass' of a district is approximated by the number of located firms. This index can easily be substituted by a more realistic proxy of firms' dimension (such as employment or sales). However the number of firms has an obvious advantage in its simplicity and is the best indicator when the inter-firms relationships at study are independent of firm's size.

[16] However, this assumption can easily be relaxed by assigning the locational benefits of the cluster a value equal to the difference between the locational benefits available outside and inside the cluster.

[17] By considering average functions we indirectly assume that some market mechanisms is at work in the cluster and makes both benefits and costs equal for each incumbent.

Figure 2. Agglomeration costs and benefits for incumbents and critical sizes of a district

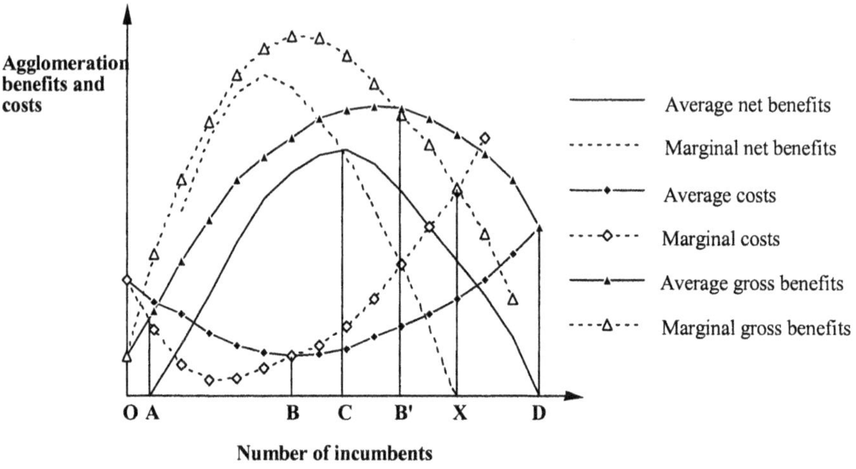

A is the minimal sustainable dimension of the district (i.e. where agglomeration net benefits start to be positive and, consequently, $\frac{dn_q(t)}{dt} > 0$). Prior to A no firm will spontaneously enter the district (because agglomeration benefits are negative). A can be called the 'critical mass' of the region. A can be reached only by a group of co-ordinated firms entering together, or by direct intervention of a public authority aimed at subsidising entries until $n(t) = A$.

B is the dimension where average agglomeration costs are minimum. B' is the district dimension which maximises gross average agglomeration benefits. B and B' underline the importance of analysing both costs and benefits of location to avoid harmful misrepresentation of the economic reality, as in some early contributions of location theory. Obviously, it could also be the case that B' < B.

C gives the maximum per firm net benefits (i.e. average net benefits). Up to C every new entrant increases (by its very entry) the average benefits of all incumbents; after C the average benefits decrease. C is therefore the optimal size of district for incumbent firms; however, it is neither the social efficient outcome (given that marginal benefits are still greater than marginal costs) nor the maximum possible dimension (average benefits are still positive). At C, several firms outside the district might still want to enter, while firms already in the district would like to deter further entries. Here we have a contrast between incumbents, outsiders and public authorities, each of them with a different view of what is the optimal outcome.

X is the economically efficient (i.e. social optimum) dimension of the industrial district. At X marginal costs equal marginal benefits, therefore the total benefits (number of firms times *per capita* benefits) are maximised. However, as the average benefits at X are still positive, some outsiders would still like to enter. Such entries would reduce the total amount of benefits available to incumbents.

D is the maximum dimension of the region (in terms of economic mass) since $B_{fq} = c_{fq}$. From D onwards no more net entry is deserved because, after this point, average benefits are negative and therefore there are no incentives to enter. However, new entries are still possible but these would be at the expense of some incumbents who would be driven out of the district[18]. Thus D identifies the ceiling level K_q of the logistic model shown in Figure 1.

By choosing specific functional forms for locational benefits and costs, one can thus endogenously determine K_q, the maximum dimension of the district, corresponding to 'size' D in figure 4.2, where net locational benefits are equal to zero. If, for example, we assume a quadratic formulation for the average net benefits function (i.e. $N_{fq} = an_q^2 + bn_q + c$, with $a < 0$, $b > 0$, and $c > 0$), then the district equilibrium level D would be equal to:

$$\mathbf{D} = \frac{-b + \sqrt{b^2 - 4ac}}{2a} \qquad [6]$$

By similar calculations every other critical dimension of the district can be found[19].

In this first formulation, the number of located firms directly generates (through agglomeration dynamics) the level of locational benefits; since the entry rate is assumed to be proportional to the level of locational benefits, it also indirectly determines the location of new firms into the district.

Such a formulation of the location process is very simple but can be used to empirically estimate key parameters of the location path of different districts[20]. These estimated parameter could also be used as dependent variables in cross-section analyses in order to assess the influence of different factors on the level of the intrinsic growth rate of a cluster or on its maximum dimension.

However this formulation can be criticised on a number of grounds. The first refers to the 'isolation hypothesis': firms' location decision are modelled as a dichotomous choice, in the sense that there is only one possible site for location and the choice variable is just the timing of the entry. The second concerns the indi-

---

[18] After D new entries thus support a turnover process without causing relevant changes to the equilibrium level.

[19] Referring to Figure 2 other main critical dimensions – in terms of the underlining parameters of the net benefit function – are as follows: $A=(-b-(b^2-4ac)^{1/2})/2a$, $C=-b/2a$, $X=(-b-(b^2-3ac)^{1/2})/3a$.

[20] Where the difference may refers to different industries in the same geographical site, or to different geographical sites in the same industry.

vidual rationality of these firms: the entry rate is modelled as proportional to the level of locational benefits and firms do not anticipate the effects of further locations on such benefits.

Since we have already dealt with expectation in other papers (Maggioni and Porro, 1994; Maggioni, 2001) we focus here on the 'isolation hypothesis' to show how this model can easily accommodate for inter-sectoral and inter-regional relationships.

2.4 Location process in a two-districts framework

Several different interactions may arise when a framework with two districts (1 and 2) is considered. The first concerns the location decisions of a set of industry-specific potential entrant firms confronted with the choice of locating in one out of two districts. This problem can be formally described through a model of inter-regional competition with a given number of potential entrant firms[21]. The entry decision is again modelled as dependent on the number of firms located in each district, reflecting the effects of agglomeration economies and diseconomies, as described in the previous section.

Formally, the multi-region situations can be described through the following system of equations which directly recalls the single district development process of equation [5a]:

$$\begin{cases} \dfrac{dn_1}{dt} = r_1 n_1(t)\left[1 - \dfrac{n_1(t) + c_{12} n_2(t)}{K_1}\right] \\ \dfrac{dn_2}{dt} = r_2 n_2(t)\left[1 - \dfrac{n_2(t) + c_{21} n_1(t)}{K_2}\right] \end{cases} \tag{7a}$$

where r_1 and K_1 are respectively district 1 incipient growth rate and equilibrium levels and c_{12}, is the 'competition coefficient' which measures the extent to which district *2* compete as an alternative location to district *1*.

However – for expositional purposes – it is more convenient to replace this formulation (which underlines the similarities between the two-district case and the model of development of one isolated district depicted in equation 5a) with an alternative formalisation which is able to model different multi-district interac-

[21] This peculiar approach to the process of multi-regions development (where the district is modelled as an agent and firms as passive industrial mass) reveals its advantages when modelling active policy pursued by different local authorities willing to foster the industrial development of their respective industrial districts.

tions through simple differences in the sign of parameters. We thus replace equation [7a] by [7b], where time references are dropped for the sake of clarity:

$$\begin{cases} \dfrac{dn_1}{dt} = \left(a_1 - a_{11}n_1 - a_{12}n_2\right)n_1 \\ \dfrac{dn_2}{dt} = \left(a_2 - a_{22}n_2 - a_{21}n_1\right)n_2 \end{cases} \qquad [7b]$$

In system [7b], n_1 and n_2 are the 'economic masses' (number of incumbents) of each district, a_1 and a_2 are the intrinsic rates of increase of each district in isolation (r_1, r_2 in system 4.7a), a_{11} and a_{22} – the parameters on the quadratic terms which reflect the concavity of the net agglomeration benefits function – i.e. that reflect the inhibiting effects that a firm's entry has on the growth rate of the same district[22] (intra-district competition parameters); a_{12} and a_{21} show the inhibiting effects that one firm – locating in a region – has on the growth of the other region[23] (inter-district competition parameters).

According to the relative value of these parameters, three main outcomes[24] are possible, as shown by phase diagrams in Figure 3.

Phase diagrams are useful tools to describe systems of differential equations. At any moment in time the state of the system is fully described by the number of firms located in each district (n_1 and n_2). Within the phase plane one can further draw for each district the locus of points (i.e. combination of n_1 and n_2), which is called the isocline, where the district industrial mass does not change (i.e. where $\frac{dn_1}{dt} = 0$ and $\frac{dn_2}{dt} = 0$). The multiple equilibria of the system are thus identified by the intersections of one isocline with either the axes or the other isocline[25]. The arrows in the phase planes indicate the direction in which the system,

---

[22] a_1 / a_{11} and a_2 / a_{22} in this formulations are the regions' maximum size or isolated equilibrium values (K_1 and K_2).

[23] Different interpretations for these parameters are possible: the simplest relates the values of inter-districtual competition parameters to the fact that – given a certain number of potential entrants – each firm that locates in district 1 does not locate in district 2. Then the inter-district coefficients should be equal, i.e. $a_{12} = a_{21}$. Another interpretation explicitly relates to the possible differences in the logistic growth path of the two districts in isolation (which reflect differences in the net benefits function). In this case inter-district competition coefficients can be different.

[24] Selected from the analysis of 13 possible consistent cases. For a more complete analysis, see Capelo (1989, pp. 165-169) and Beltrami (1987, pp. 72-73).

[25] The intersection of an isocline with an axis is an equilibrium point since one district in deserted and the other does not change its industrial mass, the intersection of two isoclines

at that point, will move; the colours and shapes of the dots show whether an equilibrium is either stable or unstable or a saddle point.

Figure 3. Inter-regional competition

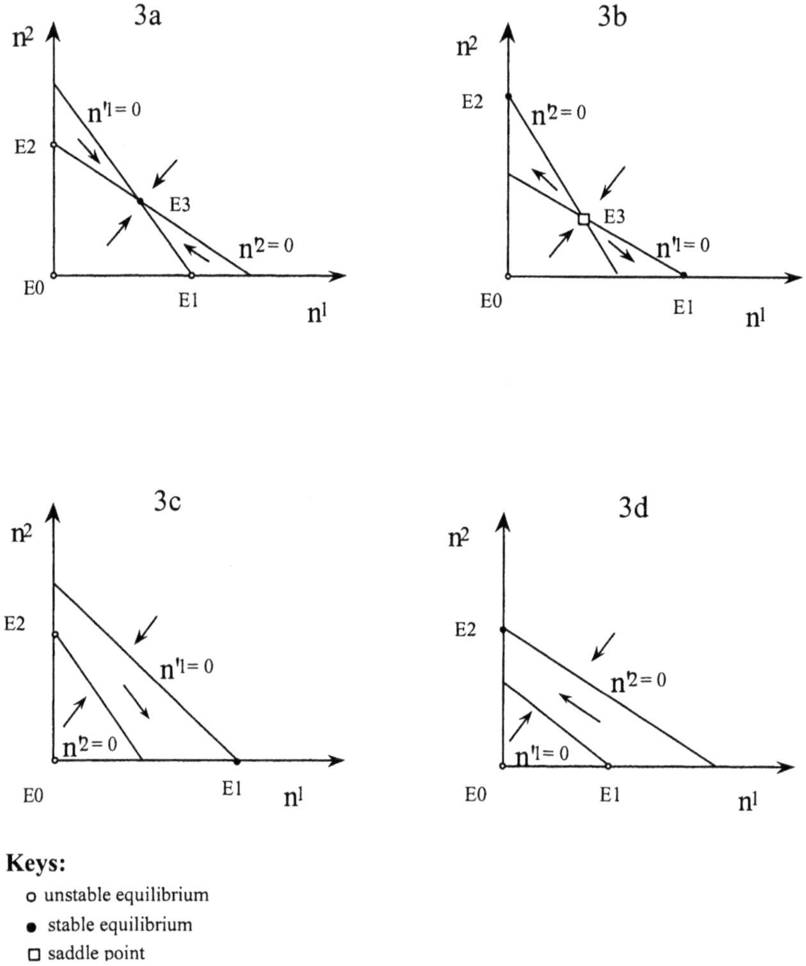

Figure 3a shows the persistent coexistence of two district (both recording a positive number of locations) as a stable equilibrium when:

---

is an equilibrium point since both districts do not register any changes in the number of located firms.

$$\frac{a_1}{a_{12}} > \frac{a_2}{a_{22}} \quad \text{and} \quad \frac{a_1}{a_{11}} < \frac{a_2}{a_{21}} \qquad [8a]$$

Figure 3b shows the coexistence of two districts as an unstable equilibrium, which is extremely vulnerable to local perturbations and initial biases[26] when:

$$\frac{a_1}{a_{12}} < \frac{a_2}{a_{22}} \quad \text{and} \quad \frac{a_1}{a_{11}} > \frac{a_2}{a_{21}} \qquad [8b]$$

Figures 3c and 3d show the total exclusion of one district from the location process that occurs only in the other district (irrespective of the initial state) which reaches its carrying capacity leaving the opponent deserted [27]. In particular, only district 1 is entered when:

$$\frac{a_1}{a_{12}} > \frac{a_2}{a_{22}} \quad \text{and} \quad \frac{a_1}{a_{11}} > \frac{a_2}{a_{21}} \qquad [8c]$$

while only district 2 is entered when:

$$\frac{a_1}{a_{12}} < \frac{a_2}{a_{22}} \quad \text{and} \quad \frac{a_1}{a_{11}} < \frac{a_2}{a_{21}} \qquad [8d]$$

Since the industrial growth of the district in isolation follows a logistic path, then the isoclines are linear and the stability of equilibrium depends on the equi-

[26] More precisely the equilibrium defining the coexistence of two districts is a saddle point; any initial state off the stable manifold, or any perturbation that move the system far from it, leads to the 'deprivation' of one out of two regions.

[27] This case includes two opposite situations of only district 1 surviving and only district 2 surviving.

librium point being above the line[28] joining the maximum dimensions of each district in isolation: $\left(\frac{a_1}{a_{11}},0\right)$ and $\left(0,\frac{a_2}{a_{22}}\right)$[29].

This presentation of the possible outcomes of firms location processes in a two-districts framework would seem to suggest that inter-district competition leads – almost necessarily – to a situation of locational monopoly, in which only one district is chosen as the preferred site by industry-specific firms while the other remains deserted. Although the economic history of industrial districts displays few cases supporting this result, the coexistence of multiple non-empty districts in the same industrial sectors is a more common phenomenon[30]. This must therefore be caused by factors still missing from the above analysis.

One way to explore the problem assumes the intrinsic growth rates of two regions to be equal (i.e. $a_1 = a_2$). Then the persistent coexistence of both districts is achieved when:

$$a_{11} > a_{21} \text{ and } a_{22} > a_{12} \qquad [8e]$$

These two inequalities state that, to obtain persistent coexistence of both districts, an increase in the numbers of incumbents in either district must inhibit its own growth more than it inhibits the growth of the other (e.g. that the degree of intra-district competition and congestion must be larger then the degree of inter-district competition). This result is likely to be achieved if the districts are, at least partially, 'attracting' different types of firms or, in other words, if firms within the same industry have significantly heterogeneous locational preferences.

This observation has been acknowledged within the population ecology literature for a long time under the name of the 'Gause principle'[31]. This principle states – in the original biological terms – that in a situation where "two species are resource-limited, [...] only one species can survive on a single resource" (Levin 1970, p. 413). Translated into an economic framework, the Gause 'principle of competitive exclusion' states that – if firms are completely homogeneous in their locational preferences – then the most likely outcome is the unanimous choice of one single district (the one ensuring at the start the highest level of geographical

---

[28] Whose equation is $n_1=(1-(n_2/(a_2/a_{22}))(a_1/a_{11})$.

[29] Alternatively, if one expresses the district size in 'equivalent numbers' (i.e. as fractions of their maximum dimension), the equilibrium is stable if the total equivalent number at the joint equilibrium is greater than unity (i.e. greater than the equivalent number at the isolated district's equilibrium).

[30] Such as: Lumezzane and Cusio for taps and fittings, Maniago, Frosolone and Premana for cutlery, Treviso and Carpi for knitwear.

[31] Being formally stated by Gause (1934), even though an earlier and slightly different formulation is ascribed to Volterra (1926).

benefits). The existence of a plurality of non-empty district inhabited by firms belonging to the same industrial sector can therefore be explained in terms of the existence of different types of firms which have different locational needs and preferences.

The degree of competition between two districts is therefore proportional to the degree of locational homogeneity of the firms population. The same concept can be expressed in a rather different way: the degree of inter-district competition is inversely proportional to each district's specificity[32]. According to this observation, each district should foster the development of specific distrectual characteristics in order to develop a 'niche' strategy able to limit the degree of inter-district competition. This observation, which has a general value, has a particular interest for any public authority willing to foster the development of a 'new' industrial district in an area where there are already other established district. The identification and the attraction of a specific subset of firms is the crucial element which determine the success, or the failure, of the regional development project.

2.5 Co-operative interactions

The analytical representation, used in equation [7b] to describe the location decisions of a set of potential entrants between two alternative locations, can be easily encompassed to describe other significant interactions which may arise during the development process of two industrial districts belonging to the same industry or the development process of two industries within the same district.

These different relationships can be modelled simply by changing the signs (or values) of the inter-district coefficients a_{12} and a_{21} in the system of equations [7b] as shown in Table 1.

Table 1. Different types of bilateral interactions

Tipe of Interaction	*Sign and values of interaction coefficients a_{12} and a_{21}*
Competition	- -
Mutualism	+ +
Commensalism	+ 0
Predation	+ -
Neutralism	0 0
Amensalism	0 -

[32] Where regional specificity means that each district is the ideal location for a different type of firm.

The regional economics literature records several contributions on predatory relationships (see for example; Dendrinos and Mullally, 1985; Maino, 1989; Nijkamp and Reggiani, 1991). This section therefore concentrates on co-operative relations that – despite their empirical relevance – have been neglected by the main stream of population ecology and, consequently, by the early pioneers of 'ecological' regional economics[33]. These relationships describe a set of co-operative interactions where the development of a certain set of firms, such as an industrial cluster in a given nation (or one industry in a given region) is not obtained at the expense of the development of another industry (or cluster) but the two processes enjoy synergistic effects.

Both ecological and economic systems record empirical evidence of mutualistic relationships between two populations over a long period of time.

In particular, we think that the model analytically described in expression [9] could well describe the behaviour of two technologically related industries within the same industrial, each receiving positive benefits from the development of the other one[34]. We think that the analysis of inter-industry relations is as important as (if not more than) the analysis of inter-district competition since one of the main findings of the most recent literature on industrial districts is that the physical proximity of firms belonging to different but interrelated industries $i = (1,2)$ act as an engine sustaining the district long-run growth[35].

The above remarks can thus be described more formally as follows:

[33] There are two different reasons why mutualistic and amensalistic relations have been ignored. The first reason is an historical one: both Lotka and Volterra – generally acknowledged as the forerunners of population ecology – limited their analyses to predation and competition. The second reason is a theoretical one: in its purest formulation, a mutualistic model does not have a stable equilibrium and one or both populations undergo "an unbounded exponential growth in an orgy of mutual benefaction" (May 1976, p. 95). A growth model without stable equilibrium seemed of no interest to regional and applied economists. However, more sophisticated models of mutualistic interaction allow for saturation in the magnitudes of reciprocal benefits and obtain – as general result – a stable equilibrium with both populations (usually districts, but they can equally be industries) developing larger than in isolation. It must be said, furthermore, that this equilibrium is less stable – in the sense that perturbations are more slowly damped – than in correspondent systems without mutualistic relations.

[34] However it is also theoretically possible that inter-industries relations, within a district, become conflicting because of competition on non industry-specific inputs such as real estates, financial services, fiscal and accounting consultants etc. The prominence of co-operative versus competitive interactions must therefore be empirically assessed.

[35] Swann defines such a process as 'convergence between technologies' and shows that "single technology clusters, while they might grow faster in the formative stages, did not have the lasting power of diversified clusters" (Swann 1998, p. 64).

$$\begin{cases} \dfrac{dn_1}{dt} = \left(a_1 - a_{11}n_1 + a_{12}n_2\right)n_1 \\ \dfrac{dn_2}{dt} = \left(a_2 - a_{22}n_2 + a_{21}n_1\right)n_2 \end{cases} \quad [9]$$

In this model each industry in isolation would follow a logistic growth pattern, but the presence of one industry in the same district has a positive influence on the rate of growth of the other one. The system's dynamics – which is described through a phase diagram in Figure 4 – always displays three trivial unstable equilibria (Figure 4a): the origin (corresponding to the underdevelopment of both industries) and each industry's isolated maximum dimension. However, under certain assumptions concerning the slope of isoclines (Figure 4b), a fourth stable equilibrium (E_3) emerges which enables both industries to reach a higher level in the district. The coexistence of two industries in the same district allows both industries to grow larger than would have been possible in the isolated case, but intra-industry competition effects prevent the system from experiencing explosive growth.

Figure 4. Inter-industry facultative mutualism

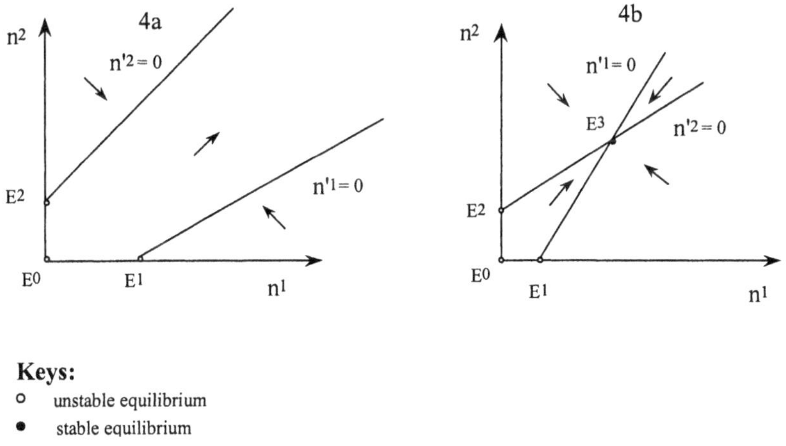

The conditions for an equilibrium solution in which both industries survive can be re-expressed as a strict positivity requirement on the co-ordinates of E_3 in the (n_1, n_2) space, formally:

$$\frac{(a_2 a_{11} + a_1 a_{21})}{(a_{11} a_{22} - a_{12} a_{21})} > 0 \qquad [10a]$$

$$\frac{(a_2 a_{12} + a_1 a_{22})}{(a_{11} a_{22} - a_{12} a_{21})} > 0 \qquad [10b]$$

Given that all coefficients are positive, the two conditions reduce as follows:

$$a_{11} a_{22} > a_{12} a_{21} \qquad [10c]$$

A stable equilibrium in a mutualistic interaction is therefore achieved when the product of the two intra-industry competition coefficients exceeds the product of the inter-industries co-operation coefficients.

The same analytical framework may describe the mutualistic growth of two neighbouring districts, specialised into complementary productions, where the driving forces of the joint development are vertical and horizontal productive linkages established between districts.

Other stimulating results are obtained when the symmetry assumption required by facultative[36] mutualism is relaxed. An interesting case concerns a situation in which the first district in isolation will grow according to a logistic path and its growth is further fostered by the presence of industrial activity in the neighbouring one, whilst the second district is strictly dependent on the existence of the industrial activity in the first cluster for its very survival. A typical example of this kind is the relationship existing between the 'core' of an industrial area (where the large companies are located) and the surrounding area, 'periphery', crowded by several small firms (i.e. sub contractors and suppliers) – specialised in accessory productions – which are vitally linked to the industrial success of the core[37]. Similar relationships (which do not imply any spatial differentiation) may well arise between two different industries in the same district. Another example (thoroughly discussed in Swann 1993) is constituted by the relationships between the manufacturing core of an high-tech cluster and the network of specialised service sectors. Since these specific business services must be 'consumed' in the very place of their production and cannot be exported to other areas, the existence of a service sector within the cluster is crucially dependent on the existence of a prosperous manufacturing core.

---

[36] Where the term 'facultative' refers to the fact that each industry, in isolation, develops and reaches its equilibrium level. The presence of an industry is therefore no necessary condition for the development of the other one.

[37] The Motor vehicle industry (which is clustered around Detroit in the USA, Coventry in the UK, Besançon in France, and Torino in Italy) displays a core-periphery spatial configurations in almost every countries.

These interactions can be formally described as follows:

$$\begin{cases} \dfrac{dn_1}{dt} = (a_1 - a_{11}n_1 + a_{12}n_2)n_1 \\ \dfrac{dn_2}{dt} = (-a_2 + a_{21}n_1)n_2 \end{cases} \quad [11]$$

where n_1 is the economic mass (number of located firms) of the manufacturing core area and n_2 is the economic mass of the specialised service sector.

With the aid of a phase diagram (see Figure 5), it is possible to see that the system has three equilibria. The first is trivial, unstable and corresponds to the origin; the second, stable, corresponding to the isolated maximum dimension of the core manufacturing industry; the third, which is a saddle point, corresponds to a balanced co-operative relationship[38], in which manufacturing and service sectors co-exist with mutual benefits. When the system lies in this unstable mutualistic equilibrium, small positive perturbations lead the system to an unlimited growth while small negative perturbations may cause the 'extinction' of the service sector. This is the only possible outcome when $\dfrac{a_2}{a_{21}} < \dfrac{a_1}{a_{11}}$ (see Figure 5b).

Figure 5. Core-periphery dynamics (inter-industry commensalism)

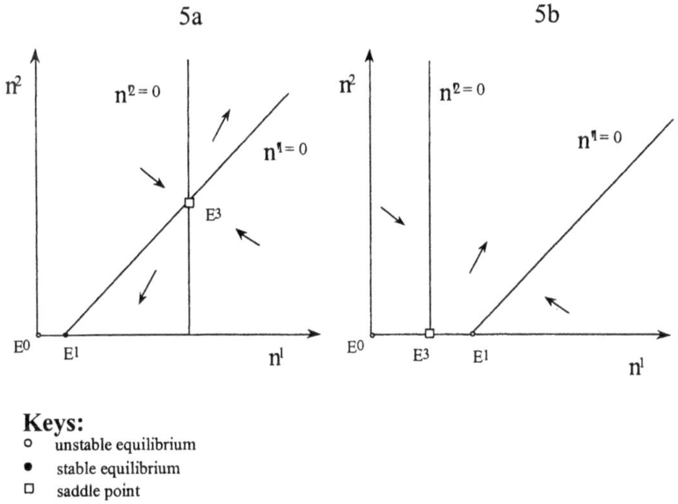

[38] Which exists if and only if $(a_2/a_{21}) > (a_1/a_{11})$ (see Figure 5a).

Various type of interactions coexist within the development dynamics of industrial districts. The qualitative identification of such interactions and the quantitative measurement of their effect on the growth rate of industries is thus left to the empirical analysis.

The following section is therefore devoted to the analysis of the governance structures of industrial districts in order to understand the inner mechanisms of districts and forecast their evolution.

3 How do industrial districts work? Emergence and evolution of governance structures

A large part of the economic literature on industrial districts is devoted to the reconstruction of the structural changes that industrial districts have experienced from the early 1970s.

Brusco (1990) divides the evolution of industrial districts into two main phases, which he calls Mark I (1970-1990) and Mark II (1990, onwards). In the first phase the district behaves as described in the "canonical literature: diffuse vertical cooperation and horizontal competition, strong interaction between market and non-market institution and cultures, very dense and a-hyerarchical structure of inter-firm relations" (Brusco, 1990, p. 16-17). In the second phase Brusco stresses the emergence of technological innovation and industrial restructuring and signals the problem arising from the "strong heavy inertia" of the governance structure of traditional industrial districts. "(N)ew markets and new technologies have developed since the beginning of the 1980s, posing a problem for both large and small firms (...) industrial districts must now face the problem of how social districts are to be endowed with the new technologies which are necessary to revive a process of creative growth" (ibid.). Brusco sees the solution in a set of public policy interventions aimed at offering small firms "what they need in kind, instead of offering them money to buy what they need" (ibid.). A direct interventions of local public authorities providing real services rather than financial help seems the only possible alternative to the crisis of industrial districts caused by dramatic changes in both commercial and technological environments.

Other authors (Dei Ottati,1994; Bramanti and Maggioni, 1994; Markusen, 1996; Varaldo and Ferrucci, 1997; Balloni *et al.*, 1998; Paniccia, 1998; Brioschi and Cainelli, 2001; Guerrieri and Pietrobelli, 2001; Withford, 2001) highlight the role played by structural changes in the governance structure of industrial districts. In their opinion[39] some industrial districts have gradually transformed their governance structure and their geographical scope in order to cope with external pressure both on the technological and on the commercial side: "only where lead

[39] Which, for the sake of simplicity, have been classified in one single group, downplaying the existing relevant differences between these works.

firms are able successfully to impose a quasi-hierarchical model will local systems thrive" (Withford, 2001, p. 52).

The main problem with most of the literature on industrial districts[40] lies in its intrinsic annedoctical character and in the lack of a general vision being mostly based on case studies. To overcome such limitation, we decided to modify a taxonomical device put forward by Storper and Harrison (1991) and to use it in order to analyse the structure and evolution of a significant number of industrial districts by classifying them into different cells of a matrix whose columns identify different types of governance structure (from the most egalitarian, *all ring no core*, to the most vertically structured, *core ring with leading firm*) and whose rows identify the geographical agglomeration/dispersion of the network of firms.

We therefore used different types of information - from direct sources (own empirical analysis and interviews with 'qualified observers'[41]), to indirect ones (other scholars papers and surveys[42]) cross-checking information whenever possible – and we ended up with the table below which at least gives you a overall perspective of the evolution of a number of Italian industrial districts.

[40] Despite some relevant exceptions.

[41] Such as: entrepreneurs, local authorities, service centres' managers, chamber of commerce analysts, etc.

[42] Such as Gandolfi, (1988); Gobbo, (1989); ICE, (1989); Pyke *et al.*, (1990); Moussanet and Paolazzi, (1992); Nuti, (1992); Onida *et al.*, (1992); Unioncamere, (1995); Belussi, (1997); Bramanti and Maggioni, (1997b); Crestanello, (1997); Storper, (1997); Paniccia, (1998); Belussi and Gottardi, (2000); Maggioni and Bramanti, (2002); Viesti, (2001); Withford (2001).

Table 2. The dynamics of Italian industrial districts (1970s-2000)

		All ring, no core	Core-ring with co-ordinating firm	Core-ring with leading firm
Local agglomerated network	Small units	CARRARA (marble) FROSOLONE (cutlery) VALDUGGIA (valves) ARZIGNANO (leather) SOLOFRA (leather) BARLETTA (footwear) VALENZA PO (jewels) SANTO STEFANO (balances) VAL VIBRATA (clothes) PALOSCO (compasses) PREMANA (scissors) VIGEVANO (footwear) LUMEZZANE (taps and fittings) MANIAGO (cutlery) ▪ BRIANZA (furniture) ↘ ▪ MIRANDOLA (biomedical instruments) →	CEREA-BOVOLONE (furniture) *Selva* CUSIO (taps and fittings) *Zucchetti – Paini* GRUMELLO (buttons) *Zaravit – Mdp* ▪ MURANO (glassware) *Venini – Murrina* ↘ ▪ CASTELFIDARDO (musical instruments) *Bontempi* ↘ ▪ GUASTALLA (agricultural machineries) ↓ ▪ CASTELFIDARDO (musical instruments) *Bontempi* ↘ ⇨▫ **MIRANDOLA**	
	Large Units	OMEGNA (pots and pans) *Alessi – Lagostina* CANNETO SULL'OGLIO (toys) *Grazioli – Peg* SETTIMO TORINESE (felt-pens) *Universal – Walker pen* VIADANA (brushes and brooms) *Tonkita Spa* ▪ UDINE (chairs) *Calligaris – Euroline* → ▪ CARPI (knitwear) *Mit – Cp Company* ↓ ▪ VARESE (car alarms) *Gemini – Elser* ↓ ▪ PRATO (textile) *Pecci* ↓	▪ CADORE (spectacle-frames) *Safilo – Luxottica* ↓ ▪ MONTEBELLUNA (boots and footwear) *Nordica – Brixia – Caber – Lotto* ↓ ▪ CALTELGOFFREDO (nylon stockings) *Golden Lady – Cps International* ↓ ▪ SASSUOLO (ceramic tiles and pottery) *Marazzi – Ires Ceramiche* ↓ ▪ BIELLA (wool) *Fila – Loropiana* ↓ ▪ COMO (silk) *Ratti – Mantero* ↓ ▪ TREVISO (knitwear) ↘ *Benetton* ⇨ BRIANZA (furniture) *Poliform, Molteni* ↘ ⇨▫ **UDINE**	CASARANO (footwear) *Filanto* AREZZO (jewels) *Uno A Erre* TOLENTINO (leather goods) *Nazareno Gabrielli* ▪ TRENTO (porphyry and stone materials) *Consorzio Albiano* ↓ ⇨▫ **MURANO (glassware) *La Murrina*** ⇨▫ **CALTELFIDARDO**
Local-Global network	Italy	⇨ CARPI ↘ ⇨ VARESE ↘ ⇨ PRATO ↘		▪ SANTERAMO (leather sofas) *Natuzzi* ↓ ⇨ TREVISO ↓ ⇨▫ **BRIANZA (furniture) *Poliform, Molteni***
	World		⇨▫ **CADORE** ⇨▫ **MONTEBELLUNA** ⇨▫ **CASTELGOFFREDO** ⇨▫ **SASSUOLO** ⇨▫ **BIELLA** ⇨▫ **COMO** ⇨▫ **CARPI** ⇨▫ **VARESE** ⇨▫ **GUASTALLA** ⇨▫ **PRATO**	⇨▫ **SANTERAMO** ⇨▫ **TRENTO** ⇨▫ **TREVISO (knitwear) *Benetton***

The Table shows the initial situation and the evolution of 45 Italian IDs (1970s-2000) through the following symbols: SMALL CAPITALS: name of the milieu; (in brackets): prevalent or dominant production; *italics:* name of the most important firms in the area. Black square ▪ initial situation; black arrows → ↘ ↓ recent dynamics of the ID (moving to a different cell); white arrows ⇨ and white square ▫ current situation (**BOLD TYPE**). Source: adapted from Bramanti - Maggioni (1997a).

Even if it is impossible to identify a unique and uncontroversial trend in the evolution of industrial districts, the table shows a diffused shifts of most of them towards the bottom right corner of the table. In other words, in order to cope with the growing international competition and the turbulent technological environment, the most successful industrial districts have changed their traditional (and canonical) governance structure – based on flat and a-hierarchical inter-firm relationships and exclusively local agglomeration of firms – and have moved towards greater hierarchy and wider geographical distribution of production processes.

3.1 The evolution of industrial districts: a simulation exercise

One may perhaps simply concludes that, as size increase and distance grows, a more structured relational form is need in order to manage any social and economic entities. Inter-firms co-ordination within early and small industrial districts was achieved thank to daily unstructured contacts between entrepreneurs or, in Brusco's own words, thanks to "continual informal interactions in cafés and bars and in the street. In this way, new ideas are formed and transmitted" (Brusco, 1990, p.16). When both number and geographical distance hindered the management of industrial districts through daily informal interactions, hierarchy and multi-layer governance structure where identified as the only viable organisational solution.

However we are convinced of a deeper relationship existing between the evolution of transaction costs related to inter-agent relations and the emergent governance structure of these relations. Transaction costs involved in establishing and using a relational channel between two agents are heavily dependent on past history and behaviours, in other words they are path dependent. Usually it is assumed that transaction costs are a decreasing function of the number of interaction between two agents since the repeated structure of the interaction games should (under certain circumstances) discourage opportunistic behaviours and support co-operation[43]. On the contrary we think that, depending on various exogenous and endogenous conditions (i.e. related both to the content of the interaction and to the environment in which these interactions are embedded) transaction cost may be either a decreasing or a convex non-monotonic function of previous interactions.

In order to study the relationship between transaction costs 'dynamics' and governance structure evolution we used the simplest possible model in which the industrial district is represented by a 5-nodes undirected[44] network (as in Figure 6).

[43] This is at least the trivial version of Axelrod's (1984) original ideas which is widely quoted in both industrial districts and OSSCs related literature.

[44] I.e. a network in which the direction of linkages is irrelevant, thus A-B is equal to B-A.

Figure 6. The initial governance structure of the system

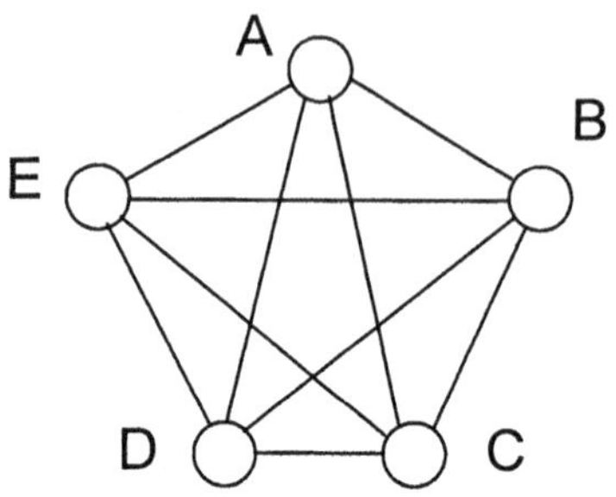

All nodes are potentially connected (i.e. the network has a maximum density) through linkages. Attached to each linkage there is a level of transaction cost which, at the beginning, is fixed as equal to 10 for all linkages. A random process chooses one out of 10 possible couplets of nodes (10 = (5*4)/2)), then the computer calculate the cheapest route to connect the two nodes forming the couplet. This exercise is repeated for 100 runs. If transaction costs are independent from previous interactions, then the situation repeats identically period after period. If, on the contrary, transaction costs are made dependent from previous interactions, then, according to different functional forms, interesting results can be obtained.

In these simulations transaction costs were assigned three different dynamic laws.

In the first case (Figure 7a), transaction costs are assumed to decrease sharply according to previous interactions. This simulates a situation in which trust is built by few repeated interactions in such an efficient way that, after few runs, one node becomes the obligate 'bridge' through which all connections must pass since linkages connecting this node with all other nodes in the network have lower transaction costs than any other linkages. Formally:

$$CT_t = CT_{t-1}\left(0.9^n\right) \qquad [12]$$

where *CT* are transaction costs, *t* and *t-1* are time periods and *n* is the number of previous periods.

Figure 7.a Transaction costs dynamics (case 1)

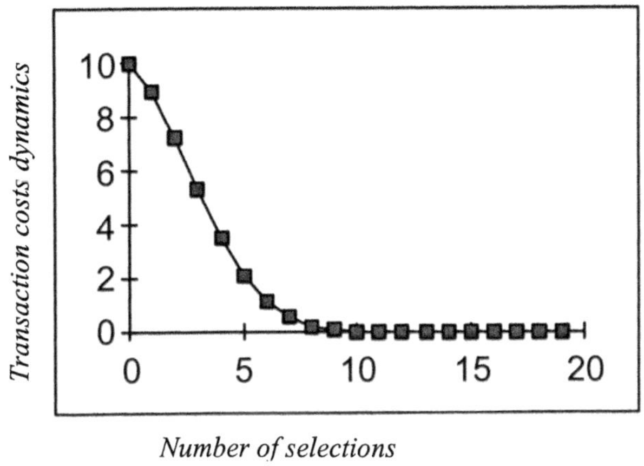

Figure 7.b Simulation results, 100 runs (case 1)

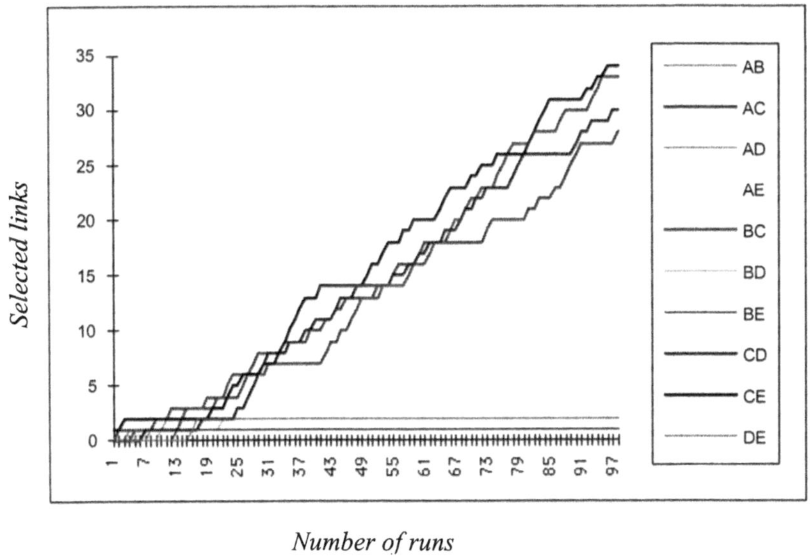

This means that one agent in the system assumes a pivotal role and establishes a relational leadership (Figure 7b). The system's governance structure evolves from a maximum dense and a-hierarchical (all ring-no core) network where each agent has the same relational power and role to a minimally dense 'star-shaped'

(core-ring leading) network where the leader (node C) controls all transactions (Figure 7c). In such a situation, where trust is established very easily and transaction costs fall sharply after few interactions, the dynamics of the system is heavily non-ergodic and the systems locks-in after few runs.

Figure 7.c Evolution of the governance structure (case 1)

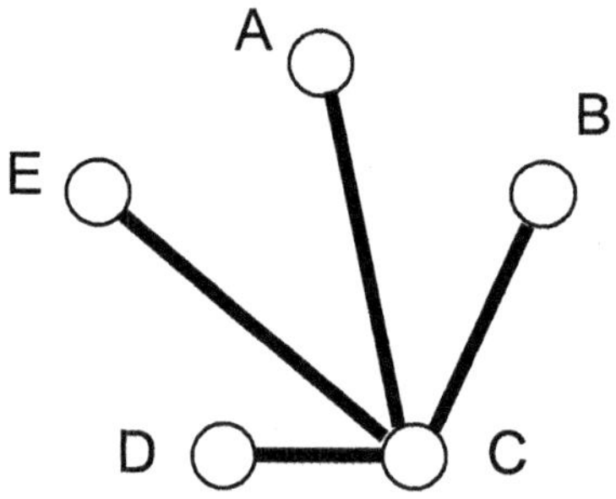

network structure after 100 runs

In a second case, (Figure 8a), transaction costs are assumed to decrease slowly according to previous interactions. This simulates a situation in which trust must be built through many repeated interactions in order to achieve a sensible reduction in transaction costs. Formally:

$$CT_t = CT_{t-1}(0.9) \qquad [13]$$

Figure 8.a Transaction costs dynamics (case 2)

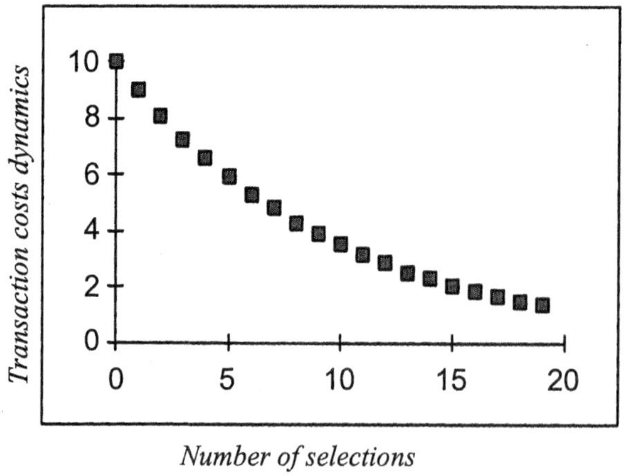

Figure 8.b Simulation results, 100 runs (case 2)

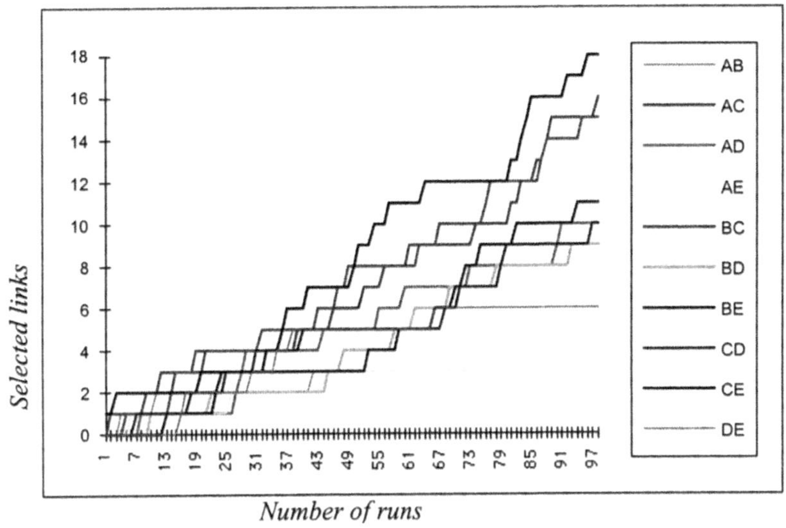

In this case a single node (for expositional convenience, C) still emerges as the network leader but the reduction in transaction costs due to repeated interactions is too weak to establish a relational monopoly. Thus (Figure 8b) the governance structure of the system displays a leading agent (node C) and two other nodes (node B and node D) which have acquired an intermediate level of relational powers. Such a governance structure can be defined as a quasi-hierarchical (core-ring coordinating) network.

Figure 8.c Evolution of the governance structure (case 2)

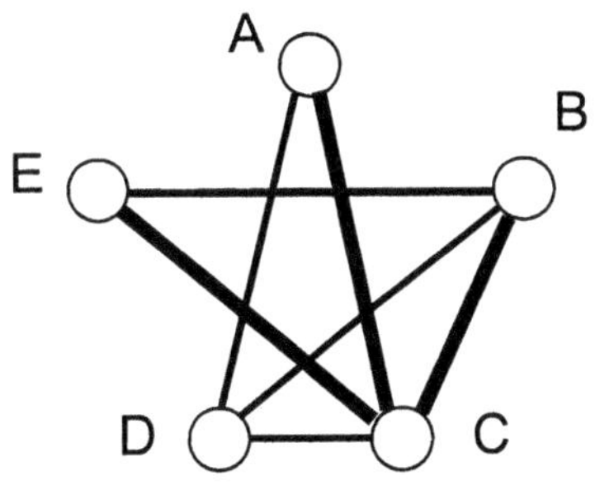

Network structure after 100 runs

In a third case, (Figure 9a), transaction costs are assumed to initially decrease and then increase as the number of previous interactions grows. This simulates a situation where agents interact mainly in order to learn something (i.e. acquire information, knowledge and know-how) from one another. In this case trust, which is achieved through repeated interactions, lowers transaction costs only up to a point in which each firm in a couplets starts loosing interest in continuing the relationships since each one has extracted all possible information from the other. The desire for new and alternative interactions, counteracting the effects of trust on transaction costs is modelled, in this simulation, by making transaction costs a convex non monotonic function of the number of previous interactions (or selections). Formally:

$$CT_t = 0.1(s)^2 - 2s + 10 \qquad [14]$$

Figure 9.a Transaction costs dynamics (case 3)

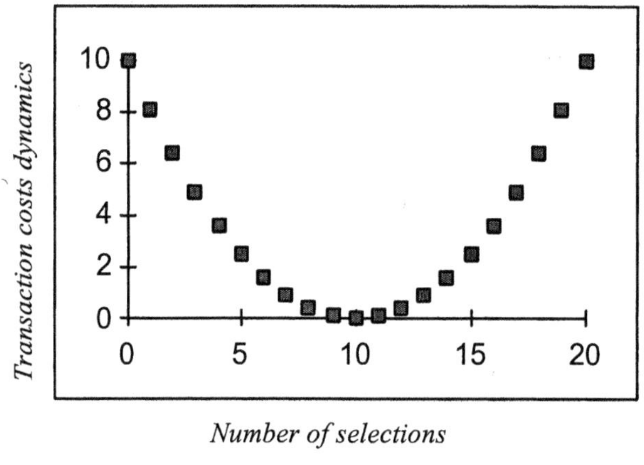

Figure 9.b Simulation results, 100 runs (case 3)

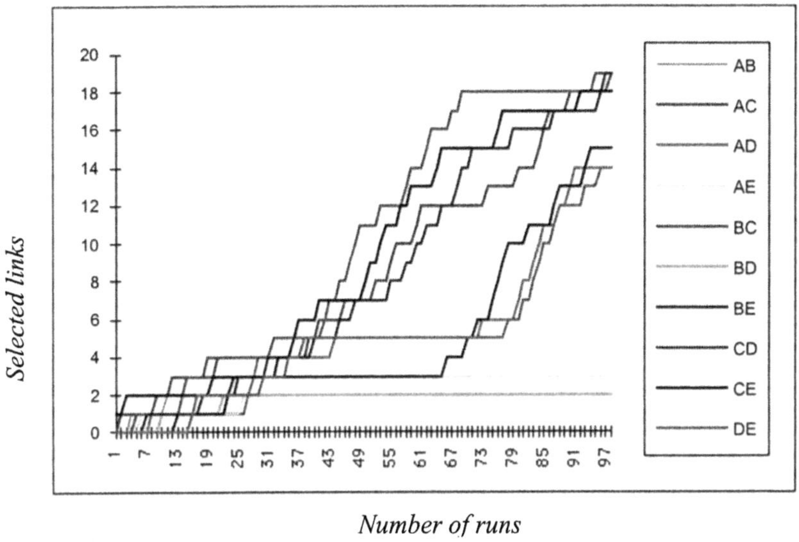

In this situation the temporary leadership of a single node (for instance node C) in the network is rapidly substituted by a new one (node D) which, in turn, loose his prominent position in favour of a new leader (probably node E) according to a dynamics of successive leadership 'waves'. The picture become brisker if one assumes that trust has an expiry date, so that relationships must be periodically renewed for transaction costs to keep low values. This may be easily achieved (in the simulation) by assuming that transaction costs for each un-chosen node follow an increasing function of the number of runs. This result is shown in figure 9c where the two networks graphically emphasises the growth and decline of central node's leadership and the emergence of a new one.

Figure 9.c Evolution of the governance structure (case 3)

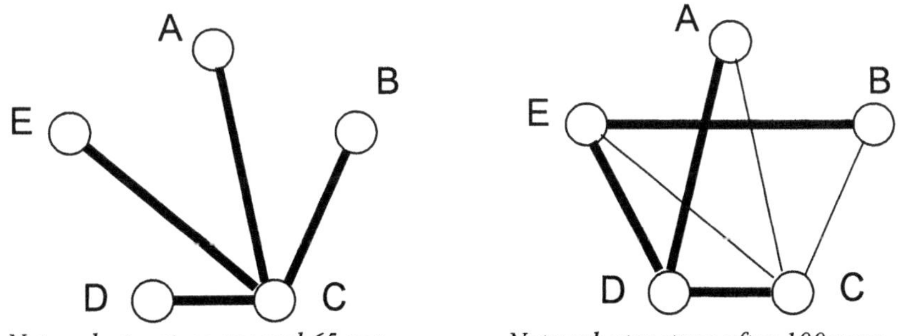

Network structure around 65 run *Network structure after 100 runs*

One may thus discuss about what functional form of transaction costs evolution describes better the relational contents and the environmental feature of the specific system which is under study.

Many successful industrial districts are currently managed by a 'central' core (which may be either a single entity or composed by a restricted number of entities) which takes care of the development of the system either through a polarised structure of hierarchy (as in Figure 2c) or through a multi-layer governance structure in which intermediate role between the core and the periphery of the system emerged.

4 Conclusions

This work has firstly focussed on the basic economic principles which may explain the existence of industrial districts: agglomeration economies, geographical features and inter-industry interactions by showing a simple analytical framework able to identify the long-run equilibrium of complex interactive systems (composed by several industries within the same districts a/o several cluster belonging to the same industry. The paper has further shown a tentative explanation of the origin and evolution of the governance structures of industrial districts based on transaction cost dynamics through a simple simulation exercise.

The recent literature on industrial district has stressed the importance of rigorous empirical analysis together with the in depth knowledge of specific cases while the theoretical side of analysis, being 'hijacked' by the new economic geography approach, has been almost neglected by the community of 'industrial districts analysts'. The role of this chapter within such a volume is therefore to show that it is still possible to use several theoretical tools to analyse such a complex entity.

In other words while we are still convinced that, while an industrial district is "a socio-territorial entity which is characterised by the active presence of both a community of people and a population of firms in one naturally and historically bounded area" (Becattini, 1990, p. 39) we do believe that economic theory and economic modelling may give a great help to anyone interested in understanding the structure and evolution of industrial districts, being he/she either an entomologist or, even, an ant.

References

ARTHUR W.B. (1988), "Urban System and Historical Path Dependency", in: AUSUBEL J.H., HERMAN R. (Eds.), *Cities and Their Vital Systems*, National Academy Press, Washington D.C..

ARTHUR W.B. (1990), "Silicon Valley Locational Clusters: When Do Increasing Returns Imply Monopoly?", *Mathematical Social Sciences*, 19, pp. 235-251.

AXELROD R. (1984), *The Evolution of Cooperation*, Basic Books, New York.

BALLONI, V, CUCCULELLI M., IACOBUCCI D. (1998), *Le politiche locali nel modello NEC*, paper presented at the "Incontri pratesi sullo sviluppo locale", Prato.

BECATTINI G. (1990), "The Marshallian District as a Socio-Economic Notion", in: PIKE F., BECATTINI G, SENGERBERGER W. (Eds.) *Industrial Districts and Inter-Firm Co-operation in Italy*, International Labor Office, Geneva.

BECATTINI G. (2002), "Industrial Sectors and Industrial Districts: Tools for Industrial Analysis", *European Planning Studies*, vol. 10, n.4, pp. 483-493.

BELTRAMI E. (1987), *Mathematics for Dynamic Modelling*, Academic Press, Boston.

BELUSSI F. (1997) "Le politiche knowledge intensive per lo sviluppo dei sistemi produttivi territoriali", in: BRAMANTI A. MAGGIONI M.A. (Eds.) *La dinamica dei sistemi produttivi teritoriali: teorie, tecniche, politiche*, FrancoAngeli, Milano.

BELUSSI F., GOTTARDI G. (2000), *Evolutionary Patterns of Local Industrial Systems*, Ashgate, Aldershot.

BIANCHI G. (1994), "Tre e più Italie: sistemi territoriali di piccola impresa e transizione post-industriale", in: BORTOLOTTI F. (a cura di) *Il mosaico e il progetto: Lavoro, imprese , regolazione nei distretti industriali della Toscana*, FrancoAngeli, Milano.

BRAMANTI A. MAGGIONI M.A. (1994), "The Dynamics of Mileus: From Governance Structure to Network Analysis", in: *Dynamis-Quaderni IDSE*, n. 5/94, IDSE-CNR, Milano.

BRIOSCHI F., CAINELLI G. (a cura di) (2001), *Diffusione e caratteristiche dei gruppi di piccole e medie imprese nelle aree distrettuali. Dove va il distretto industriale?*, Giuffrè, Milano.

BRUSCO S. (1992), "Small Firms and the Provision of Real Services", in: PIKE F., BECATTININI G, SENGERBERGER W. (Eds.), *Industrial Districts and Local Economic Regeneration*, International Labor Office, Geneva.

CAPELO A.C. (1989), *Modelli matematici in biologia*, Decibel, Padova.

CRESTANELLO P. (1997), "Le trasformazioni in 10 distretti industriali durante gli anni 80", in: VARALDO R., FERRUCCI L. (Eds.), *Il distretto industriale tra logiche di di impresa e logiche di sistema*, Franco Angeli, Milano.

DEI OTTATI G. (1994), "Cooperation and Competition in the Industrial District as an Organization Model", *European Planning Studies*, vol 2, n. 4, pp. 463-484.

DENDRINOS D.S., MULLALLY H. (1985), *Urban Evolution. Studies in the Mathematical Ecology of Cities*, Oxford University Press, Oxford.

GAMBAROTTO F., MAGGIONI M.A. (1998), "Regional Development Strategies in Changing Environments: an Ecological Approach", *Regional Studies*, 32, pp. 49-61.

GAMBAROTTO F., RANGONE M., SOLARI S. (2001), *Competenze, attori, istituzioni. Le forme dell'apprendimento collettivo nel distretto cadorino dell'occhiale e della scarpa sportiva di Montebelluna*, paper presented at the XXII AISRe Conference, Venice, 10-12 october.

GANDOLFI V. (1988) *Aree sistema: internazionalizzazione e reti telematiche*, Franco Angeli, Milano.

GAUSE G.F. (1934), *The Struggle for Existence*, Williams & Wilkins, Baltimore.

GOBBO F. (a cura di) (1989), *Distretti e sistemi produttivi alla soglia degli anni '90*, Franco Angeli, Milano.

GUERRIERI P, PIETROBELLI C. (2001), *Models of Industrial Districts' Evolution and Changes in Technological Regimes*, paper presented at the 2001 EAPE Conference, Siena, 8-11 november.

HARNOFF D, HENKEL J., VON HIPPEL E. (2000), *Profiting from Voluntary Information Spillovers: How User Benefits by Freely Revealing Their Innovation*, opensource.mit.edu/online_papers.php.

HOOVER E.M. (1971), *An Introduction to Regional Economics*, Knopf, NewYork.

ICE (1989), *Il sistema Italia e le economie locali*, Marsilio, Venezia.

ISARD W. (1956), *Location and Space-economy*, MIT Press, Cambridge (Mass.).

MAGGIONI (2002c) *Open Source Software Communities and Industrial Districts: A Useful Comparison?*, DISEIS W.P. n. 6, Università Cattolica, Milano.

MAGGIONI M.A. (1993), *Ecological Dynamics and Critical Mass Processes in the Location of High Tech Firms*, paper presented at the 40th RSAI (Regional Sciences Association International) Conference, North American Section, Houston, 11-14 november.

MAGGIONI M.A. (2002a), "The Development of High-Tech Clusters: Theoretical Insights and Policy Implications" in: FELDMAN M, MASSARD N., (Eds.) *Institutions and Systems in the Geography of Innovation*, Kluwer, Dordrecht.

MAGGIONI M.A. (2002b), *Clustering Dynamics and the location of High-Tech Firms*, Springer, Heidelberg.

MAGGIONI M.A., BRAMANTI A. (2002), "Local and Global Networks in the Economics of SMEs. Is Proximity the Only Thing that Matters?" in: MCNAUGHTON R., GREEN M., (Eds.) *Global Competition and Local Networks*, Ashgate, London.

MAGGIONI M.A., PORRO G. (1994), "Dinamiche di crescita regionale: il ruolo delle aspettative in modelli di tipo ecologico", *Quaderno della ricerca di base "Modelli di sviluppo e regional competition"*, 6, Università Bocconi, Milano.

MARKUSEN A. (1996), "Sticky Places in Slippery Spaces: A Typology of Industrial Districts", *Economic Geography*, n. 72, pp. 293-313.

MARSHALL A. (1921), *Industry and Trade*, McMillan, London.

MAY R.M. (1976), *Theoretical Ecology, Principles and Applications*, Saunders, Philadelphia.

MIYAO T., KANEMOTO Y. (1987), *Urban Dynamics and Urban Externalities*, Harwood Academic Publishers, London.

MOUSSANET M. , PAOLAZZI L. (1992), *Gioielli, bambole, coltelli. Viaggio de 'Il Sole24Ore' nei distretti produttivi italiani*, Edizioni Il Sole24Ore, Milano.

NUTI F. (1992), *I distretti dell'industria manifatturiera in Italia*, Franco Angeli, Milano.

ONIDA F., VIESTI G., FALZONI A.M. (a cura di (1992) *I distretti industriali: crisi o evoluzione?*, Egea, Milano.

PANICCIA I. (1998), "One, a Hundred, Thousend of Industrial Districts. Organizational Variety in Local Network of Small and Medium-sized Enterprises", *Organization Studies*, vol 19, n. 4, pp. 667-699.

PAPAGEORGIOU Y.Y. (1979), "Agglomerations", *Regional Science and Urban Economics*, 9, pp. 41-49.

PEARL R., REED L.J. (1925), "Skew-Growth Curves", *Proceeding of the National Academy of Natural Sciences of the USA*, 11, pp. 16-22.

PIKE F., BECATTININI G, SENGERBERGER W. (Eds.) (1990) *Industrial Districts and Inter-Firm Co-operation in Italy*, International Labor Office, Geneva.

PIKE F., BECATTININI G, SENGERBERGER W. (Eds.) (1992), *Industrial Districts and Local Economic Regeneration*, ILO, Geneva.

RICHARDSON H. W. (1978), *Regional and Urban Economics*, Penguin Books, Harmondsworth.

ROUGHGARDEN J. (1979), *Theory of Population Genetics and Evolutionary Ecology: an Introduction*, MacMillan, New York-London.

STORPER M. (1997), *Regional World*, The Guilford Press, New York.

SWANN G.M.P., PREVEZER M., STOUT D. (Eds.) (1998), *The Dynamics of Industrial Clustering*, Oxford University Press, Oxford.

SWANN G.M.P. (1998), "Towards a Model of Clustering in High-Technology Industries", in: SWANN G.M.P., PREVEZER M., STOUT D. (Eds.), *The Dynamics of Industrial Clustering*, Oxford University Press, Oxford.

SWANN G.M.P., PREVEZER M., STOUT D. (Eds.) (1998), *The Dynamics of Industrial Clustering*. Oxford University Press, Oxford

TAUCHEN H., WITTE A.D. (1983), "An Equilibrium Model of Office Location and Contact Patterns", *Environment and Planning A*, 15, pp. 1311-1326.

UNIONCAMERE (1995), *Imprese e istituzioni nei distretti industriali che cambiano*, Franco Angeli, Milano.

VARALDO R., FERRUCCI L. (a cura di) (1997), *Il distretto industriale tra logiche di di impresa e logiche di sistema*, Franco Angeli, Milano.

VERHULST P.F. (1845) Recherches Mathématique sur la loi d'accroissement de la population, *Nouveaux Mémoires de l'Académie Royale des Sciences et Belles-Lettres de Bruxelles*, vol. 18, n.2, pp. 3-38.

VIESTI G. (2001), *Come nascono i distretti industriali*, Laterza, Bari.

VOLTERRA V. (1926), "Variazioni e fluttuazioni del numero di individui in specie di animali conviventi", *Memorie Accademia dei Lincei*, 2, pp. 31-113.

WEBER (1929), *Theory of the Location of Industry*, Chicago, Chicago University Press.

WHITFORD J. (2001), "The Decline of a Model? Challenge and Response in the Italian Industrial Districts", *Economy and Society*, vol. 30, n. 1, pp. 38-65.

Part II. Firm Ownership and District Organisation

Institutional innovations in industrial districts

Riccardo Varaldo[1] and Luca Ferrucci [2]

[1]Sant'Anna School, Via Carducci 40, 56100 Pisa, Italy
[2]Faculty of Economics, University of Perugia, Via Pascoli 20, 06123 Perugia, Italy

1 Introduction

The present essay aims to analyse the nature and the evolution of industrial district firms in terms of their institutional model. It is divided into four parts. In the first part, we observe the presence of certain barriers preventing institutional change in each model of firm. In the second part, some specific institutional and historical features in industrial district firms are observed which confirm the presence of high barriers preventing change in ownership. In the third part, the main paths of institutional evolution of district capitalism are outlined, especially with regard to the internal local system. Finally, in the last part, we suggest some possible implications regarding industrial policy supporting institutional innovations to the ownership of these firms.

2 Barriers to institutional changes inside the firms

The change in the institutional model in the firms involves a lot of organisational problems[1]. This is true for all kinds of firms, whether they be small or large, managerial or entrepreneurial, capitalistic or co-operative. These problems constitute barriers to the implementation of innovation, because they are linked to change in ownership and, consequently, in the management and in the organisational model.

The presence of these barriers arises from specific features of this innovation: (i) institutional innovation does not allow experimentation. This change can be studied and analysed *ex ante*, but it cannot be experimented. Owners have a limited knowledge of possible new partners, especially with regard to their real expectations, which results in a highly uncertain context for decision making, well known adverse selection and moral hazard problems; (ii) this change represents an irreversible organisational process. Once it has been implemented, it involves a great number of sunk costs. Therefore, it becomes difficult and expensive to change the decisions taken. The cost of an institutional mistake is very high. If innovation has been structured and managed in an incoherent way (for instance, the choice of a wrong partner or the lack of shared objectives between partners),

[1] For the meaning of institutional firm models, see Hansmann (1996).

decisional paralysis and loss of qualified personnel may result; (iii) the benefits arising from innovation are uncertain both in terms of their nature and intensity. This institutional innovation has many effects on the technological, organisational and market behaviour of the firm. Therefore, it is not possible to foresee every change of strategy and their consequent economic benefits within the firms.

Given these characteristics, institutional innovation is very rare in the life of a firm. Normally, it happens during a period characterised by economic, financial or organisational problems inside the firm. In this context, the owners may have divergent opinions on strategies, consolidated routines are not adequate and new entrepreneurial and managerial approaches are needed. The barriers to the implementation of this institutional innovation are, however, very different inside the firms. In other words, there are firms with higher or lower propensity for innovation. This depends on their institutional models and on their different business histories. Some firms, such as public companies or producers' cooperatives, are characterised by a high degree of institutional flexibility. On the other hand, other firms have high barriers to change in their institutional models, as in the case of industrial district firms.

3 The industrial district firm between institutional specificities and barriers to changes

The reluctance of industrial district firms to implement institutional innovation arises from some economic and social relationships within the local system. There are three different factors to be noted:

The first one is related to the ownership model. The industrial district firm is typically a small family company. Its small size is a structural feature of this model of firm (Varaldo, 1993). As indicated in Table 1, the average size of the industrial district firms in Tuscany is very small (we have analysed only the most important district in this region: the Prato textile district). There is only a small number of firms with more than 49 employees in the district. A large majority of firms have up to 9 employees, (these, in 1996, constituted 83% of total firms in Prato textile district). This historical evidence suggests that the small size is a structural and not a contingent feature. The district entrepreneurs believe that their small size constitutes a competitive advantage. In the minds of these entrepreneurs, flexibility, quality of the products, artisans' skills and creativity are strong linked to the small size of the firms. They are not, therefore, in favour of a strategy of expansion based on size changes.

Moreover, the industrial district firm is often a family company (Bull and Corner, 1993). The strong ties of family and friendship between owners are important socially and culturally for organisational and decisional activities. Inside the firm, there are consolidated routines for supporting decision processes, with three main characteristics: (i) well defined roles with informal hierarchy; (ii) informal meeting places such as those outside the firm for sports or tourism; (iii) shared belief held by the owners regarding the competitive advantage given by the district paradigm (such as small size, high flexibility, high level of outsourcing activities and so on).

As a consequence, the small family company in the district can take decisions in a short time with a high degree of agreement (Kepner, 1991). Therefore, there are high barriers against pursuing and favouring the entry of new entrepreneurs or managers to decision making processes. These institutional innovations can modify the consolidated decision making routines, both in terms of the process and in the results.

Table 1. Total number of firms and employees in Prato textile district

	Firms			*Employees*		
Range size	1-9	10-49	More	1-9	10-49	More
1971	7,415	734	114	14,430	14,622	13,540
1981	9,817	1,078	96	21,351	19,680	8,551
1991	5,906	858	62	15,315	15,196	4,490
1996	4,596	887	71	13,089	15,990	5,599

Source: ISTAT, Census of Industry, different years

The second factor *is related to the presence of a large number of local suppliers and customers*. This means that the district firms can develop and exploit these vertical relationships both in manufacturing and information flow (Goodman *et al.*, 1989). These links are very strong with a high degree of trust among entrepreneurs (Spagnolo, 1999). The industrial atmosphere can support a quasi-vertical integration strategy. In certain circumstances, one district firm can buy new specific technology for the production of a new component ordered by another district firm. On the other hand, without any consolidated trust relationship, these technological investments have not been realised due to high transaction costs. If the firms, however, do not have financial resources or adequate guarantees to buy this technology, customers can grant them a specific loan to be paid off over time. These financial relationships are frequent and permit investments for facilities or technology with risk sharing divided between local suppliers and their customer network. This generates a closed network between district firms (Dei Ottati, 1994). These facts confirm that the small family district firm can utilise these vertical relationships as a substitute for expansion based on new partners. As a consequence, the district firm can avoid institutional innovations.

The third factor is related to local banks. In a large majority of industrial districts in Italy, there is a developed banking system with local or regional banks (cooperatives or savings banks), which is historically linked to these manufacturing firms. Bank management has often been influenced by district entrepreneurs.

This fact reinforces the co-dependency of these two institutions (manufacturing firms and banks) in the evolution of the district (Conti and Ferri, 1997). As Becattini (1991) asserts, the local bank is an institution which has been established and grown in the district. It has a strong link with local entrepreneurs and knows all their activities. As a consequence, these local banks can analyse claims for credit by taking personal characteristics of entrepreneurs into consideration and not only the economic guarantees they can offer.

In this way, the local bank has developed a specific *modus operandi* based on personal trust and social reputation. The proliferation of new small firms inside the industrial districts over the last decades depends on these local banking activities (Ferri and Messori, 2000). They have supported the growth of the district by stimulating the setting up of new small firms. They act as complementary financial integrators and, as a consequence, the district firm can avoid institutional innovations. As an empirical research has shown, the Prato textile firms ask for a loan more often than non-district textile firms (Signorini, 1994). Therefore, in Italy, given the historical lack of innovative financial institutions, such as merchant banks or venture capitalists, the industrial districts have contributed to the establishment of their local financial institutions with specific routines with regard to credit standing and granting. This co-evolutionary path has generated a strong hybridization with the risk capital of the firms by local banks. Moreover, the manufacturing specialisation at district level has contributed to the upgrading of the economic risk of these local banks during periods of recession in this industry, generating many economic problems inside these financial institutions in the long run, as the Cassa di Risparmio di Prato's bankruptcy illustrates (Filippi, 1999; Beretta *et al.*, 2000).

Given these three characteristics, the district firm has an high aversion to growth via institutional innovations. It can achieve the same objective or it can solve a problem avoiding changes in ownership and using manufacturing or banking relationships inside the industrial districts.

4 New competitive strategies and institutional innovation

Institutional innovation is necessary in order to implement new competitive strategies. Today the district firm has at least two strategic problems.

On the one hand, this firm has often a generational problem. The founder entrepreneur is old and it is not easy to guarantee continuation of the firm. The risk for industrial district is to lose the tacit and immaterial competence and knowledge owned by the entrepreneurs and by their firms. In evolutionary theory, this means that the industrial district has to acquire the immaterial and tacit knowledge possessed by these firms again but by trial and error. The costs and risks in repeating the learning processes are high and can lead to a reduction in district competitiveness in the long run. In the past, this problem was not significant since the industrial birth and death rate was high. Since the eighties however, the industrial birth rate has decreased and the duration of the firm has become important. As is indicated in Tables 2, there are a lot of small firms in Prato textile district that are likely to close their businesses in a short period[2]. More than 55% of the entrepreneurs interviewed want to give up business within four years and the most important reasons regard both the market and problems connected to age. There may be a certain pessimism in the answers, but this shows the existence of a widespread problem among these small district firms.

On the other hand, the district competitiveness needs new immaterial competence to face the international markets and technological changes. Traditional manufacturing competence is not adequate and there is a strategic need for new immaterial competence, such as in marketing, information technology and product development within district firms (Ferrucci, 1996). These skills are possessed by specialists in human resources, who often do not come from local labour market. Their intellectual knowledge and their decisional routines are very different from the entrepreneur's capabilities. Therefore, the employer-employee relationship reveals a typical agency problem with adverse selection and moral hazard situations. As a consequence, there is a need to draw up an efficient labour agreement between these two subjects, with stock options and a share in the profits for instance. Therefore, the institutional innovation is a proper way of acquiring new immaterial competence within district firms.

Table 2. Time horizon for ceasing business by entrepreneur in Prato textile district

Number of years	*Number of entrepreneurs (%)*
within 2 years	40.3
2-4 years	17.3
4-6 years	6.1
More	36.3

5 Towards new institutional paths for districts firms

Since the nineties, the industrial districts have had to face increasing competition. Radical changes in technology (such as electronic manufacturing systems, CAD design technology or ICT), in the market (such as new international markets like Eastern European countries or China) and in competition (such as the growing concentration in the retail chain in Western countries or the emergence of new

[2] The empirical research based on telephone interview realised in 2001 regards 544 small firms with at least one entrepreneur older than 55 years.

competitors in South Eastern Asiatic countries) have influenced past game rules and traditional strategies played by these district firms. As a consequence, the internal organisation and the institutional structure of the firm has changed in order to support new innovative strategies and constitute dynamic competition (Arcangeli *et al.*, 1988).

There are four principal ways of transforming institutional district firms each with different repercussions in terms of their organisational model and policy.

The first way is based on the entry of *entrepreneur's family members* into the firm management. This new entrepreneurial structure does not change the decisional making process and does not involve any changes in management or policy. On the contrary, sometimes the organisational impact can be hard with destabilising effects on the entrepreneurship, if the new member of the family in question presents new cultural attitudes and innovative visions. Therefore, this evolution can in the first hypothesis result in high dependence on existing policy or in the second case highly innovative processes. In both cases, this institutional change is easier than in the following ones because there are family preferences and a common problem-solving language.

The second way is based on the *entry of employees into the entrepreneur structure*. Some of them are highly specific investments for the firm (as Williamson suggests in 1986), based on their managerial competence and their tacit knowledge. Therefore, the district firm can build strong ties with these employees through profit sharing, stock option and other incentive pay. In a certain way, the employee becomes an entrepreneur, with broadening entrepreneurship inside the firm. The existence of trust between these two subjects gives rise to this institutional evolution (Solinas, 1982). The historical, social and economic link between them constitutes a shared knowledge of manufacturing problems, opinions about strategies and decision making processes. As a consequence, this institutional change generates incremental innovations in organisational settings: it is only a new combination of local and tacit internal knowledge within the district firm. Persisting difficulties in the implementation of this institutional solution are characterised by employees' age (if they are old, they do not have any entreprenerial propensity) and on their limited financial resources (if the employee has to pay to enter the entrepreneurship). Therefore, this institutional solution can require specific contractual agreements (such as instalments, payment by profits or other incentives) or the presence of an innovative financial intermediary (such as a merchant bank) in the industrial district.

The third way consists in *mergers and acquisitions between district firms*. If there is an aged entrepreneur and the continuation of business is not possible via the two aforementioned solutions, the merger or acquisition on the part of another district firm may be the only solution. Usually, the potential buyer is linked with the firm to be sold by customers, suppliers or competitor relationships. As a consequence, the buyer – given their competence in the industry and location inside the district – knows this firm, its business history and its strategic assets and relationships and it is able to appreciate its economic and manufacture values. The barriers to transfer tacit knowledge from the firm to be sold to a new buyer are limited and they can be reduced by adequate contractual solutions (ie. the presence of the seller in the management activities of new firm for some years and bonus payment based on profit performance indicators).

Moreover, these merger and acquisition processes can promote the growth of formal groups in industrial districts, exceeding the traditional small size limit. As a consequence, the typical industrial organisation of the district can change, with the emergence of a leader firm and vertical and horizontal integration strategies.

The fourth way consists in *merger and acquisition realised by firms localised outside the industrial district*. This strategy is difficult to pursue, especially if the buyers are not already customers or suppliers of the firm to be sold. There are two kind of difficulties: information and assessment problems.

The problems regarding information depend on the typical rules and mechanisms of how the firm market works. In order to sell or buy a firm shared knowledge and transparency in the information flows is required. Each firm needs to know about the existence of a potential seller and of a potential buyer, but if this information become public, the seller firm can incur damages in economic terms and in terms of its reputation with regard to its customer, supplier, financial and employee relationships: some customers may search for new possible sellers; most banks may reduce their financing plans; some qualified employees might try to find a new job and so on. The geographical distance between the district firm to be sold and potential buyers increases these problems because, on the one hand, the willingness to sell or acquire needs information flows and, on the other hand, these diffused information flows result in economic damages. In other words, there is a clear paradox: if the information on the firm to be sold becomes public, even outside the industrial district, the chances of finding a potential buyer increases but, at the same time, the economic value of the firm decreases.

The problems of assessment derive from structural characteristics of district firms which are essentially based on tacit knowledge and competence of the entrepreneur. He has specific artisan skills and social and manufacturing reputation which is historically consolidated into supplier, customer and financial relationships (Varaldo and Ferrucci, 1995). Therefore, the economic value of the firm to be sold depends essentially on this immaterial asset, owned by the entrepreneur. In other words, the reputation belongs to the entrepreneur and, only in an indirect way, to the firm. Therefore, in merger and acquisition processes, especially with non-district firms, it is important to preserve this immaterial capital by using contractual instruments such as the former entrepreneur continuing to take part in managerial activities and bonus payments based on financial or economic indicators.

Information and assessment problems give rise to a high level of uncertainty both in the decision making process and in the contracts resulting from the negotiations between the firm to be sold and the potential buyer. Their geographical distance contributes to increasing this uncertainty. Indeed, if the two subjects are both located in the same industrial district, there are fewer information problems and the immaterial and social reputation of the firm to be sold can be transferred more easily to a district buyer than to a non-district buyer. For all these reasons, this fourth institutional solution is the most difficult to pursue and a specific financial and industrial policy is required to promote a real and efficient market.

In spite of all these problems, this institutional solution encourages innovative strategies. With new partners in the entrepreneurship, the district firm can modify its paradigmatic model of development, once strong ties with new international markets, new technological suppliers or new foreign customers have been established (Grandinetti and Rullani, 1994; Rullani,1997). Cultural and economic cross fertilisation furthers organisational innovation due to a critical revision of different consolidated routines (Nonaka and Takeuchi, 1995). In this dynamic process, the district firm learns new routines and explores new organisational models and strategic goals. As a consequence, in the present competitive situation which is characterised by radical changes, this institutional solution allows firms to learn more than with previous solutions that are based only on a new combination of district knowledge.

In the Prato textile district, there is an interesting dichotomy between entrepreneurs without any realistic chances of continuing and others who believe in the the survival of the firm in their absence (as indicated in Table 3). Almost 40% of the entrepreneurs affirm that survival is impossible; this shows the existence of high barriers preventing the implementation of efficient and realistic institutional solutions. As a consequence, if the industrial district wants to preserve the immaterial asset possessed by these small firms, it requires efficient financial and industrial instruments.

Of the four different institutional solutions analysed previously, a large majority of entrepreneurs prefers to transfer the small firm inside the family, especially to their heirs (as indicated in Table 4). The possibility of transferring firms to employees or of effecting mergers and acquisitions is low. It confirms the varying degrees of difficulty to implement these institutional innovations: the easier the solutions are, the higher the number of preferences it obtains.

Table 3. Different options for firm continuity in Prato textile district

Options	*Frequency*
Business cease	39.2
Willing to continue	41.7
High uncertainty on options	19.6

Table 4. Entrepreneur preference about firm continuity in Prato textile district

Institutional solutions	*Number of preferencies*
Sons	67.1
Relatives	10.8
Employees	7.0
Merger&Acquisition	13.3
Others	3.1

6 Industrial and financial policy in supporting institutional changes in districts firms

As shown in Table 3 and 4, preferences for the merger and acquisition and the employee solutions are not so considerable as desired. If the industrial districts want to have a policy to further the survival of the firm, it is necessary to develop specific instruments to realise these two institutional solutions.

A first possible policy can improve the efficiency of the market for firms. Mergers and acquisitions are not frequent, particularly with non-district firms and the problems of meeting firm demand and firm supply are numerous. Therefore, new managerial activities in industrial districts, such as firm market advisors, is needed to further these specific transactions in terms of negotiations, contractual arrangements and economic assessment.

Moreover, the industrial district needs new local financial intermediaries, that are able to uphold continuity of the firm with their employees. Venture capital can manage a transitional period and helps the transfer of the firm from entrepreneur to employees. Maybe some tax incentives are needed in support of this institutional change, as happens in Italy for producer cooperatives. Specific management training programmes designed for employees can decrease the economic and organisational risk associated with this transfer.

7 Conclusions

The institutional change of district capitalism is becoming a fundamental factor for strategic innovation of these small firms. Moreover, the historical and economic characteristics of district firms represent today strong barriers preventing the implementation of radical institutional innovation. The paradigmatic model of district development prefers incremental institutional changes based on relatives or employees solutions. But they combine only local knowledge possesed by different individuals (Maskell, 1999; Senker, 1995). As a consequence, this institutional innovation is marginal and their effects on organisation and strategies are weak (Grabher, 1993).

District renewal needs different institutional solutions able to reinforce local knowledge with external competences possessed by non-district firms. To avoid the high barriers to the entry of non-district subjects within the firm, it is useful to develop a specific industrial policy, such as tax incentives, new financial intermediaries and new advisors for this market.

References

ARCANGELI F., BELUSSI F., GOTTARDI G. (Eds.) (1988), *Evolutionary Pattners of Local Industrial Systems*, Gower, Ashgate.

BECATTINI G. (1991), "Il distretto industriale marshalliano come concetto socio-economico", in: PYKE F., BECATTINI G., SENGERBERGER W., (a cura di) *Distretti industriali e cooperazione fra imprese in Italia*, Studi e Informazioni, Quaderni Banca Toscana, Firenze.

BERETTA E., OMICCIOLI M., TORRINI R. (2000), "Banche locali e amplificazione degli shocks economici attraverso il canale creditizio", in: SIGNORINI L.F. (a cura di), *Lo sviluppo locale. Un'indagine della Banca d'Italia sui distretti industriali*, Meridiana, Roma.

BULL A., CORNER P. (1993), *From Peasant to Entrepreneur: the Survival of the Family Economy in Italy*, Berg, Oxford.

CONTI G., FERRI G. (1997), "Banche locali e sviluppo economico decentrato", in: BARCA F. (a cura di), *Storia del capitalismo italiano*, Donzelli, Roma.

DEI OTTATI G. (1994), "Trust, Interlinking Transactions and Credit in Industrial Districts", *Cambridge Journal of Economics*, vol. XVIII, n.6.

FERRI G., MESSORI M. (2000), "Bank-Firm Relationships and Allocative Efficiency in Northeastern and Central Italy and in the South", *Journal of Banking and Finance*, vol. XXIV, n. 6.

FERRUCCI L. (1996), *I processi evolutivi nei sistemi di piccole imprese - Il distretto tessile pratese*, Guerini e Associati, Milano.

FILIPPI E. (1999), "Credito e finanza nei sistemi locali: alcuni commenti", GRUPPO DI ANCONA (a cura di), *Trasformazioni dell'economia e della società italiana*, Il Mulino, Bologna.

GOODMAN E., BANFORD J., SAYNOR P. (1989), *Small Firms and Industrial Districts in Italy*, Routledge, London.

GRABHER G. (1993), *The Embedded Firm: on the Socioeconomics of Industrial Network*, Routledge, London.

GRANDINETTI R., RULLANI E. (1994), "Sunk Internalization: Small Firms and Global Knowledge", *Revue d'Economie Industrielle*, n.67.

HANSMANN H. (1996), *The Ownership of Enterprise*, Harvard University Press, Cambridge.

KEPNER E. (1991), "The Family and the Firm: a Co-evolution Perspective", *Family Business Review*, vol. 4, n.4.

MASKELL P. (1999), "Localised Learning and Industrial Competitiveness", *Cambridge Journal of Economics*, n.23.

NONAKA I., TAKEUCHI H. (1995), *The Knowledge-Creating Company*, Oxford University Press, Oxford.

RULLANI E. (1997), "L'evoluzione dei distretti industriali: un percorso tra decostruzione e internazionalizzazione", in: VARALDO R., FERRUCCI L.(EDS.), *Il distretto industriale tra logiche di sistema e logiche di impresa*, F.Angeli, Milano.

SENKER J. (1995), "Tacit Knowledge and Model of Innovations", *Journal of Corporate Change*, vol. 4, n.2

SIGNORINI L.F. (1994), "The Price of Prato, or Measuring the Industrial District Effect", *Papers in Regional Science*, vol. 73, pp. 369-392.

SOLINAS G. (1982), "Labor Market Segmentation and Worker's Careers", *Cambridge Journal of Economics*, n.6.

SPAGNOLO G. (1999), "Social Relations and Cooperation in Organizations", *Journal of Economic Behaviour and Organization*, n.1

VARALDO R. (1993), "Artist, Artisan and Development of Small-Medium Sized Enterprises in Italy", *Small Business*, vol. IV, n.2

VARALDO R., FERRUCCI L. (1995), "The Evolutionary Nature of the Firms within Industrial Districts", *European Planning Studies*, vol.4, n.1

Groups of small and medium-sized firms in industrial districts in Italy

Donato Iacobucci

Computer Science Institute, University of Ancona, Via Brecce Bianche, 60131, Ancona, Italy

1 Introduction

Literature on industrial districts has emphasised the 'systemic' dimension of such productive organisation. Two 'levels' have drawn the attention of scholars: i) the 'interfirm network', i.e. the set of firms exchanging goods and services within a production chain (Staber, 2001; Whitford, 2001); ii) the 'local system', which is composed not only of firms located in a specific area but also of the social and political institutions that characterise the area[1].

The impact of these two levels on the efficiency of firms has been widely analysed, both at a theoretical and an empirical level. Some authors have emphasised the former aspect, some the latter; all have recognised the importance of the relationships between the two. Indeed they both contribute to the *external economies* which are at the heart of the competitiveness of the firms located in industrial districts. Given the fact that these firms are mostly small and medium sized enterprises (SMEs), researchers have normally taken for granted that the 'elementary unit' forming the system - either the network or the local system - is a 'single plant – single company' firm, owned and managed by an entrepreneur (or a family).

Recent empirical investigations regarding the Italian industrial districts of the so-called NEC (North-Eastern-Central) regions have demonstrated that this is not always the case, because of the increasing presence of business groups (Balloni and Iacobucci, 1997; Brioschi and Cainelli, 2001). Both the group and the network are made up of legally independent firms. Moreover, there is a clear distinction between the two: the companies in a network are owned and managed by different people, while the companies forming a group (as defined here) are bound under a common ownership. It is this element, i.e. ownership, and the control rights attached to it, that distinguishes a group from other forms of contractual arrangements between legally independent firms (Gulati and Singh, 1998). This does not automatically imply that the 'tightness' of control is greater for the companies in a group than for those in a network: a company in a group could enjoy greater autonomy in its strategic and operative activities than a company in a network. Still, there is a fundamental difference: in the case of a

[1] The second aspect has been emphasised by Italian scholars (Brusco, 1982; Becattini, 1991, 2000; Pyke and Sengenberger, 1992). See also Harrison (1992).

network the 'ties' between firms are restricted to the aspects pertaining to the contractual arrangement (formal or informal); in the case of a group, ownership allows the owner to discretionally intervene in the major decisions of the controlled companies[2].

In order to avoid any misunderstanding in the terms we are going to use, from now on we will adopt the following definitions:

- company: a legal entity, normally a partnership or a limited company;
- group: a set of companies under common ownership and control[3];
- ultimate owner(s): the person(s) at the head of the controlling chain, where there are several layers of controlled companies;
- network: a set of companies belonging to different owners cooperating towards a common end[4].
- entrepreneur: a person controlling a company (ultimate owner) and directly involved in the management of it.

Depending on the definition of the firm and on the kind of analysis one is interested in, the firm (i.e. the economic unit) can be associated with the legal entity (the company) or with the group. Official statistics normally consider only the technical and the legal unit (the plant and the company respectively). This is one of the reasons why researchers normally associate the firm with the legal entity. At the same time, scholars interested in the phenomenon of groups generally take for granted that the economic unit is the group rather than the single company. Because ownership and legal autonomy are both important elements for the definition of the firm, the association of the firm either with the group or with the single company is not straightforward.

Starting from this premise, the aims of this paper are the following:

- To discuss whether and to what extent we can consider the group rather than the single company as the economic unit (the firm);
- To present some data regarding the presence of groups in the Italian manufacturing sector and its industrial districts;
- To review the reasons put forward by the literature to explain the presence of groups among SMEs;

---

[2] Both groups and networks can be thought of as mechanisms for solving problems of contractual incompleteness (Grossman and Hart, 1986); however, the scope of ownership is much larger than that of the contractual arrangement in a network.

[3] We refer here to *effective* control rather than to *formal* (legal) control (Hansmann, 1996). The latter, which typically gives the right to elect the firm's board of directors, is by law associated with the owners of a company, although not always exercised (as in the case of a public company). Effective control means not only the exercise of formal rights but also the direct involvement in the strategic decision-making process as a member of the board of directors of the company.

[4] The relationships between companies engaged in a network are mostly as part of a production chain: i.e. they exchange products or services that are the output of one firm and the input of another. Horizontal cooperation is rather unusual in industrial districts.

- To discuss the consequences of the spread of business groups on the evolution of industrial districts, taking into account the case of the Marche region.

These aims are the contents of sections 2, 3, 4 and 5 respectively. Section 6 draws some conclusions and discusses their implications for further research.

2 The unit of analysis: companies or groups?

Some papers in this volume (see, for example, Brioschi *et al.*) argue that the degree of economic concentration (however measured) within Italian industrial districts is higher than it appears to be from official statistics which ignore the presence of business groups: sets of companies legally distinct but controlled by the same owner(s). In this assertion it is understood that the group rather than the single company should be considered the appropriate economic unit (i.e the firm).

The question of whether the group or the single companies should be considered as the elementary unit in economic analysis cannot be answered in a general way. Groups come in a wide variety of types, according to their internal organisation, the activities of the member companies, the structure of ownership, the operating mechanism, etc. (Brioschi and Cainelli, 2001). At the same time, the answer depends on the aim of the analysis and on the related way of defining the firm.

If the emphasis is laid on the firm as a decisional unit for resource allocation then it is the group as a whole rather than the single companies that should be taken into account. Whatever the strategic and operative autonomy of the single companies, the ultimate owner retains the right to selectively intervene in the resource allocation process of the controlled companies. In this respect the group can be assimilated to a multidivisional company (M-form) where the role of the top management (the ultimate owner in the case of a group) is that of deciding the resources to be allocated to the existing divisions (companies) and whether divisions should be opened (set up or acquired) or closed (liquidated or sold).

If in defining the firm the emphasis is laid on its contractual autonomy with third parties, then the single companies rather than the group become pre-eminent. In the Italian civil code, as well as in that of other countries, the group does not count as a legal entity (except for marginal aspects). It is the single companies that are responsible for obligations with third parties and for the most important aspects of corporate governance. Notwithstanding the analogy of the companies with the divisions of an M-form, some authors have stressed the importance of the legal regime in setting the boundaries of the firm (Masten 1991). Although the ultimate owner can discretionally intervene in the main decisions of the controlled companies, the kind of hierarchical relationship between an owner and the directors is different from that of a CEO with the managers (heads of divisions). The analogy between the group and the M-form (Chandler, 1982) overlooks the importance of the legal framework of the group and of its consequences on the internal relationships between the companies of a group and with third parties.

From this brief discussion it emerges that, whatever the theoretical perspective adopted in defining the firm, the group remains an 'hybrid' form, that shows some features of the economic firm but that does not correspond to it[5].

For this reason, the question of whether the group or the single companies should be considered as the elementary unit must take into account the individual characteristics of the groups. Two aspects seem pre-eminent: the existence of economic synergy among the companies (beside the common ownership) and the degree of association between ownership and control[6].

In the case of large groups we can identify two main types of groups: the *financial group* and the *industrial group*. In the financial group companies do not show any economic relationships apart from the reference to a common ownership: it is what Williamson (1975) called the holding company (H form) and can be associated with a conglomerate multidivisional. On the contrary, an industrial group is made up of companies that show economic relationships between each other, based on horizontal synergy (like the sharing of common resources) or the exchange of goods and services. In very large groups we often observe both forms given the presence of several layers of controlled companies: the first layer, made up of sub-holdings operating in very different sectors, can be associated with a financial group; the second layer, formed of the companies controlled by each sub-holding, could be associated with an industrial group[7].

In some cases, both the holding company and the sub-holdings are listed on the stock exchange; this allows the ultimate owner to exploit the mechanism of 'stock leverage' that has been identified as one of the main reasons for the existence of pyramidal groups in Italy (Brioschi *et al.*, 1990; Barca *et al.*, 1994).

In the case of groups composed of SMEs one of their main features is that the companies forming the group are completely owned by the entrepreneur or his/her family (Brioschi and Cainelli, 2001; Iacobucci, 2002). In addition to ownership concentration (and clearly linked to that), the main companies of the group are directly controlled by the entrepreneur or by the members of his/her family. In this case, according to the degree of economic synergy between the

---

[5] According to Gerlach (1997), the spread of the group form is part of a more general tendency towards the loosening of organisational boundaries.

[6] There are other important aspects regarding the internal organisation of the groups. Some, like the degree of strategic and operative autonomy of the companies, are clearly related to the two aspects considered in the analysis.

[7] Most of the major Italian groups are organized in this way. An emblematic case is that of the Benetton group: the holding company (Edizione Holding) owned by the members of the Benetton family controls several sub-holdings operating in disparate sectors: clothing, catering, telecommunications, etc. Each sub-holding is organised as a group of companies operating in the same sector and showing a high degree of synergy in their activities. A similar structure is found by Shimotani (1997) in the largest Japanese groups.

activities performed by the companies, we can distinguish between *entrepreneurial groups* and *family groups* (Figure 1).

Figure 1. A typology of groups

Ownership concentration and association between ownership and control		Horizontal synergy and/or vertical relationships between the companies: Yes	Horizontal synergy and/or vertical relationships between the companies: No
	No	a) Industrial group	b) Financial group
	Yes	c) Entrepreneurial group	d) Family group

The vertical dimension of Figure 1 is clearly related to the size of the group, although this is not the only relevant variable to explain the degree of ownership concentration and the association between ownership and control.

For most of the economic analysis, types a) and c) should be considered the appropriate unit: i.e the firm. In these cases, in fact, the group can be considered as a specific organisational form, whatever the reasons are that justify its adoption instead of other organisational forms (specifically the M-form). This definition of the firms adheres to the one proposed by Penrose: "... we define the industrial firm with reference to its administrative framework within which industrial activities are co-ordinated ... It is the 'area of co-ordination' – the area of 'authoritative communication' – which must define the boundaries of the firm"; (Penrose, 1959, p. 20-21). She excluded from her definition the cases in which the companies were bound together by a mere element of common ownership (the 'financial' and 'family' groups in our typology).

Of course, there are cases in which either the 'financial group' or its member companies can also be the appropriate unit. In general, however, the choice is not between the single *company* and the *group* but should consider the appropriate bunch of companies that fit the economic definition of the firm. It is important to stress that the typology presented in Figure 1 does not necessarily refer to the group as a whole; in some cases we can have more that one type according to the *layers* considered in the analysis. This is often the case in large groups, whose activities are spread over different sectors and hundreds of companies organised in several layers. It is less relevant in the case of groups of SMEs, where the entrepreneurial group seems to be the dominant form.

3 The presence of business groups in the Italian manufacturing industry

Several surveys conducted in Italy during the nineties revealed the importance of the phenomenon of business groups and the fact that it was not restricted to large firms (Barca *et al.*, 1994; Barbetta *et al.*, 1995).

In 1997, for the first time in the collection of census statistics in Italy, manufacturing companies were asked whether they belonged to a group or not[8]. The result is shown in Table 1, where a comparison is also made with the same information collected by Mediocredito Centrale on a sample of Italian manufacturing firms with more than 10 employees.

Table 1. Companies belonging to a group by class of employees in the Italian manufacturing sector (1997)

ISTAT		*Mediocredito Centrale*	
Class of employees	% of companies	Class of employees	% of companies
1-9	0,8	11-20	9,1
10-19	3,8	21-50	14,2
20-49	12,3	51-100	28,2
50-99	23,2	101-250	45,1
100-249	43,9	251-500	73,7
250+	72,7	501 -	84,1

Sources: ISTAT, Intermediate census; Mediocredito Centrale (1998)

The two surveys show a remarkable similarity in the entity of the phenomenon, except for the smaller firms (10 to 20 employees) for which the Mediocredito survey shows a larger percentage of firms belonging to a group. Overall they demonstrate that the phenomenon of groups is practically absent in companies with less than 10 employees but grows steadily with company size and is remarkable also in medium sized firms: indeed, about half the companies in the class 100-249 employees declared they belonged to a group. In the case of firms with less then 20 employees, it should be taken into account that in Italy the majority of them are artisan firms (craftmen) whose owner, by law, is supposed to be directly involved in the operative activity of the firm. For this reason it is difficult that the same person can own and manage more than one company.

Information regarding firms belonging to a group does not on its own allow for an assessment of the impact of the phenomenon on the size structure of the

[8] The definition of group adopted by ISTAT is similar to that adopted here and in other survey studies: group is defined as a set of companies under a common control.

manufacturing sector or on industrial districts. In order to construct a new size distribution of firms based on groups rather than single companies, one should have information regarding the companies of the individual groups[9]. The Mediocredito Centrale survey gives some information about this; specifically, any company belonging to a group was asked to provide information about the number of companies belonging to the group and its overall employees (Table 2). The average number of companies per group is 37; this figure is strongly affected by a few very high values: indeed, the median is 6 companies and 450 employees per group. The average size of the group in terms of employees is 8 to 50 times the average number of employees of its member companies (Table 2).

Table 2. Number of companies and employees of groups by class of employees of the firms declaring they belong to a group, 1998.

	Companies per group		*Total employees of the group*		*Average employees per company of the group*
Class of Employees	Average	Median	Average	Median	
11-20	6.6	4	165	100	31
21-50	17.9	5	663	150	51
51-100	16.9	5	1,568	300	79
101-250	36.2	6	2,376	450	124
251-500	63.7	10	5,216	1,000	188
501-	76.9	20	17,148	3,551	276
Total	36.9	6	4,223	450	124

Source: Elaboration from Mediocredito Centrale database

Given the fact that the presence of groups is negligible in firms with less than 20 employees, the conclusion that can be reached from the data presented in Table 2 is as follows: considering the group, the structure of the Italian manufacturing sector is more *dichotomic* than it appears when considering the companies (Barca, 1996). This is because it leaves the size of very small firms unchanged while it determines a significant increase in the average size of medium sized and large firms. This conclusion is also valid in the case of the industrial districts, where the presence of very small firms is still relevant, but where the importance of medium sized firms is probably underestimated when we consider the single companies rather than the groups.

Up to now we do not have census data regarding the presence of business groups within the Italian industrial districts. Studies regarding specific regions or districts (Brioschi and Cainelli, 2001; Balloni and Iacobucci, 1997; Caprio, 2002)

[9] One can imagine two extreme cases: the minimum in which all the groups are formed of only one manufacturing company (in this case there would be no consequence on the size structure of sectors or districts); the maximum in which all the manufacturing firms declaring they belong to a group are part of the same group.

demonstrate that the presence of groups in industrial districts by size of firms is similar to that observed in the overall Italian manufacturing system. Relying on data at company level for analysing the structural characteristics of industrial districts can overshadow three important aspects.

The first is that several companies within the industrial district may be part of the same group; in this case the concentration within the district is higher that it appears to be when the single companies are considered. The second is that district groups can hold companies outside the district; this does not change the concentration within the district but underestimates the effective size of the largest companies. The third is the fact that single companies within the district can be part of major groups at a national or international level.

The first two cases are the result of the growth process of medium sized firms within industrial districts, while the third is the result of the acquisition of district firms by Italian or foreign firms. Although in some districts the penetration of external (even foreign) firms has been substantial, we think that the growth process of district firms is the phenomenon that is exerting the greatest influence on the district organisation. The reasons why district firms grow through the group form and its consequences on the organisation of industrial districts is the subject of the next section.

4 The causes explaining the presence of business groups among SMEs in industrial districts

There are two important strands of literature regarding business groups: economic literature and entrepreneurial literature[10]. The former has focused its attention on large groups, while the latter is mainly concerned with SMEs. The common element of these two strands of literature is that they regard the business group as a way of managing the organisational complexity arising from the diversification of activities.

Economic literature has put forward several explanations of why large firms are organized as a group, in developing as well as in most developed countries. There are different ways of classifying this literature. One is that of separating the authors depending on whether they consider the group as an efficient or an inefficient form of business organisation (Khanna, 2000).

Several authors have interpreted the group as an efficient form of organisation of business activities compared with alternative organisational forms. The most common theoretical approach of this strand of literature is that of transaction

[10] There is also sociological literature, that we are not going to review here. It focuses its attention mainly on personal ties between the owners and directors of companies in order to assess the degree of power concentration in economic activities. This is another reason why sociological authors normally give a definition of a group which is larger than that adopted here (see Granovetter 1994 for a review of this literature).

costs, according to which the group is identified as an intermediate form between market and hierarchy. The efficiency of the group form stems from the fact that it offers a solution to problems created by market imperfections: mitigating the opportunism and the information problems between firms exchanging goods and services (Goto, 1982; Kester, 1993); enhancing the incentive for managers running the companies (Williamson, 1975; Collin, 1998); reducing the information asymmetries in allocating financial resources to the companies. Within this strand of literature specific attention has been dedicated to the Japanese groups (Aoki, 1984; Kester, 1993; Shiba and Shimotani, 1997; Gerlach, 1997) whose organisation, based on cross ownership between industrial and financial firms, is rather different from that of the 'pyramidal' groups prevailing in continental Europe.

Authors studying the organisation of business activities in developing countries and transitional economies have seen the group as an organisational device that is able to overcome the market imperfections and the lack of economic institutions (like a developed financial market) that are typical of these economies (Chang and Choi, 1988; Feenstra *et al.*, 1997; Ghemawat and Khanna, 1998; Keister, 1998; Khanna and Palepu, 1999; Khanna and Palepu, 2000; Khanna and Rivkin, 2001)[11]. In general this literature sees a beneficial role of groups in fostering efficiency in product and capital markets and in promoting the development of business activity in these countries (Leff, 1978).

Some of the same authors have also underlined the potential welfare loss determined by the presence of groups, both in developed and developing countries. The inefficiencies can originate from several factors: the exploitation of minority shareholders in pyramidal groups (Brioschi *et al.*, 1989; Bianco and Casavola, 1999); the rent seeking behaviour of group owners in less developed countries; the exercise of market power (Encaoua and Jacquemin, 1982).

The empirical evidence available up to now does not allow for a clear discrimination between these two competing views (Khanna, 2000). Overall, business groups are considered as playing a positive role in developing countries while they are seen with more 'suspicion' in developed countries, where their presence is less justified by efficiency reasons. This position is prevalent among Italian scholars who consider the (pyramidal) group as a financial device devoted to maximising the activities controlled by a family, while minimising the amount of capital invested by the same family[12]. In this respect the 'pyramidal' group, which is common among the large Italian firms, is interpreted as a way of separating ownership and control in a situation of under-development of both

---

[11] See also n. 4, vol. XXXI, 1993 of Developing Economies, dedicated to the phenomenon of business groups in developing countries.

[12] "By spreading the voting rights of minority shareholders out over a large number of firms, and concentrating those of the entrepreneur in the company at the top of the pyramid, this model allows the latter to obtain control over the greatest possible amount of other people's capital with the smallest possible amount of his own." Barca (1996), p. 14.

capital markets and other institutions of corporate governance (Brioschi *et al.*, 1989; Barca, 1996).

Although economic literature offers a rich set of hypotheses in order to understand the causes and the consequences of the phenomenon of business groups, its main limitation, from the point of view of our analysis, is that it considers only large groups. In several cases the cited authors take into consideration just a few of the major groups within a country. In general, they consider the group as a 'special' form of business organisation, efficient under certain circumstances. This approach seems in contrast with the empirical evidence presented in the previous section, according to which the group should be considered, at least in the Italian case, as the 'normal' form of organising the activities of the firm as it grows beyond a certain size.

As in the case of the multidivisional form (M-form), the setting up or the acquisition of new companies is the result of a growth process achieved by diversifying the original activity[13]. It is normally taken for granted that growth by diversification and the associated need for organisational differentiation is characteristic of large firms. For this reason, some Italian scholars consider the presence of groups among SMEs as induced by the specific Italian institutional context rather than by the need for organisational differentiation. Lorenzoni 1990 (p. 103-104) sees the group of SMEs as the result of the lack of managerial skills on the part of entrepreneurs: as the original company grows in size and adds new activities, the entrepreneur prefers to repeat the same entrepreneurial formula, i.e. the elementary unitary-form, rather than develop a more complex organisation. Other scholars argue that the spread of groups can be explained by the desire to take advantage of some benefits associated with the small size of its legal units. Among these advantages, those most often cited are: the possibility of circumventing labour or fiscal regulations, the possibility of being entitled to financial benefits reserved for small firms and the greater flexibility allowed in the use of labour in small companies (Barca *et al.*, 1994).

These explanations do not seem convincing for several reasons. One of them is that the presence of business groups among SMEs is not peculiar to the Italian situation. Indeed, entrepreneurship scholars have found that there is a significant percentage of entrepreneurs who found, and simultaneously own and manage, a series of businesses. Although we lack a clear picture of how far this phenomenon occurs across different countries, it can be said that the impact of 'multiple business ownership', i.e. of groups, in small firms seems relevant in all industrialised countries (Donckels *et al.*, 1987, Birley and Westhead, 1993, Scott and Rosa, 1996; Rosa and Scott, 1999; see also issue n. 4, vol. 22 1998 of *Entrepreneurship Theory and Practice* dedicated to the subject)[14].

---

[13] According to Chandler 1982, the group is the European counterpart of the American M-form, justified by the differences in the history of the industrialisation process and in the institutional context.

[14] Entrepreneurial literature uses several definitions for the phenomenon: habitual entrepreneurs; portfolio entrepreneurs; multiple business owners; serial entrepreneurs.

Three main explanations for the phenomenon of business groups among SMEs have been put forward by entrepreneurial literature.

The first considers the group as a specific organisational form, especially suited to accommodate a *growth process through diversification* (Rosa, 1998). The idea that firms grow mainly through a diversification process, and that this process generates organisational differentiation, is not new (Penrose, 1959; Chandler, 1962; Rumelt, 1974). What seems new is that the prevailing demand and technology conditions have reduced the size at which the phenomenon occurs (Robson et al. 1993).

A second way of looking at business groups is to consider them the result of *entrepreneurial dynamics* (Rosa, 1998). Once the original company set up by the entrepreneur has reached success (or 'maturity'), he/she delegates the management of the company to professional managers (while retaining company ownership) and concentrates his/her efforts on the development of new businesses. This perspective recognises that entrepreneurship is not a "single-event" action (Wright *et al.*, 1998) and that some entrepreneurs, especially successful ones, tend to replicate their experience in new businesses[15].

It has also been suggested that the presence of groups among SMEs can be explained considering the process of *capital accumulation* by the entrepreneur or his/her family (Scott and Rosa, 1996). In this case it is the concentration of ownership, typical of SMEs, which can favour a portfolio strategy in allocating financial resources.

In another work (Iacobucci, 2002) we have tried to discriminate between these three explanations, using data regarding the groups of SMEs in the Italian manufacturing sector. The internal organisation of these groups and the activities performed by the companies favour the first hypothesis, confirming that the spread of groups among SMEs is the result of a growth process through diversification.

While in the case of large groups, the setting up or acquisition of new companies has been associated with a policy of conglomerate diversification (Williamson, 1975), this does not seem to be the case for groups of SMEs. The data available show a low degree of diversification in their activities: indeed, growth is mainly achieved through horizontal expansion in *similar* activities (Brioschi and Cainelli, 2001; Balloni and Iacobucci, 2001). The setting up of new companies is mainly aimed at ensuring operative autonomy in these new activities, although they are clearly part of a unitary strategy. The group form allows the autonomy of the individual units to be enhanced while the group as a

---

Except for the last definition, that refers to entrepreneurs who own one business after another but effectively own only one business at a time, the other definitions correspond to our definition of group.

[15] In the Italian experience it is also common that companies are set up in order to accommodate the entrepreneurial aspirations of other members of the entrepreneur's family (typically his/her children).

whole can take advantage of economies of scale in some activities (such as marketing, finance, R&D, etc.).

5 The role of groups of SMEs in the evolution of industrial districts

The increasing presence of business groups within the Italian industrial districts is mainly linked to the growth process of some firms within the district. Recent empirical analysis regarding the largest firms in industrial districts showed that the formation of groups is the result of a growth process that follows two main directions (Brioschi and Cainelli, 2001; Balloni and Iacobucci, 2001): i) the setting up or the acquisition of companies operating in activities similar to the original one (differentiation); ii) the expansion of manufacturing and distribution activities abroad (internationalisation).

While strategy i) is followed even by small groups, strategy ii) is strictly associated with size (Mariotti and Mutinelli, 2001; Balloni and Iacobucci, 2001).

This confirms that the setting up of new companies is not only induced by diversification but also by the development of similar activities aimed at penetrating different segments of the same market (in some cases companies are set up or acquired in order to manage different brands). Less important within group strategy is the vertical integration of companies operating along the production chain[16]. The presence of business groups within industrial districts has two consequences: firstly it raises the concentration of the district output; secondly it enhances the ability of the largest firms to control the final markets of district products.

In order to analyse the growth strategies followed by groups of small and medium sized firms and some of their structural characteristics we have taken advantage of a data-set concerning the largest manufacturing firms of the Marche region (see Balloni and Iacobucci in this volume). For each district identified in the region we isolated the 8 largest companies and studied their group organisation.

Only in the case of the largest districts (the furniture district of Pesaro, the household appliance district of Fabriano and the footwear district of Civitanova-Fermo) were we able to identify 8 groups for each; the other cases showed that only one or a few groups dominate the local system (Table 3).

[16] Although the largest companies do not tend to integrate their suppliers, there is a tendency to reduce their number and to set-up more collaborative relationships with them.

Table 3. Largest firms in the industrial districts of the Marche region by type of ownership

	*Formal Group**	*Informal Group***	*Subsidiary of foreign groups*	*Total*
Furniture (Pesaro)	2	5	1	8
Textiles and clothing (Urbania)		2		2
Household appliances (Fabriano)	4	3	1	8
Furniture (Treia)		1		1
Textiles and clothing (Urbisaglia)	1			1
Leather (Tolentino)	1	1		2
Footwear (Civitanova-Fermo)	4	4		8
	12	16	2	30

* Groups with a holding company and a consolidated balance sheet
** Groups whose companies are owned directly by the same people

At the end of 1999 the 30 groups examined controlled 252 companies, with an average of 8 companies per group. About 60% of these companies are manufacturing companies, most of them located within the same district[17].

5.1 Furniture groups

In the case of the furniture district of Pesaro, the largest companies are kitchen producers (Table 4). Their group structure is rather different, reflecting the different attitudes of the founding entrepreneurs, who still control the groups. In the case of the Scavolini group (the largest kitchen producer in Italy) the group is formed of two companies, both operating as kitchen producers although serving different market segments. In the case of the second largest kitchen producer (the Berloni group) the founding brothers (and their wives and children) control several companies operating in the furniture sector (kitchens, office furniture, home furniture) and electric appliances (small refrigerators for hotels and boats).

Furniture is mainly sold on the domestic market. This is reflected in the type and location of the companies belonging to the largest groups; they are concentrated in manufacturing activities and are mainly located within the district (Table 5).

[17] Some general features of these districts and the methodology to identify them are discussed by Balloni and Iacobucci in this volume.

Table 4. Largest groups in the furniture district of Pesaro

Group	*Type**	*Number of companies*	*Consolidated turnover (million €)*	*Consolidated employees*	*Turnover of the original company (million €)*	*Employees of the original company*
Scavolini	F	2	129	449	117	362
Berloni	I	9	101**	554**	56	269
L.C. Mobili	S	6	n.a.	n.a.	37	148
Febal	F	3	32	161	28	146
Curvet	I	6	n.a.	n.a.	28	155
I.F.I.	I	4	30**	180**	19	117
Imab	I	2	34**	216**	13	105
S.I.F.A.	I	2	n.a.	n.a.	9	75

* F=formal; I=informal; S=subsidiary of foreign group
** Sum of data regarding the main companies

Overall the furniture groups show a low level of diversification and internationalisation. This is evident from the fact that most of their companies are furniture manufacturing companies located within the district (Table 5).

Table 5. Companies belonging to furniture groups by type and location

	District	*Same region*	*Abroad*	*Total*
Distribution			5	5
Service	1	1		2
Manufacturing	24	2	1	27
Total	25	3	6	34

An interesting case of furniture groups is the Berloni group; it is the largest in terms of companies controlled and the most diversified. The original company was founded by the two Berloni brothers in 1960 as a kitchen maker. At first the firm expanded its activities by setting up companies in other segments of the furniture market: home and office furniture. All these companies are located within the furniture district. Subsequently the group took the opportunity to diversify its activities acquiring some companies which produce small fridges for hotels and boats. These companies are located near the furniture district. Two (Indel Marine and Bartech) are operated as joint ventures with foreign partners.

Until 2001 the group operated as an informal group, with all the companies owned directly by members of the Berloni family. In 2001 a holding company was created to concentrate the ownership of all the companies. At the same time some activities of the group (finance, human resource development, etc.) were centralised.

Table 6. The Berloni group

Company	*Activity*	*Relationship with the original company*	*Location*
Mobili Berloni Spa	Kitchens	Original	District
Iterby Srl	Office furniture	Similar	District
Betacinque Srl	Home furniture	Similar	District
Giemmetre Srl	Office furniture	Similar	District
Indel B Srl	DC powered fridges and fridge-bars	Concentric diversification	Same region
Condor B Srl	Components for refrigeration	Concentric diversification	Same region
Indel Marine Srl	Marine fridges	Concentric diversification	Same region
Bartech Sa	Fridge-bars	Concentric diversification	Abroad
Terme Di Carignano Srl	Thermal spa	Conglomerate diversification	Same region

5.2 Footwear groups

In the case of the footwear district most of the groups are informal groups and only the largest ones show a formal structure. They are all entrepreneurial groups, still owned and controlled by the founding entrepreneurs or his/her children (Table 7).

Table 7. Largest groups in the footwear district of Civitanova-Fermo

Group	*Type**	*Number of companies*	*Consolidated turnover (million €)*	*Consolidated employees*	*Turnover of the original company (million €)*	*Employees of the original company*
Della Valle	F	5	220	883	207	718
Fornari	F	10	61	204	60	152
Falc	I	3	n.a.	n.a.	38	174
Due Esse	I	3	n.a.	n.a.	31	175
Pizzuti	F	8	30	313	29	164
Ruggeri	I	8	85**	502**	27	116
Melania	I	7	n.a.	n.a.	25	54
R.P.M.	F	3	31	237	13	103

* F=formal; I=informal; S=subsidiary of foreign group
** Sum of data regarding the main companies.

Because the controlled companies operate along the same production chain (especially in the distribution phase) the difference between the consolidated and the original company is larger in terms of employees than in terms of sales. The degree of diversification is very low; all the companies belonging to the groups operate in the footwear sector. Most of the manufacturing companies are located within the district, while the distribution and service companies (including financial companies) are located abroad (Table 8).

Table 8. Companies belonging to the footwear groups by type and location

	District	*Other regions*	*Abroad*	*Total*
Distribution	1	3	7	11
Service	3		4	5
Manufacturing	23	4	2	29
Total	27	7	13	47

The low degree of internationalisation of manufacturing activities can be explained in two ways: firstly the small size of the groups; secondly the fact that in most cases foreign firms are owned directly by the entrepreneurs and for this reason they are difficult to detect.

One of the most internationalised groups is the Ruggeri group; it is an interesting case as it shows the main features of footwear groups (Figure 1).

The original company was set up in 1961 as a producer of women's shoes. Given the success of the original company in 1971 the four brothers set up a new company producing men's shoes. A 20% stake of the new company was given to a former employee who took management responsibility in it. In 1977 the entrepreneurs set up the IPR to enter the market of synthetic soles; the new company serves not only the companies of the group but also other footwear firms in the district and abroad. Also in this case the entrepreneurs gave a stake of the company to former employees that took on management responsibility in the new company. The growing demand for synthetic soles induced the entrepreneurs to set up two new companies in this sector in order to serve different market segments.

In 1990 the group acquired a company in Hungary to produce shoes for that market. In 1993 the group set up a holding company in the Czech Republic controlling several companies producing and selling shoes in East European countries (Table 9).

The group is a combination of a formal and informal group. The three main companies are owned by the four Ruggeri brothers with some managers of the companies having minority stakes. One of these companies, the IPR, is a formal group as it controls several companies in Italy and abroad. For this reason we have a consolidated balance sheet for IPR but not for the group as a whole.

Figure 2. The structure of the Ruggeri group

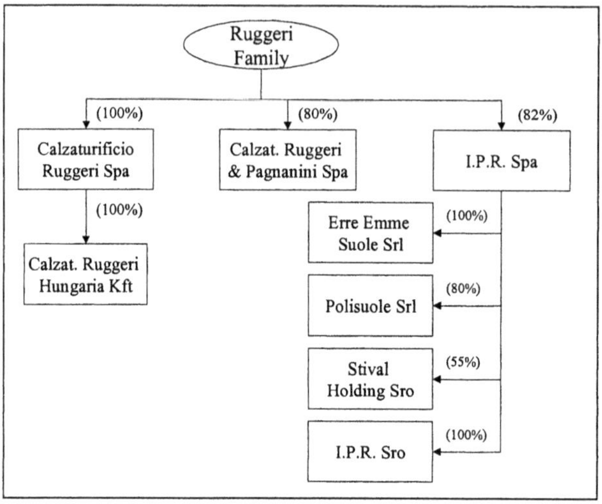

Table 9. The Ruggeri group

Company	*Activity*	*Relationship with the original company*	*Location*
Calzaturificio Ruggeri Spa	Footwear	Original (women's footwear)	District
Calzaturificio Ruggeri & Pagnanini Spa	Footwear	Similar (men's footwear)	District
IPR Spa	Synthetic soles	Complementary	District
Emme Erre Suole Srl	Synthetic soles	Complementary	District
Polisuole Srl	Synthetic soles	Complementary	District
Calzaturificio Ruggeri Hungaria Kft	Footwear	Similar	Hungary
Stival Holding Sro	Holding of several footwear companies in the Czech Republic	Similar	Czech Republic
IPR Sro	Synthetic soles	Complementary	Czech Republic

5.3 Household appliance groups

The household appliance district of Fabriano (table 10) is dominated by the presence of three large groups (the Merloni groups). Although originating from the same family they are owned and managed autonomously by the three sons of the founding entrepreneur. They are multinational groups with commercial and manufacturing subsidiaries in all the main European countries and in other parts of the world (Table 11).

Table 10. Largest groups in the household appliance district of Fabriano

Group	*Type**	*Number of companies*	*Consolidated turnover (million €)*	*Consolidated employees*	*Turnover of the original company (million €)*	*Employees of the original company*
Merloni Elettrodomestici	F	41	1,420	7,793	1,026	4,771
A. Merloni	F	7	681	4,636	537	3,624
Merloni Termosanitari	F	33	432	2,845	283	1,417
Elica	F	4	169	987	139	783
Galassi	F	16	101	359	64	191
Best	S	4	82	658	61	462
Turboair	F	5	59	503	47	272
Tecnowind	I	3	n.a.	n.a.	27	163

* F=formal; I=informal; S=subsidiary of foreign group

Table 11. Companies belonging to the household appliance groups by type and location

	District	*Same region*	*Other regions*	*Abroad*	*Total*
Distribution				35	35
Service	6	1	5	19	31
Manufacturing	19	4	4	20	47
Total	25	5	9	74	113

The medium sized groups are all producers of range hoods; in fact, Fabriano has the largest concentration of range hood production in Europe. During the nineties one of these companies (Best) was acquired by a USA multinational.

Despite the large size of these groups, their degree of diversification is still low. Except for a few companies, all of them operate within the production chain of the household appliance sector.

However, the degree of diversification is strongly influence by the 'level' at which it is considered. Data on ownership are normally available at the level of industrial or entrepreneurial groups. It is much more difficult to obtain data at the family level, unless the holdings of the same family are concentrated in a holding company. This is the case of one of the Merloni brothers who controls the Merloni Elettrodomestici group. Although this group shows a high coherence in the activities performed by its companies, the holding company that owns Merloni Elettrodomestici (Fineldo) has important stakes in groups operating in other sectors: motor vehicle production, energy production, paper, etc. If we consider the Merloni Elettrodomestici group, it is an industrial groups where all the companies are related to its main business (household appliances). If on the contrary we consider Fineldo it shows a high level of diversification of its activities. It is a family group where the members of the family have important positions within the boards of the controlled companies[18].

5.4 Groups operating in other districts

The other groups examined operate in industrial districts that show an absolute size far lower than that of the three districts previously examined. For this reason they normally represent a large quota of the output of the district they belong to[19].

Overall the groups located in these districts show the same characteristics as those located in the main districts previously mentioned. Most of their companies are manufacturing companies located within the district, while the remaining ones are the result of production and distribution internationalisation.

An interesting case in the Eugenio Tombolini group, a clothing company located in the local system of Urbisaglia.

---

[18] Family groups are probably much more widespread than they appear to be from the data examined. This is the result of the difficulty in gathering information on the holdings at a personal level.

[19] Given these two features it is questionable whether these local systems can be considered as industrial districts. This point is discussed by Balloni and Iacobucci in this volume.

Table 12. Companies belonging to the groups by type and location

District	Type of company	Location				Total
		Same district	Same region	Other regions	Abroad	
Textiles and clothing (Urbania)	Distribution				1	1
	Service					
	Manufacturing	4			1	5
	Total	4			2	6
Furniture (Treia)	Distribution					
	Service					
	Manufacturing	4			2	6
	Total	4			2	6
Textiles and clothing (Urbisaglia)	Distribution	5		1	5	11
	Service	1				1
	Manufacturing	6	1	2		9
	Total	12	1	3	5	21
Leather (Tolentino)	Distribution	2				2
	Service	1				1
	Manufacturing	5			1	6
	Total	8			1	9

The group is made up of 19 companies, 8 manufacturing and 11 distribution companies. Out of the 11 distribution companies 6 are retail outlets in Italy while the others are commercial subsidiaries located in the main export markets. Most of the manufacturing companies are located within the district and perform complementary activities for the original company (Figure 3).

The overall employees in the manufacturing companies located in the district are more than half the overall employees of the district[20]. The setting up of complementary companies is justified by the need to control the production chain in order to guarantee the quality of the final products.

---

[20] In spite of this, the local system where the Tombolini group is located is classified as a system of small and medium sized firms because the companies of the group are considered as independent units.

Figure 3. The structure of the Tombolini group

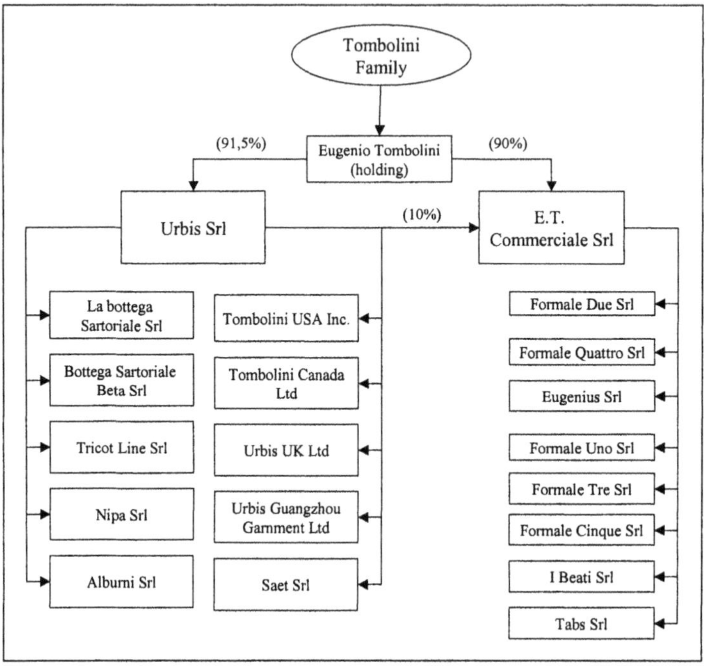

6 Conclusions

This chapter has discussed three main questions regarding groups of SMEs: whether the group or its member companies should be considered as the appropriate unit of analysis; the reasons why entrepreneurs develop their activities by setting up or acquiring new companies rather than by expanding the existing ones; the characteristics of the groups operating within industrial districts. The empirical analysis has taken into consideration mainly the groups operating in the Marche region.

We have reached the conclusion that the first question cannot be answered at a general level and that some internal features of the group should be considered in order to isolate the appropriate economic unit. For this reason a typology of groups has been proposed based on the existence of economic synergy between the companies and the degree of association between ownership and control. It has been suggested that the *industrial group* and the *entrepreneurial group* should be considered as a specific form of firm organisation and, as such, the appropriate unit in most of the industrial analyses. On the contrary, the *financial group* and the *family group* are of importance when the focus of the analysis is on the capital accumulation process of the family or on the functioning of capital markets.

Regarding the presence of groups among manufacturing firms, up-to-date evidence concerning Italy shows that the group form is not peculiar to large firms and is largely present even among medium sized firms. Indeed, it can be considered the prevailing organisational form adopted by firms in their growth process. For this reason the choice of the appropriate unit of analysis is of critical importance in any study regarding the performance and growth of firms.

Both economic and entrepreneurial literature associate the set-up (or the acquisition) of new companies with a growth process based on the diversification of activities. This seems to contrast with the fact that the degree of diversification of activities in SMEs is generally low. Based on a data-set regarding the main groups located in the industrial districts of the Marche region, it has been demonstrated that the manufacturing companies of the groups are principally engaged in similar activities: i.e. they serve different segments within the same market (horizontal or vertical differentiation). Despite the many similarities between these activities and the original one, entrepreneurs prefer to manage them through independent companies. The autonomy of the companies offers both internal and external advantages: in the first case it helps to enhance the incentives for the managers running a specific unit (even by giving them a quota in the stock of the company); in the second case it allows for a more effective adaptation of marketing policy and of contractual relationships with suppliers to the specific needs of every line of business.

In the case of small groups the contractual autonomy of the companies seems to be the most important advantage of the group form. Each company is able to adapt its marketing policy and its relationships with the suppliers to the specific needs of its line of business[21]. This aspect is particularly important for groups that rely on a network of external suppliers, as is the case of groups located in industrial districts.

The second main reason for the formation of groups is the foreign expansion of distribution and production processes. In this case the setting up of new companies is made necessary in order to adhere to the legal framework of the foreign countries.

The study of business groups is made difficult by the absence of large data sets regarding the phenomenon. This is especially true in the case of groups of small and medium sized firms. Small groups are mostly "informal" groups: i.e. groups formed by companies owned by the same persons but where there is no holding company at the head of the group; this makes it difficult to trace a complete picture of the companies belonging to the group and of the relationships among them (Brioschi *et al.* in this volume).

The relationship between diversification and business group formation depends not only on the size of the group but also on the unit of analysis. In the case of

---

[21] There are frequent cases in which several companies are set up or acquired in order to manage different brands within the same product category.

industrial or entrepreneurial groups the activities controlled by the companies normally show a high degree of coherence and, conversely, a low degree of diversification. On the contrary when the analysis is extended to include all the companies controlled by a holding company or a family the degree of diversification rises substantially.

Overall the results of this study confirm that the group of companies is the preferred organisational form not only in the case of large groups, where the diversification is of a conglomerate nature, but also in the case of SMEs whose activities are normally concentrated within the same sector. In the latter case the need for organisational differentiation is explained by the fact that firms operate in a market context characterised by a high degree of demand variety (horizontal differentiation) and variability (Best, 1990; Audretsch 1995). In these circumstances growth by vertical integration would reduce the flexibility of the firm (Carlsson, 1989) while growth by horizontal integration is possible only through product differentiation.

Three main reasons can be identified to explain the growing presence of groups within the industrial districts.

The first is a historical one: groups are the result of the growth process achieved by a limited number of firms within industrial districts. The internal learning processes and the changes in external conditions stimulate the growth of some firms that lead to the formation of groups.

The second reason is linked to the characteristics of industrial district specialisation. Firms in industrial districts operate in highly segmented markets that require a high degree of organisational differentiation even when the diversification of activities (traditionally measured) is rather low. Firms prefer to set up or acquire new companies even when they operate in different segments of the same market (men's and women's footwear; high quality and medium quality furniture; etc.). This is because these segments often require a completely different approach to marketing and production activities. Firms also prefer to maintain the legal autonomy of the companies when they perform complementary activities within the same production chain (e.g. sole production in the case of footwear). Legal autonomy makes it easier for the complementary company to operate on the market and not just for the group.

The third reason explaining the formation of groups within industrial districts is the process of internationalisation carried out by the largest firms. This process seems to follow a clear pattern that starts with the setting up of distribution companies in the export markets and develops through the establishment of manufacturing units abroad. Although the companies located abroad have no impact on the concentration of district activities, the presence of groups operating at an international level increases the disparity between large and small firms within the district. The former concentrate their investment in activities that are increasingly important for the competitiveness of their products: R&D, marketing, internationalisation, etc. The latter continue to concentrate on manufacturing activities and operate mainly as suppliers for the former.

Whether and to what extent this will undermine the organisation of industrial districts is still a question of debate, and cannot be answered on a general level, without taking into account the specific features of the districts being analysed.

References

AOKI M. (1994), "The Japanese Firm as a System of Attributes: A Survey and Research Agenda", in: AOKI M., DORE R. (Eds.), *The Japanese Firm*, Oxford University Press, Oxford.

AUDRETSCH D.B. (1995), "Innovation, Growth and Survival", *International Journal of Industrial Organization*, vol. 13, n. 4, pp. 441-457.

BALLONI V., IACOBUCCI D. (1997), "Cambiamenti in atto nell'organizzazione dell'industria marchigiana", *Economia Marche*, n. 1, pp. 29-66.

BALLONI V., IACOBUCCI D. (2001), "I "Nuovi Protagonisti" dell'Industria Italiana", *L'Industria*, vol. 22, n. 4, pp. 633-675.

BARBETTA G.P., PIGA C., VIVARELLI M. (1995), "Il fenomeno dei gruppi di imprese in Italia, *Osservatorio sulle piccole e medie imprese. Quaderni di politica industriale, Mediocredito Centrale*, Roma.

BARCA F. BIANCO M., CANNARI L., CESARI R., GOLA C., MANITTA G., SALVO G., SIGNORINI L.F. (1994), *Assetti proprietari e mercato delle imprese. Vol. I. Proprietà, modelli di controllo e riallocazione delle imprese industriali italiane*, Il Mulino, Bologna.

BARCA F. (1996), "On Corporate Governance in Italy: Issue, Facts and Agenda", *Fondazione ENI Enrico Mattei Working Paper Series* (Nota di lavoro 10.96).

BECATTINI G. (1991), "The Industrial District as a Creative Milieu", in: BENKO G., DUNFORD M. (Eds.), *Industrial Change and Regional Development: The Trasformation of new Industrial Spaces*, Belhaven Press/Pinter, London.

BECATTINI G. (2000), *Il distretto industriale*, Rosenberg & Sellier, Torino.

BEST M. (1990), *The New Competition: Institution of Industrial Restructuring*, Polity Press, Cambridge.

BIANCO M., CASAVOLA P. (1999), "Italian Corporate Governance: Effects on Financial Structure and Firm Performance", *European Economic Review*, vol. 43, pp. 1057-1069.

BIRLEY S., WESTHEAD P. (1994), "A Comparison of New Businesses Established by 'Novice' and 'Habitual' Founders in Great Britain", *International Small Business Journal*, vol. 12, n. 1, pp. 38-60.

BRIOSCHI F., BUZZACCHI L., COLOMBO M.G. (1990), *Gruppi di imprese e mercato finanziario. La struttura di potere nell'industria italiana*, La Nuova Italia Scientifica, Roma.

BRIOSCHI F., CAINELLI G. (2001), *Diffusione e caratteristiche dei gruppi di piccole e medie imprese nelle aree distrettuali*, Giuffrè Editore, Milano.

BRUSCO S. (1982), "The Emilian Model: Productive Decentralization and Social Integration", *Cambridge Journal of Economics*, vol. 6 , pp. 167-184.

CARLSSON B. (1989), "Flexibility and the Theory of the Firm", *International Journal of Industrial Economics*, vol. 7, pp. 179-203.

CHANDLER A.D. (1982), "M-Form: Industrial Groups, American Style", *European Economic Review*, vol. 19, pp. 3-23.

CHANG S.J., CHOI U. (1988), "Strategy, Structure and Performance of Korean Business Groups: A Transactions Cost Approach", *Journal of Industrial Economics*, vol. 37, n. 2, pp. 141-158.

COLLIN S. (1998), "Why Are These Islands of Conscious Power Found in the Ocean of Ownership? Institutional and Governance Hypotheses Explaining the Existence of Business Groups in Sweden", *Journal of Management Studies*, vol. 35 , pp. 719-746.

DENIS D.J., DENIS D.K., SARIN A. (1997), "Agency Problems, Equity Ownership, and Corporate Diversification", *Journal of Finance*, vol. 52, n. 1, pp. 135-160.

DONCKELS R., DUPON B., MICHEL P. (1987), "Multiple Business Starters. Who? Why? What?", *Journal of Small Business and Entrepreneurship*, vol. 5 , pp. 48-63.

ENCAOUA D., JACQUEMIN A. (1982), "Organisational Efficiency and Monopoly Power: The Case of French Industrial Groups", *European Economic Review*, vol. 19 , pp. 25-51.

FEENSTRA R.C., YANG M., HAMILTON G.C. (1997), "Business Groups and Trade in East Asia: Part 2, Product Variety, *NBER Working Paper* n. 5887.

GERLACH M.L. (1997), "The Organisational Logic of Business Groups: Evidence from the Zaibatsu", in: SHIBA T., SHIMOTANI M. (Eds.), *Beyond the Firm. Business groups in international and historical perspective*, Oxford University Press, Oxford.

GHEMAWAT P., KHANNA T. (1998), "The Nature of Diversified Business Groups: A Research Design and Two Case Studies", *The Journal of Industrial Economics*, vol. XLVI, pp. 35-61.

GOTO A. (1982), "Business Groups in a Market Economy", *European Economic Review*, vol. 19, pp. 53-70.

GULATI R., SINGH H. (1998), "The Architecture of Cooperation: Managing Coordination Costs and Appropriation Concerns in Strategic Alliances", *Administrative Science Quarterly*, vol. 43, pp. 781-814.

GRANOVETTER M. (1994), "Business Groups", in: SMELSER N.J., SWEDBERG R. (Eds.), *The Handbook of Economic Sociology*, Princeton University Press, Princeton.

GROSSMAN S., HART O. (1986), "The Costs and Benefits of Ownership: A Theory of Vertical and Lateral Integration", *Journal of Political Economy*, vol. 94, n. 4, pp. 691-719.

HALL E.H., ST. JOHN C.H. (1994), "A Methodological Note on Diversity Measurement", *Strategic Management Journal*, vol. 15, pp. 153-168.

HANSMANN H. (1996), *The Ownership of Enterprise*, Harvard University Press, Cambridge (Mass.).

HARRISON B. (1992), "Industrial Districts: Old Wine in New Bottles?", *Regional Studies*, vol. 26, n. 5, pp. 469-483.

IACOBUCCI D. (2002), "Explaining Business Groups Started by Habitual Entrepreneurs in the Italian Manufacturing Sector", *Entrepreneurship and Regional Development*, vol. 14 , pp. 31-47.

JACQUEMIN A.P., BERRY C.H. (1979), "Entropy Measure of Diversification and Corporate Growth", *Journal of Industrial Economics*, vol. 27, n. 4.

KEISTER L.A. (1998), "Engineering Growth: Business Group Structure and Firm Performance in China's Transition Economy", *American Journal of Sociology*, vol. 104, n. 2, pp. 404-440.

KESTER W.C. (1993), "Industrial Groups as Systems of Contractual Governance", *Oxford Review of Economic Policy*, vol. 8, n. 3, pp. 24-44.

KHANNA T. (2000), "Business Groups and Social Welfare in Emerging Markets: Existing Evidence and Unanswered Questions", *European Economic Review*, vol. 44, pp. 748-761.

KHANNA T., PALEPU K. (1999), "Policy Shocks, Market Intermediaries, and Corporate Strategy: The Evolution of Business Groups in Chile and India", *Journal of Economics and Management Strategy*, vol. 8, n. 2, pp. 271-310.

KHANNA T., PALEPU K. (2000), "Is Group Affiliation Profitable in Emerging Markets? An Analysis of Diversified Indian Business Groups", *Journal of Finance*, vol. 55, n. 2, pp. 867-891.

KHANNA T., RIVKIN J.W. (2001), "Estimating the Performance Effects of Business Groups in Emerging Markets", *Strategic Management Journal*, vol. 22, pp. 45-74.

LANCASTER K. (1979), *Variety, Equity and Efficiency. Product Variety in an Industrial Society*, Blackwell, Oxford.

LEFF N.H. (1978), "Industrial Organisation and Entrepreneurship in the Developing Countries: The Economic Groups", *Economic Development and Culturale Change*, vol. 27.

LORENZONI G. (1990), *L'architettura di sviluppo delle imprese minori. Costellazioni e piccoli gruppi*, Il Mulino, Bologna.

MARIOTTI S., MUTINELLI M. (2001), "La formazione dei gruppi multinazionali nei distretti: interpretazioni generali ed evidenze empiriche per il caso della meccanica italiana", in: BRIOSCHI F., CAINELLI G. (a cura di), *Diffusione e caratteristiche dei gruppi di piccole e medie imprese nelle aree distrettuali*, Giuffrè Editore, Milano.

MASTEN S.E. (1991), "A Legal Basis for the Firm", in: WILLIAMSON O.E., WINTER S. (Eds.), *The Nature of the Firm. Origins, Evolution, and Development*, Oxford University Press, Oxford.

MONTGOMERY C.A. (1994), "Corporate Diversification", *Journal of Economic Perspectives*, vol. 8, n. 3, pp. 163-178.

PENROSE E. (1959), *The Theory of the Growth of the Firm*, Basil Blackwell, Oxford

PYKE F.,SENGENBERGER W. (1992), *Industrial Districts and Local Economic Regeneration*, International Labour Office, Geneva.

ROBSON G., GALLAGHER C. DALY M. (1993), "Diversification Strategy and Practice in Small firms, *International Small Business Journal*, vol. 11, pp. 37-52.

ROSA P. (1998), "Entrepreneurial Processes of Business Cluster Formation and Growth by 'Habitual' Entrepreneurs", *Entrepreneurship Theory and Practice*, 22 (4, Summer), pp. 43-61.

ROSA P., SCOTT M.G. (1999), "The Prevalence of Multiple Owners and Directors in the SME Sector: Implication for Our Understanding of Start-Up and Growth", *Entrepreneurship and Regional Development*, vol. 11, pp. 21-37.

RUMELT R.P. (1974), *Strategy, Structure and Economic Performance*, Harvard Business School Press, Boston.

SCOTT M., ROSA P. (1996), "Has Firm Level Analysis Reached its Limits? Time for a Rethink", *International Small Business Journal*, vol. 14, n. 4, pp 81-89.

SHIBA T, SHIMOTANI M. (1997), *Beyond the Firm. Business Groups in International and Historical Perspective*, Oxford University Press, Oxford.

STABER U. (2001) "The Structure of Networks in Industrial Districts", *International Journal of Urban and Regional Research*, vol. 25, n. 3, pp. 537-552.

WHITFORD, J. (2001), "The Decline of a Model? Challenge and Response in the Italian Industrial Districts", *Economy and Society*, vol. 30, n. 1, pp. 38-65.

WILLIAMSON O.E. (1975), *Markets and Hierarchies: Analysis and Antitrust Implications*, The Free Press, New York.

WRIGHT M., WESTHEAD P., SOHL J. (1998), "Habitual Entrepreneurs and Angel Investors", *Entrepreneurship Theory and Practice*, vol. 22 n. 4.

Ownership linkages and business groups in industrial districts. The case of Emilia Romagna

Francesco Brioschi[1], Maria Sole Brioschi[2] and Giulio Cainelli[3]

[1]Polytechnic of Milan, Via Giuseppe Colombo 40, 20133, Milan, Italy
[2]University of Bergamo, Viale Marconi 5, 24044, Dalmine (BG), Italy
[3]University of Bari and CERIS-DSE, National Research Council, Via Bassini 15, 20133, Milan, Italy

1 Introduction

Italy's model of industrial development has always been viewed as anomalous in the panorama of the industrialised economies. Satisfactory long-run performance, on the whole, despite a number of structural weaknesses, has been achieved by an industrial system consisting largely of small and medium size enterprises not generally operating on a large enough scale to withstand international competition.

This essay examines two of the most convincing among the many explanations suggested for this anomaly, in their relation to the phenomenon of industrial districts. The first explanation lies on the presence of long-run informal linkages among firms, which tend to generate agglomeration economies and, more generally, Marshallian externalities that take the place of economies of scale internal to the individual firm. This explanation (which has a general validity) has been proffered to account for the success of Italy's industrial districts, constituted by territorial agglomerations of small and medium size businesses interacting in a single production sector (Brusco, 1982; Becattini, 1979, 1989, 1990). The second explanation essentially disputes the significance of the statistics on firm size in the Italian manufacturing industry. If instead of the single firm one considers the business group – if, that is, one discards the legal definition in favour of the economic substance of the concept of enterprise – then firm size rises substantially (Barca *et al.*, 1994). This is because, as it is well known in the business groups literature, a set of firms under the control of a single owner (or a single coalition of shareholders) may in many respects (central handling of various business functions; extended power towards the banking system, suppliers and customers; exploitation of other synergies) be considered as a larger-size single enterprise. It is perfectly plausible that such a phenomenon be found also within industrial districts with the gradual replacement of informal relations between firms by mutual ownership linkages.

Despite the potential importance of inter-locking shareholdings among firms within Italy's industrial districts, with the significant exception of some recent empirical research (reviewed in section 2) the vast literature on industrial districts has shown surprisingly little interest in this issue. One possible reason is that "the precise prevalence of groups in industrial districts is extremely hard to identify with available statistical data" (Whitford, 2001, p. 48). In this chapter we show that: (i) the phenomenon of business groups is extremely widespread in the industrial districts of Emilia Romagna; (ii) if groups instead of single firms are taken into account the degree of industrial concentration turns out to be extremely high. It has to be noted, however, that also neglecting groups the degree of industrial concentration of the districts of Emilia Romagna is still high, a fact which is clearly in contrast with the traditional view of the district literature; (iii) a large number of groups detected in the analysis are 'district groups' namely are composed by firms belonging to the district. A significant trend towards internationalisation has also been detected.

As already said this paper is empirical; the conclusions derive from an interpretation of observed facts. The empirical analysis is based on a survey that is carried out on thirteen historical industrial districts of Emilia Romagna with the aim to bring to light the features of the business groups and the reasons for their formation and development (section 3). Finally, a number of theoretical and policy implications are offered with regards to the effects of industrial concentration and corporate grouping on the district organisation (section 4).

The choice of Emilia Romagna as our field of study is motivated by the fact that in a number of ways the region's industrial system represents a paradigmatic model of local capitalism, combining the large-scale presence of industrial districts with a marked entrepreneurial spirit, strong social cohesion, and an exceptionally efficient system of local institutions and intermediate organisations (see, among others, Brusco, 1982; Bellini, 1990; Bianchi and Gualtieri, 1990). Moreover, even though we start from the empirical analysis of one case (albeit a significant one like Emilia Romagna), our contribution also has the more general aim of analysing the ways and forms in which a particular form of local capitalism marked by industrial districts has evolved and been transformed over the years, while at the same time signalling the need to revise the theoretical concepts and methods of empirical inquiry used to analyse and interpret the new forms of Italian local capitalism.

To conclude, it is worth mentioning that our study represents the most comprehensive empirical research ever carried out in Italy on corporate grouping. On the other hand, while recognising the relevance of our results, we are fully aware that solely considering the firms with disclosed ownership structure allows us to 'capture' the behaviour of only a portion of the firms of a district, typically the largest size ones.

2 Industrial districts, ownership linkages and business group structures

The debate over the structural, organisational and developmental specificities taken on by the Italian industrial system in the course of the 1990s has grown and been articulated into at least two different directions of inquiry. One has led to the recognition of the pervasiveness of the business group form not only among large firms (listed and unlisted alike) but also among small and medium size enterprises. The work in this strand – above all the studies of the Bank of Italy (Barca *et al.*, 1994) and of Mediocredito Centrale (Barbetta *et al.*, 1996) – has emphasised that in many cases the business group[1] is small Italian enterprises' preferred way to cope with changes in market conditions and rules of competition and to respond to the need to expand in size. The second strand of research focuses on the evolution of the main industrial districts (Pyke *et al.*, 1990; Cossentino *et al.*, 1996). This line of research has adopted the theoretical paradigm and the analytical approaches first developed, in the Italian literature on industrial districts, by Brusco (1982) and Becattini (1979, 1989, 1990). As it is well known, these two scholars returned to the Marshallian concept of industrial district[2] as a unit of analysis offering a most useful conceptual framework in the interpretation of the structure and industrial organisation of a large number of Italian local systems of small and very small businesses.

Despite the importance of business groups and industrial districts in the interpretation of the specificities of the Italian economy, only recently have the two strands of inquiry been integrated (Bianchi and Gualtieri, 1990; Viesti, 1992; Dei Ottati, 1996; Brusco *et al.*, 1996; Cainelli and Nuti,1996; Balloni and Iacobucci, 1997; Becattini, 1998; Bianchi *et al.*, 1999; Brioschi and Cainelli, 2001). Bianchi and Gualtieri (1990) clear-sightedly understand the on-going evolution of industrial districts already a decade ago, spotlighting the process of concentration via the formation of business groups. Viesti (1992) shows how the formation of groups within industrial districts identifies for these local production systems a new evolutionary model. In her analysis of the main Tuscan industrial districts, Dei Ottati (1996) shows that district firms tend to organise themselves in groups. This tendency apparently dates back to the early 1970s, though it is only in the subsequent decade that she finds that district firms essentially grow "by the creation of new units [...] and by the acquisition of new companies which were already operational but possibly found themselves in financial difficulties, owing to a lack of orders because of changes in demand" (Dei Ottati, 1996, p. 52). Analysing the same industrial districts under investigation in this chapter, Brusco *et al.* (1996) suggest that in Emilia Romagna there is a strong tendency towards an increase in industrial concentration. The authors show, in particular, that this process has been primarily carried out through the creation of business group

---

[1] A business group can be defined as a set of legally independent companies connected by reciprocal shareholding linkages which, taken as a whole, permit unitary control of all their activities (Brioschi et *al.*, 1989, 1991).

[2] An industrial district can be defined as "a socio-territorial entity which is characterised by the active presence of both a community of people and a population of firms in one naturally and historically bounded area" (Becattini, 1990, p. 38). It follows from this definition that industrial districts have a two-fold scope, one referring to the social environment and one related to the economic environment (Dei Ottati, 1994).

linkages rather than through mergers and acquisitions resulting in single juridical firms. Cainelli and Nuti (1996), in a paper published within a three article symposium in the *Journal of Industry Studies* on the development of Italian industrial districts in the late 1980s and early 1990s, suggest that these production systems reorganise themselves in hierarchies because predictable relationships among partners can be an effective weapon against unpredictable market relationships, "privileging voice over exit" (Whitford, 2001, p.48).

A common feature of these papers is that they do not try to measure the extent of corporate grouping within the industrial districts under investigation. An attempt in this direction has been provided, for the first time, by Balloni and Iacobucci (1997), who present the findings of a survey on groups of small firms in the leading industrial districts in the Marche region. This study brings out several interesting features of the group form within districts: (i) close relationship between the activities of the group firms and those of the 'originator' firm; (ii) growth of groups through formation of new companies engaged in activities similar (especially in the 1970s) or complementary (especially in the 1980s) to those of the originator firm; (iii) low degree of separation between ownership and control: the equity held by shareholders outside the controlling core (typically, the family of the founder of the originator firm) is unlikely to reach 20 percent. Becattini (1998) documents the 'district groupification' with reference to a number of Tuscany's industrial districts. Bianchi *et al.* (1999) and Brioschi and Cainelli (2001) are the first contributions to the research presented in this chapter. Their main achievement is to show that "at least for some industries, the phenomenon of industrial districts can be traced back not only to the sort of informal links among firms largely highlighted in the literature but also to the presence of formal, equity arrangements" (Bianchi *et al.*, 1999, p. 281, our translation).

Finally, it has to be noted that the present contribution draws heavily from another paper of the same authors (Brioschi *et al.*, 2002) which presents the same methodology and reaches the same conclusions but, however, reports the empirical evidence only for a smaller set of industrial districts (4 instead of 13).

3 The corporate grouping within the industrial districts of Emilia Romagna: the empirical evidence

In this section we present the results of an extensive survey undertaken during the period 2000-2001 in thirteen industrial districts of Emilia Romagna with the aim of identifying the main features of business groups and understanding the reasons for their formation and development. The industrial districts examined are those displayed and listed in Figure 1. The research project has been carried out in two different stages. The first stage – of statistical nature – has concerned, by means of the Impero database, all the Emilia Romagna firms. The second one has been carried out through a high number of interwiews directed to the firms of the industrial districts under investigation. The main features of both stages of the research project are described in Brioschi *et al.*, 2002.

Before showing the main results of our empirical survey it is useful to present, by using two different statistical sources (ISTAT and Impero), some characteristics of these districts in terms of size and industrial concentration. These findings are shown in Table 1 and 2. In particular, Table 1 suggests, first of all, that the districts taken into account in our analysis are characterised by a strong size eterogeneity. In fact it is possiblc to find large district in terms of employees such as those of Sassuolo and Parma, and smaller districts such as, for example, the wood processing machinery district of Carpi. From this Table turns also out that Impero – the database used for stratifying the sample of firms utilised in the empirical survey – shows a quite good degree of fitting with respect to the ISTAT data which comes from the intermediate industrial census of 1996.

The Table 2 shows another aspect characterising the thirteen industrial districts examined: i.e., the high industrial concentration.

Figure 1. The thirteen industrial districts of Emilia Romagna

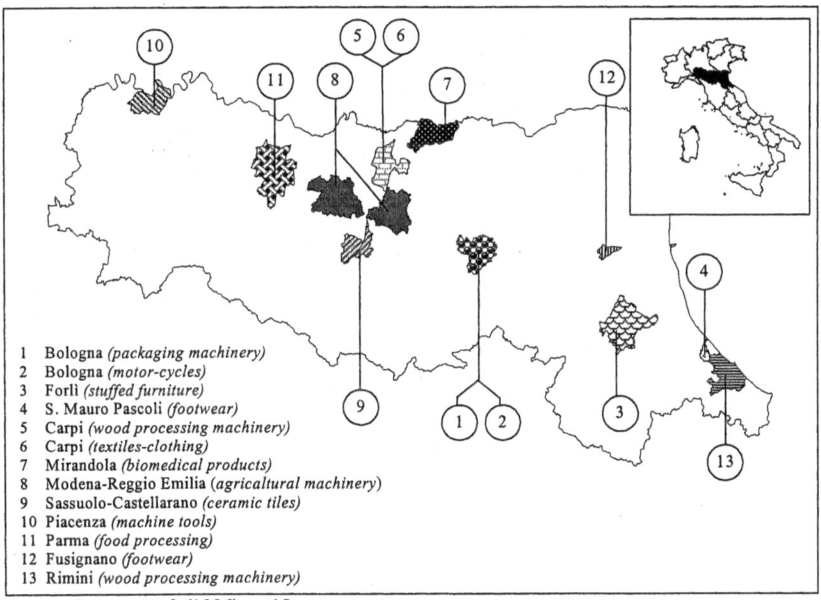

Source: Brusco *et al.* (1996), p. 18

Table 1. The thirteen industrial districts of Emilia Romagna: firms and employees

Industrial districts	*ISTAT 1996*		*Impero 1997*	
	Firms No.	Empl. No.	Firms No.	Empl. No.
Motor-cycles (Bologna)	50	3,567	29	2,093
Wood processing machinery (Carpi)	31	962	34	945
Stuffed furniture (Forlì)	544	3,638	380	2,671
Biomedical products (Mirandola)	296	3,490	296	3,017
Ceramic tiles (Sassuolo-Castellarano)	341	23,200	347	19,459
Machine tools (Piacenza)	46	1,797	64	1,238
Food processing (Parma)	1,313	17,802	1,271	11,733
Footwear (Fusignano)	160	1,794	158	1,657
Wood processing machinery (Rimini)	30	1,527	29	1,440
Packaging machinery (Bologna)	177	7,006	249	7,251
Footwear (San Mauro Pascoli)	127	2,233	137	2,520
Textiles-clothing (Carpi)	2,325	13,907	2,343	11,027
Agricaltural machinery (Modena-Reggio Emilia)	363	8,947	275	9,951
Total	5,803	89,870	5,612	75,002

Table 2. Concentration in the thirteen industrial districts of Emilia Romagna

Industrial districts	*ISTAT 1996*		*Impero 1997*	
	C4	C8	C4	C8
Motor-cycles (Bologna)	69.13	81.65	86.00	92.26
Wood processing machinery (Carpi)	52.03	77.11	49.31	76.93
Stuffed furniture (Forlì)	9.92	13.18	14.34	20.85
Biomedical products (Mirandola)	48.90	60.48	46.70	55.45
Ceramic tiles (Sassuolo-Castellarano)	15.97	29.30	19.00	31.36
Machine tools (Piacenza)	46.91	69.45	57.51	65.58
Food processing (Parma)	47.60	51.47	31.35	38.35
Footwear (Fusignano)	13.36	18.78	16.17	25.83
Wood processing machinery (Rimini)	83.05	90.70	85.49	92.01
Packaging machinery (Bologna)	50.60	60.46	48.08	61.12
Footwear (San Mauro Pascoli)	33.18	53.10	31.55	45.32
Textiles-clothing (Carpi)	6.19	8.84	4.81	7.56
Agricaltural machinery (Modena-Reggio Emilia)	61.60	68.80	64.39	71.76

C4 and C8 are respectively the share of the employees of the first four firms and of the first eight firms over total districts employees

In fact, by adopting two different measures of industrial concentration (C4 and C8) and two different statistical sources, from Table 2 turns out that the industrial districts under investigation are characterised, with the exception of Carpi and Forlì, by production structures highly concentrated. This finding contrasts the traditional view of the industrial district that generally assumes this industrial organisation as composed of a great number of small size firms. On the contrary our evidence suggests a significant presence within these production structures of big firms.

Let us now pass to the analysis of the diffusion of the group form within the industrial districts. The main quantitative results of the field survey are set out in Tables 3 to 7. A total of 211 groups were identified in the districts under inquiry, with an average of 6.3 firms per group. This overall average conceals a substantial difference between the 'traditional' districts (for example, textiles and footwear) and the 'higher-tech' districts (for example agricultural and packaging machinery) in that the former tend to exhibit corporate groups composed of a significantly lower number of firms.

For a closer reading of this evidence, we developed a taxonomy of business groups, according to which the group structures found within the industrial districts under investigation could be classified in four types: (*i*) 'pseudo-groups', (*ii*) 'conglomerate groups', (*iii*) 'district groups', (*iv*) 'international groups'. The term 'pseudo-group' refers to a structure consisting of just one production company plus one or more financial and/or real estate firms, the latter often employing very little staff or none at all. It is generally employed as a mean to separate the owners' corporate activities from their additional properties (real estate or other). 'Conglomerate groups' are arrangements of two or more firms operating in separate industries and generally include one or more financial/real estate companies. These structures highlight the entrepreneurs' business dynamism and ability to diversify across investment types. A 'district group' is a cluster of firms headquartered in the district that are engaged in the district reference production sector, whatever their compartment or level in the product chain. They may include one or more financial or real estate companies. The presence of this type of group significantly affects the district organisational structure since it enhances the district industrial concentration and/or degree of 'hierarchisation'. Finally, an 'international group' is a highly evolved and diversified structure with one or more companies within the district (generally, production companies) and one or more companies abroad (typically, marketing and financial arms).

Applying this classification to the 211 groups identified in the field survey, we obtained the results shown in Table 3. The most frequent type of business group detected in the industrial districts of Emilia Romagna is the 'district group', testifying the importance of this particular organisational form in local production systems such as those under investigation. This architecture is pre-eminent in the districts of Reggio Emilia, San Mauro Pascoli, Parma, Mirandola, and Carpi, whilst the packaging machinery district in Bologna, the ceramic tiles district of Sassuolo, and the biomedical products district of Mirandola seem to be distinguished by the presence of group structures with a pronounced international orientation.

Table 3. Types of business groups in the industrial districts

Industrial districts	*All groups*	*Firms per group*	*'Pseudogroup'*	*'Conglomerate'*	*'District'*	*'International'*
	No.	Average	No.	No.	No.	No.
Motor-cycles (Bologna)	12	4.0	1	0	9	2
Wood processing machinery (Carpi)	10	4.1	3	3	1	3
Stuffed furniture (Forlì)	11	2.9	2	2	7	0
Biomedical products (Mirandola)	25	3.1	4	1	13	7
Ceramic tiles (Sassuolo-Castellarano)	30	7.6	7	2	13	8
Machine tools (Piacenza)	12	2.8	3	0	7	2
Food processing (Parma)	21	15.5	1	1	14	5
Footwear (Fusignano)	7	3.1	1	0	6	0
Wood processing machinery (Rimini)	4	8.5	0	0	0	4
Packaging machinery (Bologna)	20	10.6	3	0	8	9
Footwear (San Mauro Pascoli)	9	3.3	0	1	6	2
Textiles-clothing (Carpi)	30	3.3	10	4	15	1
Farm machinery (Modena-Reggio Emilia)	20	7.3	2	5	12	1
Total	211	6.3	37	19	112	43

Table 4. Industrial diversification of business groups in the industrial districts

Industrial districts	Number of groups	Corresp. number of group firms	Industrial within district sector (*)	Com-mercial	Financial	Real estate	Industrial out of district sector
	No.	No.	No.	No.	No.	No.	No.
Motor-cycles (Bologna)	12	48	33	6	4	5	0
Wood processing machinery (Carpi)	10	41	14	10	1	8	8
Stuffed furniture (Forlì)	11	32	17	8	0	5	2
Biomedical products (Mirandola)	25	77	53	5	8	7	4
Ceramic tiles (Sassuolo-Castellarano)	30	228	98	43	28	17	42
Machine tools (Piacenza)	12	34	23	5	3	1	2
Food processing (Parma)	21	325	220	7	31	6	61
Footwear (Fusignano)	7	22	11	5	0	6	0
Wood processing machinery (Rimini)	4	34	9	10	10	0	5
Packaging machinery (Bologna)	20	212	98	60	15	2	37
Footwear (San Mauro Pascoli)	9	30	17	5	3	4	1
Textiles-clothing (Carpi)	30	100	57	10	13	13	7
Farm machinery (Modena-Reggio Emilia)	20	146	71	34	11	11	19
Total	211	1,329	721	208	127	85	188

(*) Industrial firms, operating in the district sector, are located in most cases, but not necessarily, within the district boundaries

Table 5. Geographical diversification of business groups in the industrial districts

Industrial districts	*Number of groups*	*Number of group firms*	*Within the district*	*In the rest of Italy*	*Abroad*
	No.	No.	No.	No.	No.
Motor-cycles (Bologna)	12	48	37	2	9
Wood processing machinery (Carpi)	10	41	29	3	9
Stuffed furniture (Forli)	11	32	32	0	0
Biomedical products (Mirandola)	25	77	58	12	7
Ceramic tiles (Sassuolo-Castellarano)	30	228	111	66	51
Machine tools (Piacenza)	12	34	21	2	11
Food processing (Parma)	21	325	77	27	221
Footwear (Fusignano)	7	22	21	1	0
Wood processing machinery (Rimini)	4	34	10	4	20
Packaging machinery (Bologna)	20	212	62	20	130
Footwear (San Mauro Pascoli)	9	30	24	2	4
Textiles-clothing (Carpi)	30	100	80	17	3
Farm machinery (Modena-Reggio Emilia)	20	146	101	13	32
TOTAL	211	1,329	663	169	497

Table 6. Business strategies of 'district groups' in the four industrial districts

Industrial districts	*Number of 'district groups'*	*Prevailing strategy*		*Main advantages of business grouping*		
		Horizontal diversification	Vertical integration	Economies of scale in distribution	Economies of scale in industrial planning	Quality/ Reliability of sub-contractors
Motor-cylces (Bologna)	9	6	3	3		3
Wood processing machinery (Carpi)	1	1	0			
Stuffed forniture (Forlì)	7	1	6	3		3
Biomedical products (Mirandola)	13	8	5		3	
Ceramic tiles (Sassuolo)	13	10	3	3	3	
Machine tools (Piacenza)	7	1	6	3	3	
Food processing (Parma)	14	2	12	3		3
Footwear (Fusignano)	7	2	5	3		3
Wood processing machinery (Rimini)	0	0	0			
Packaging machinery (Bologna)	8	7	1	3	3	
Footwear (San Mauro Pascoli)	6	2	4			3
Textile-clothing (Carpi)	15	5	11			3
Farm machinery (Reggio Emilia)	12	9	3	3		

The importance of the 'district group' strongly emerges also from Tables 4 and 5, respectively exhibiting the industrial and the geographical diversification of the 211 groups detected in our districts. These two tables confirm that a large number of the 1,329 firms belonging to the 211 groups under investigation are industrial firms operating in the district reference industry and located within the district boundaries.

Beyond recognising the numeric relevance of 'district groups', the field survey also allowed us to explore the business strategies of these local group structures. Such strategies, along with the attached main advantages of business grouping, are shown in Table 6. In the farm machinery district of Reggio Emilia (9 out of 12), in the ceramic tiles district of Sassuolo (10 out of 13) and in the biomedical products district of Sassuolo (8 out of 13), the majority of 'district groups' adopt a horizontal diversification strategy. 'District groups' entrepreneurs take advantage from the central handling of the commercial/distribution function, allowing them to supply several different customers with an articulated range of products often under a single brand. Similar results emerge from the analysis of the packaging machinery district of Bologna, where the prevailing strategy is again horizontal diversification for 7 'district groups' out of 8. In a district where the technological content of the goods produced is high, 'district groups' exploit economies of scale both in industrial planning and in the distribution phase, the latter often implying after-market presence abroad. A different story emerges in the other districts under investigation, where the prevailing strategy of the 'district groups' detected appears to be vertical integration. In particular, in the knitwear and garments district of Carpi the majority of 'district groups' is integrated upward, whereas in the footwear district of San Mauro Pascoli 'district groups' are also integrated downward. To explain why 'district groups' in Carpi resort to upward vertical integration, we shall recall that in the last ten years this district has reacted to the aggressive competition of the low labour cost countries by repositioning on a higher-quality segment of the market. To guarantee quality and delivery times in a sector where quality standards and speed of execution are crucial, many final firms acquired their sub-contractors. Also in San Mauro's footwear district the majority of 'district groups' is vertically integrated. However, as here 'district groups' are shoe factories of world-wide renown, local final firms have integrated both upward to better control for quality and downward to monitor distribution and manage sales points. The above analysis suggests that district corporate grouping alters the traditional organisation of the district: in districts mainly comprising groups with a vertical integration strategy (Carpi, Parma and San Mauro Pascoli), spontaneous co-operation along the production *filiere* is replaced by hierarchical relationships; by contrast, in districts where groups have a prevailing horizontal diversification strategy, we observe a growing concentration of the district output, while the degree of 'hierarchisation' along the production *filiere* remains unchanged.

In order to complete our analysis, we believe it is important to give a measure of the of the weight of business groups in our industrial districts. In particular, taking the view that the group is the relevant unit of analysis, we can assess the degree of industrial concentration generated by the district 'groupification' simply by relating the employees of the largest groups (in terms of employees) of each industrial district to the total number of district employees. Of course, in order to obtain a correct measure of industrial concentration of the groups considered only the firms operating in the district reference sector and located within the districts boundaries have to be taken into account.

Table 7. Concentration in the thirteen industrial districts of Emilia Romagna

Industrial districts	*Impero 1997 (Firms)*		*Impero 1997 (Business Groups)*	
	C4	C8	C4	C8
Motor-cycles (Bologna)	86.00	92.26	88.77	93.17
Wood processing machinery (Carpi)	49.31	76.93	49.52	77.35
Stuffed furniture (Forlì)	14.34	20.85	15.09	21.60
Biomedical products (Mirandola)	46.70	55.45	46.70	55.49
Ceramic tiles (Sassuolo-Castellarano)	19.00	31.36	28.16	42.45
Machine tools (Piacenza)	57.51	69.00	57.51	69.70
Food processing (Parma)	31.35	38.35	37.79	43.40
Footwear (Fusignano)	16.17	25.83	16.42	27.10
Wood processing machinery (Rimini)	85.49	92.01	85.70	92.22
Packaging machinery (Bologna)	48.08	61.12	56.25	67.10
Footwear (San Mauro Pascoli)	31.55	45.32	37.46	53.81
Textiles-clothing (Carpi)	4.81	7.56	5.37	8.16
Farm machinery (Modena-Reggio Emilia)	64.39	71.76	66.47	74.18

C4 and C8 are respectively the share of the employees of the first four firms and of the first eight firms over total districts employees

Table 7 shows the industrial concentration indexes built using the largest four groups in the industrial districts of interest. With the only exception of Carpi and partially of Forlì the data show that the districts are highly concentrated, spotlighting a significant departure from the original district nature. In the farm machinery district of Modena and Reggio Emilia, for instance, the employees of the first four groups account for about two thirds of the district employees, those of the first eight groups almost reach three quarters of total employees.

4 The evolution of the industrial district organisation

In the last few years district firms, just like all other enterprises, had to reorganise and grow in order to face the increasingly intensive transformation of rules and arrangements marking international competition. In the industrial districts taken into account in our study, this process of growth has led both to greater product differentiation – hence more horizontal integration – in order to broaden the product range and to an increased attention to quality, which has often meant vertical integration: upstream for better quality control on semi-finished products and components and downstream to ensure better control of sales channels[3]. In other words, in order to be competitive at international level district firms rearranged operations so as to operate on a larger scale and with stronger integration of production processes.

The interviews that we carried out during the field study highlighted the choice on the part of district businessmen to spur growth mainly via the business group device and revealed the motives behind this decision. In particular, the choice to expand through the group model has depended and depends both on general factors and on other factors linked with the specifities of the industrial district as such. The former include labour law (trade union representation rights are subject to a size threshold of 15 workers), tax considerations (such as the constitution of a holding company in a country offering tax breaks), the 'accommodation' of the need to bring relatives and employees into the ownership structure, and the opportunity to separate corporate activities from personal (often real estate) properties.

The field survey found other, more specific motivations as well. The enlargement towards the group form can take place either by the formation of new companies or by the acquisition of existing firms. Both avenues were found during the field study. In some cases, having decided to expand (generally in the same product line or in related products, either upstream or downstream), the firms of a district founded one or more new companies in the same district rather than simply expanding the size of the existing firms. The entrepreneurs interviewed explained that this option was taken in order to repeat a tested organisational model, i.e. that of the small enterprise. However, it was in the case of external expansion, i.e. expansion by acquisition (which we found to be the most common), that firms' membership in a district had the highest impact on their way of growth. Thanks to thorough familiarity with the other firms in the district (competitors, suppliers and customers), the district entrepreneurs could acquire businesses with sure knowledge of their characteristics, with no need for the intercession of an investment banker (not always reliable and always costly). The mutual familiarity of the firms within each district and the relationships of trust built up between businessmen – which the literature (correctly) considers to be the fundamental element in the very formation and development of the district (Lazerson and Lorenzoni, 1999) – turned out be decisive factors in many of the corporate finance operations undertaken by district firms. In most cases, the acquisition was not followed by the merger of the purchaser with the acquired company. In order to preserve a brand name or to leave the management of the acquired company intact, or in the desire (or necessity) to leave a stake in the company to the seller, the acquired company was left in existence – albeit with a totally or significantly renovated ownership – and became part of the group headed by the purchaser. The fact that external growth strategies were often carried out within the district led to the formation of 'district groups'.

---

[3] Similar patterns of evolution for district firms are observed by Innocenti (1994) and Paniccia (1998).

This empirical analysis has shown that the extent of corporate grouping within the industrial districts under investigation is very significant. The analysis has further ascertained that the nature of such groups is by no means homogeneous and that their formation can be ascribed to a number of different factors. In this respect, the taxonomy introduced in section 3 is helpful for understanding both the nature and the role of business groups in the evolution of the district organisation. The 'pseudo-group', which comprises one single industrial firm, is essentially employed to separate the owners' main economic activity from their additional wealth (real estate or other), showing in any case the financial sophistication achieved by district entrepreneurs. The 'conglomerate group' highlights the entrepreneurs' activism and ability to diversify investments, possibly enhancing their bargaining power within the district competitive process. From the point of view of the evolution of the organisational structure of the district, the most important type of group identified during the field study is the 'district group', composed of a plurality of firms located within the district and operating in the district reference sector. The field study has spotlighted its widespread presence, showing that the 'district group' can be considered as the main organisational form chosen by district firms in their process of growth. Finally, the 'international group' is a group with a strong international orientation. In many cases, it is simply a former 'district group' which has expanded abroad and often it can be regarded as a true small multinational enterprise. It is important to note that corporate grouping and, more specifically, district grouping is widespread not only in industrial districts characterised by the presence of large firms (Bologna, Modena-Reggio Emilia, San Mauro Pascoli) but also in the textile district of Carpi, still characterised by a low degree of industrial concentration.

While all group forms affect the district configuration, the presence of 'district groups' directly changes the internal organisation and thus the governance structure of the industrial district since it affects the degree of industrial concentration and the degree of 'hierarchisation' of the district. The prevalence of one of these two effects depends on the type of development strategy adopted by 'district groups', i.e. vertical integration or horizontal diversification. Our analysis has brought to light that within 'traditional' districts such as Carpi and San Mauro Pascoli 'groupification' produces an increase in the degree of 'hierarchisation' of the district via the substitution of the traditional co-operative relationships between sub-contractors and final firms – so well described in the literature on industrial districts – with ownership relationships which are by definition hierarchical. By contrast, in the industrial districts of Bologna and Modena-Reggio Emilia where groups mainly adopt a horizontal diversification strategy, the prevailing outcome of 'groupification' seems to be an increase in the level of industrial concentration.

The empirical identification of the 'district group' as a key form of business organisation in industrial districts raises several theoretical issues of considerable importance. First, it is worth noting that the concept of industrial district generally employed in the Italian literature is slightly different from that commonly attributed to the tradition drawing back to Marshall (1920). As a matter of fact, this term applied to the Third Italy describes a somewhat different type of organisational structure. In particular, this term indicates "a higher degree of co-operative co-ordination than would be present in a Marshallian industrial district. [...] Product differentiation rather than price is the dominant competitive characteristic. Tacit idiosyncratic knowledge is central to firm performance and competition is limited to certain spheres of activity in which firms might be expected to develop distinctive competencies" (Langlois and Robertson, 1995, p. 125). Moreover, the relations within the district are not only deemed to be informal but also based on equal rights, thus precluding the possibility of hierarchical relations. Co-ordination of the activities among different district firms is achieved through market transactions, but even so interaction mechanisms (including social interactions) based on trust, reputation and co-operation play a major role (Dei Ottati, 1994). In such a theoretical framework, the industrial district shapes up as a production arrangement characterised by a high degree of co-ordination and little (if any) ownership integration. The emergence of organisational forms based on ownership linkages, hence greater concentration of ownership and control, is altering these traditional arrangements, gradually replacing the old mechanisms of competition and co-operation with others based on more formal and stable relations.

Secondly, our results on the relevance of district 'groupification' allow us to take part in a theoretical discussion lively debated within the district literature: that on the prevalence of the systemic nature of the district or of the centrality of the single firm as unit of analysis. In this literature two different approaches can be identified: a system-centred view and a firm-centred view[4]. Behind the first approach, there is the idea that the district works as a complex evolutionary system, where the integration between firms is mainly achieved by means of a mix of automatic mechanisms based on both market competition between firms belonging to the same stage of the production process and co-operative behaviours between firms belonging to different stages of the *filiere*. In this perspective, mainly developed by Becattini and Brusco within the traditional theory of industrial district drawing back to the seminal contribution of Marshall, little room for the strategic behaviour of firms is left. Recently, Ferrucci and Varaldo (1993) have proposed an alternative approach to analyse the pattern of evolution of industrial districts. They suggest that the appropriate unit of analysis of a district is the district firm and that the district firm strategic behaviour is the key variable for the analysis of an industrial district and for the understanding of its pattern of evolution. Our point of view is close to Ferrucci and Varaldo's. We have emphasised the role of business strategies and hence of the strategic behaviour of 'district groups' in the process of evolution and transformation of an industrial district. This does not mean that the district as a system does not matter. On the contrary, the district constitutes a favourable environment stimulating the formation and development of 'district groups'. In this sense, the 'district group' represents both an element of rupture with respect to the traditional organisation

---

4 See Whitford (2001) for an extensive description of the two approaches.

of the district and a factor of continuity with the districts' traditional *modus operandi*.

Finally, it is worth analysing the policy implications of our investigation. It is useful to start from the actual experience of industrial policy in Emilia Romagna since the late 1980s. As we know, regional policy makers have generally designed industrial policy having the industrial district or the local system of small and medium size firms as the reference organisational model. Just consider the Ervet system centres, conceived from the outset as centres to support small businesses with the provision of real services (Bellini *et al.*, 1990; Mazzonis, 1996). Our findings show that the small juridical size of individual firms making up these local systems is only one of the variables to consider in designing the industrial policy programs. The high extent of corporate grouping (even detected among the smallest size firms) certainly suggests that firm ownership is concentrated in a lower number of controlling owners. One possible implication is that the fragmentation of organically unified enterprises into a large number of legally independent firms may result in a concentration of policy benefits on a lower number of actual beneficiaries. In addition, as juridical size is often an essential factor in determining who is eligible for a given benefit, the 'district group' form may permit circumvention of size ceilings. That is, corporate grouping could distort the allocation of public resources within a district to the disadvantage of small independent firms.

5 Conclusions

The industrial district is traditionally viewed both as one of the most interesting peculiarities of the Italian industrial system and one of its main strengths. Yet, in recent years faith in the role of industrial districts has wavered somewhat. While some scholars continue to 'believe in the district', many others wonder whether this form can cope with the challenges of globalisation, or whether it is doomed to decline. Our paper has revealed that the question is ill-put. District firms are still operating successfully in Italian and international markets. Nevertheless, over time industrial districts have been subject to a transformation process which radically altered their characteristics. Within industrial districts today we find the coexistence – albeit with relative weights that differ from case to case – of traditional arrangements with other, hierarchical mechanisms depending in many cases on the presence of business groups. During our field study we have identified 211 corporate groups, of which several are medium or large size. Significantly, many of these groups are the product of a process of growth involving district firms. That is, many of the districts' large enterprises (in the economic acceptation) started out twenty or more years ago as small firms. This result offers a new interpretation to the issue – typical of the literature on industrial districts (Tessieri, 2000) – of the interaction (fruitful or otherwise) between small district firms and large firms, the latter always being considered as 'external'. As we have seen, many of the large enterprises operating today within districts are actually the result of an on-going growth on the part of district firms themselves and should thus properly by considered 'internal'. In our view, the decade-old observation of Bianchi and Gualtieri (1990) on the Emilian model is perfectly suitable: that take-overs have been the growth device for local enterprises and that this has been "accompanied by the creation of group configurations that tend to remain unaltered in legal and operative standing, whilst still guaranteeing the control of functions considered to be strategic" (Bianchi and Gualtieri, 1990, p. 100).

Next to these findings, the paper has further shown that one particular form of business group specific to industrial districts – the 'district group' – is very common and plays a key role in the economy of district areas. From one point of view, their existence deeply alters the nature of the district, but from another standpoint they are a powerful element of continuity. On the 'change' side, the substantial presence of corporate grouping suggests the existence of a process of 'hierarchisation' of the economic relationships among district firms. On the 'continuity' side, the function of the industrial district, and of the forms of local capitalism shaped by industrial districts, stands confirmed and highlighted as the source of skills and abilities producing a competitive edge for district firms and prompting entrepreneurs to expand within their district.

References

BALLONI V., IACOBUCCI D. (1997), "Cambiamenti in atto nell'organizzazione dell'industria marchigiana", *Economia Marche*, vol. 16, n. 1, pp. 29-66.

BARBETTA G.P., PIGA C., VIVARELLI M. (1996), *Il fenomeno dei gruppi di imprese in Italia*, Quaderni di Politica Industriale, Mediocredito Centrale, Roma.

BARCA F., BIANCO M., CANNARI L., CESARI R., GOLA G., MANITTA G., SALVO G., SIGNORINI L.F. (1994), *Assetti proprietari e mercato delle imprese. Proprietà, modelli di controllo e riallocazione nelle imprese industriali italiane*, Vol. I, Il Mulino, Bologna.

BECATTINI G. (1979), "Dal 'settore' industriale al 'distretto' industriale: alcune considerazioni sull'unità di indagine dell'economia industriale", *Rivista di Economia e Politica Industriale*, vol. 5, n. 1.

BECATTINI G. (1989), "Sectors and/or Districts: Some Remark on the Conceptual Foundation of Industrial Economics", in: GOODMAN E., BAMFORD J. (Eds.), *Small Firms and Industrial Districts in Italy*, Routledge, London.

BECATTINI G. (1990), "The Marshallian Industrial District as a Socio-Economic Notion", in: PYKE F., BECATTINI G., SENGENBERGER W. (Eds.), *Industrial Districts and Inter-firm Co-operation in Italy*, International Labour Office, Geneve.

BECATTINI G. (1998), *Distretti industriali e Made in Italy. Le basi socioculturali del nostro sviluppo economico*, Bollati Boringhieri, Torino.

BELLINI N. (1990), "The Management of the Economy in Emilia-Romagna: the PCI and the Regional Experience", in: LEONARDI R., NANETTI R.Y. (Eds.), *The Regions and European Integration*, Pinter Publishers, London.

BELLINI N., GIORDANI M.G., PASQUINI F. (1990), "The Industrial Policy of Emilia Romagna: the Business Service Centers", in: LEONARDI R., NANETTI R.Y. (Eds.), *The Regions and European Integration*, Pinter Publishers, London.

BIANCHI P., GUALTIERI G. (1990), "Emilia-Romagna and Its Industrial Districts: the Evolution of a Model", in: LEONARDI R., NANETTI R.Y. (Eds.), *The Regions and European Integration*, Pinter Publishers, London.

BIANCHI R., BRIOSCHI M.S., CAINELLI G. (1999), "Struttura e localizzazione dei gruppi di imprese in Emilia Romagna", in: IDSE-CNR, *Trasformazioni strutturali e competitività dei sistemi locali di produzione, Rapporto IDSE-CNR sul cambiamento strutturale dell'economia italiana*, Franco Angeli, Milano.

BRIOSCHI F., BUZZACCHI, L,. COLOMBO M.G. (1989), "Risk Capital Financing and the Separation of Ownership and Control in Business Groups", *Journal of Banking and Finance*, n. 13, pp. 747-772.

BRIOSCHI F., BUZZACCHI, L., COLOMBO M.G. (1991), "More on Stock Market Value with Reciprocal Ownership", *Financial Analysts Journal*, May-June, pp. 76-78.

BRIOSCHI F., CAINELLI G. (2001), *Diffusione e caratteristiche dei gruppi di piccole e medie imprese nelle aree distrettuali. Dove va il distretto industriale?*, Giuffrè Editore, Milano.

BRIOSCHI F., BRIOSCHI M.S., CAINELLI G. (2002), "From the Industrial District to the District Group. An Insight Into the Evolution of Local Capitalism in Italy", *Regional Studies*, vol. 39, n. , pp. 1037-1052.

BRUSCO S. (1982), "The Emilian Model: Production Decentralization and Social Integration", *Cambridge Journal of Economics*, n. 2, pp. 167-184.

BRUSCO S., CAINELLI G., FORNI F., FRANCHI M., MALUSARDI A., RIGHETTI R. (1996), "The Evolution of Industrial Districts in Emilia-Romagna", in: COSSENTINO F., PYKE F., SENGENBERGER W. (Eds.), *Local and Regional Response to Global Pressure: the Case of Italy and Its Industrial Districts*, International Labour Office, Geneve.

BRUSCO S., PABA S. (1997), "Per una storia dei distretti industriali italiani dal secondo dopoguerra agli anni Novanta", in: BARCA F. (a cura di), *Storia del capitalismo italiano dal dopoguerra ad oggi*, Donzelli, Roma.

CAINELLI G., NUTI F. (1996), "Changing Directions in Italy's Manufacturing Industrial Districts: the Case of the Emilian Footwear Districts of Fusignano and San Mauro Pascoli", *Journal of Industry Studies*, vol. 3, n. 2, pp. 105-118.

COSSENTINO F., PYKE F. AND SENGENBERGER W. (Eds.) (1996), *Local and Regional Response to Global Pressure: the Case of Italy and Its Industrial Districts*, International Labour Office, Geneve.

DEI OTTATI G. (1994), "Trust, Interlinking Transactions and Credit in the Industrial District", *Cambridge Journal of Economics*, vol. 18, n. 6, pp. 529-546.

DEI OTTATI G. (1996), "The Remarkable Resilience of the Industrial Districts of Tuscany", in: COSSENTINO F., PYKE F., SENGENBERGER W. (Eds.), *Local and*

Regional Response to Global Pressure: the Case of Italy and Its Industrial Districts, International Labour Office, Geneve.

FERRUCCI L. AND VARALDO R. (1993), "La natura e la dinamica dell'impresa distrettuale", *Economia e Politica Industriale*, n. 80, pp. 73-97.

INNOCENTI A. (1994), "I distretti industriali: una categoria da ripensare. Una riflessione sugli studi di caso", in: BORTOLOTTI F. (a cura di), *Il mosaico e il progetto: lavoro, imprese, regolazione nei distretti industriali della Toscana*, Franco Angeli, Milano.

LANGLOIS R., ROBERTSON P.L. (1995), *Firms, Markets and Economic Change. A Dynamic Theory of Business Institutions*, Routledge, London.

LAZERSON M.H., LORENZONI G. (1999), "The Firms that Feed Industrial Districts: A Return to the Italian Source", *Industrial and Corporate Change*, vol. 8, n. 2, pp. 235-265.

LEONARDI R., NANETTI R.Y. (Eds.) (1990), *The Regions and European Integration*, Pinter Publishers, London.

MARSHALL A. (1920), *Principles of Economics*, eighth edition, Macmillan and Co, London.

MAZZONIS D. (1996), "The Changing Role of Ervet in Emilia Romagna", in: COSSENTINO F., PYKE F., SENGENBERGER W. (Eds.), *Local and Regional Response to Global Pressure: the Case of Italy and Its Industrial Districts*, International Labour Office, Geneve.

PANICCIA I. (1998), "One, a Hundred, Thousands of Industrial Districts: Organizational Variety in Local Networks of Small and Medium-sized Enterprises", *Organization Studies*, vol. 19, n. 4.

PYKE F., BECATTINI G., SENGENBERGER W. (Eds.) (1990), *Local and Regional Response to Global Pressure: the Case of Italy and Its Industrial Districts*, International Labour Office, Geneve.

SFORZI F. (1990), "The Quantitative Importance of Marshallian Industrial Districts in the Italian Economy", in: PYKE F., BECATTINI G., SENGENBERGER W. (Eds.), *Local and Regional Response to Global Pressure: the Case of Italy and Its Industrial Districts*, International Labour Office, Geneve.

TESSIERI N. (2000), "Multinazionali e distretti industriali in Italia", *Sviluppo Locale*, n. 13, pp. 71-99.

VIESTI G. (1992), "Crisi ed evoluzione dei distretti industriali", in: ONIDA F., VIESTI G., FALZONI A. (a cura di), *I distretti industriali: crisi o evoluzione*, EGEA, Milano.

WHITFORD J. (2001), "The Decline of a Model? Challenge and Response in the Italian Industrial Districts", *Economy and Society*, vol. 30, n. 1, pp. 38-65.

The role of medium-sized and large firms in the evolution of industrial districts. The case of Marche

Valeriano Balloni[1] and Donato Iacobucci[2]

[1]Faculty of Economics, University of Ancona, Piazzale Martelli 8, 60121, Ancona, Italy
[2]Computer Science Institute, University of Ancona, Via Brecce Bianche, 60131, Ancona, Italy

1 Introduction

The aim of this work is to examine the recent changes in the industrial districts of the Marche region, with specific reference to the role of medium sized and large firms.

The Marche region is an interesting case: among the so-called NEC (North Eastern and Central) regions of Italy (also known as the 'Third Italy') it is the one with the most relevant presence of industrial districts (Cannari and Signorini, 2000); although, up to now it has received less attention from scholars of industrial districts compared with other NEC regions (Tuscany, Emilia-Romagna, Veneto).

Recent empirical evidence on the evolution of Italian industrial districts has revealed that one of the main changes is the growing concentration of their output in the largest firms (Brusco *et al.*, 1996). To simplify the debate on this question, we can say that there are two prevailing views. The first holds that the Italian manufacturing districts are a specific way of organising some manufacturing activities – the so-called 'made in Italy' products (Fortis, 1996) – and that the evolution of such districts will not modify its most characteristic aspect; i.e. the pre-eminent role of small firms (Becattini, 1997 and 2000).

The second refers to the interpretation of the industrialisation process of the NEC regions made by Giorgio Fuà (Fuà, 1983 and 1988); an interpretation that has been somewhat overlooked compared with the 'industrial district' literature. Fuà stated that some of the characteristic features of the local systems of the NEC regions, like the sectors of specialisation and the small size of firms, are explained by the fact that these regions began their industrialisation process later than the other industrialised countries. In this perspective, the prevalence of small firms in the local systems of the NEC regions should be associated not only with the specific organisation of these systems but also with the poor quality of the 'entrepreneurial and managerial factor' available in the area and with the lack of institutions and services needed to sustain the development of large firms (financial markets, business schools, professional services, etc.). According to this hypothesis the constraints to firms growth should lessen over time as a result of the learning process carried out by existing firms and by a general enhancement of the social and economic environment of the local systems.

Together with the internal evolution of the local systems and of their firms, there have also been profound changes in the external conditions affecting the competitive advantages of firms. One of the most important is the increasing need for investment in R&D, marketing, and information systems induced by improvements in the quality of products and in their manufacturing and trading conditions. The investments in these activities show economies of scale and threshold effects, so that they can be efficiently carried out only by medium sized and large firms. All these changes are increasing the disparities between firms of different size within the industrial districts (Whitford, 2001 reviews the debate on these aspects).

Although we do not underestimate the importance of spatial agglomeration in enhancing the efficiency and the innovation capacity of firms, we believe that the largest firms (that are still medium sized firms when compared at an international level) are gaining more and more importance for the competitiveness of the industrial districts.

Given this premise this paper focuses its attention on the following points: i) analysing some structural characteristics of the industrial districts of the Marche region, with particular reference to the role of the largest firms; ii) measuring the importance of these firms in controlling the output of the districts and in influencing their evolution over time.

The chapter is organised as follows: section 2 discusses some aspects of the current debate on the evolution of industrial districts in Italy; section 3 examines the industrialisation process of the Marche region and discusses the methodology used to delimit its industrial districts; section 4 analyses the role and growth strategies of the main firms within the industrial districts of the Marche region; section 5 draws some conclusions.

2 The role of small and medium-sized firms in industrial districts

The internal and external changes affecting the Italian industrial districts are increasing the disparities between firms of different sizes. These disparities have been somewhat overlooked by scholars of industrial districts, who have emphasised the non-hierarchical nature of such productive organisation[1]. Together with the non-hierarchical nature of the industrial districts this literature has emphasised the role of small firms in the theoretical definition and in the empirical identification of industrial districts (Piore and Sable, 1984; Becattini, 1991).

---

[1] Indeed, it is often stressed that the collaborative network, typical of industrial districts, behaves quite well in the case of incremental changes, while it is unable to cope with radical technological or market changes because of the absence of a centralized decision-making structure. On the contrary the disparities between firms in industrial districts have been underlined by Rabellotti (1997); Rabellotti and Schmitz (1999).

As far as the theoretical level is concerned, we can note that the main advantages attributed to the spatial agglomeration of specialised firms are not necessarily associated with the prominence of small firms.

In the industrial district model the prevalence of small and medium sized firms is justified by their specialisation along the production chain; this specialisation is made possible by the collaborative relationships between the firms belonging to the network[2]. At the same time it is understood that the firms operating at the final stages of the production chain hold a small percentage of the district output.

Cross section observation reveals that the internal heterogeneity of industrial districts, in terms of size and vertical integration of firms, is rather high and increases over time: "Small firm industrial districts which are successful are unlikely to remain populated by small firms only. Reinvested profits lead to expansion and increasing differentiation by size and performance" (Rabellotti and Schmitz, 1999 pp. 105).

Apart from the theoretical justification for the importance of small firms in industrial districts, the empirical question remains as to what the size limit is for a small firm. The choice of one limit or another (50, 100 or 200 employees) can influence the number of districts identified (ISTAT, 1997; Brusco and Paba, 1998; Cannari and Signorini, 2000). Whatever limit is used to identify a small firm, an important aspect of the Italian situation is that the presence of small firms is not characteristic of industrial districts but of the manufacturing system as a whole (Traù, 1999).

We believe that the large presence of small firms (sometimes very small indeed) in the Italian industrial districts can be explained not only by the advantages of the district organisation but also by other factors characterising the Italian institutional context. The industrialisation process of the NEC regions is a recent phenomenon; indeed, most of the growth of its industrial districts came about during the sixties and seventies. At that time these regions lacked the infrastructures needed to sustain the development of large firms: a market for experienced managers and technicians, developed financial markets, the presence of business schools, etc. Above all, the quality of the entrepreneurs and the sectors of activity they chose were such as to impede the rapid growth of firms. Moreover the market segmentation that characterises the sectors in which industrial districts operate (so-called 'made in Italy') does not allow firms to exploit economies of scale in the manufacturing activities[3].

Together with the recentness of the manufacturing system that characterises the NEC regions, there are two other factors that help in explaining the persistence of a large share of small firms in industrial districts. The first is the special status accorded to craft firms (artisans) in Italy, in terms of tax scheme, access to

[2] Harrison (1992) demonstrated that the recent theory of industrial district involves more than the traditional agglomeration economies, because of the importance of non-economic institutions sustaining the collaboration between firms within the districts.

[3] In a recent survey regarding the fast growing manufacturing firms of the NEC regions, it emerged that it took at least 30 years for a firm to reach the threshold of 100 million Euro turnover (Balloni and Iacobucci, 2001).

finance, labour law, etc. The craft firm is defined as a firm with less than 15 employees, regardless of the activity performed. As a result of this definition there are a large numbers of craft firms operating as sub-contractors of other firms; they do not produce a final product (as one would expect of craft production) but perform a specialised activity along the production chain. The presence of craft firms in the NEC industrial districts is quite high; referring to the Marche region, in 1996 76.3% of the manufacturing firms were artisan firms (16,643 out of 21816) with 37.7% of the employees (70,922 out of 188,059). More than half of these craft firms had one or two employees; i.e. they could be best defined as self-employed rather than firms. Indeed one of the typical features of these firms is that they show a low propensity to growth in order to retain the advantages of craft firm status (Accornero 1999).

The second factor that helps in explaining the persistence of small firms in industrial districts (and in the Italian manufacturing sector in general) is the high level of 'underground economy' that still characterises the activities of firms. It takes the form of tax evasion (through the under estimation of production) and the non-observation of labour law (ISTAT, 1999)[4]. Underground economy is widespread especially among small firms; risks associated with these forms of illegality are such that they can be managed only in small, family run, firms. Together with artisan status, the cost advantages associated with underground economy are another factor discouraging the growth of firms.

Notwithstanding the importance of smaller firms within the industrial districts, the mere observation of the size structure of firms does not account for the relative importance of the firms in controlling some of the key factors for district competitiveness. Firms of different sizes perform different kinds of activities and concentrate their investment in different phases of the production chain: small firms concentrate on manufacturing activities while larger firms devote most of their investment to product development, marketing and finance (Guelpa, 1999). Differences in the activities performed imply differences in the type of employees and in their cost and productivity. From a recent survey conducted by the Italian Statistical Institute (ISTAT, 2001) it emerged that in the manufacturing sector the value added per employees in the firms with less than 20 employees is 53% that of firms with 20 employees or more (25,823 Euro and 48,547 Euro, respectively). Almost the same differences are observed in the cost of labour, while in the case of investment per employee the value of the larger firms is more than double that of the smaller ones (8,986 Euro and 4,390 Euro, respectively). These differences persist even when we consider the same sector: for example, in the textile and clothing industry the value added per employee in firms with less than 20 employees is 58% of that of firms with 20 employees or more.

---

[4] The new European Account System distinguishes between three forms of 'non-observed economy': illegal activities, underground economy, the informal sector. The underground economy is characterised by: a) the evasion of taxes on value added and on profits; b) the evasion of social contributions on labour; c) the non observance of the labour law for minimum wages, working hours, safety regulations, etc. Among the EU countries Italy is second to Greece for level of 'underground economy'.

Furthermore, looking at the size structure of manufacturing firms in terms of employees conceals the fact that in most cases the largest firms, that normally hold the final stages of the production chain, control a large part of the output of the district. Moreover, the largest firms are almost all organised as a group, i.e. a set of companies legally distinct but belonging to the same entrepreneur. This phenomenon is ignored by official statistics, whose unit of analysis remains the legal unit. This fact underestimates the effective size of the largest firms that not only own several companies within the district but also own production and distribution companies in other districts and abroad.

Taking the Marche region as a reference, in the following section we will analyse the recent changes in its industrial districts considering in particular the role of the largest firms.

3 The Marche region and its industrial districts

The Marche region is a small region in central Italy with a high industrial intensity. It belongs to the set of the North-Eastern-Central (NEC) regions of Italy (Friuli, Veneto, Emilia Romagna, Toscana, Umbria) which after the Second World War faced an intensive process of industrialisation, mostly based on small firms, concentrated in specialised districts. Given the analogies in the industrialisation process of these regions, the changes in the industrial districts of the Marche region should present close similarities with the changes that other NEC regions are presently undergoing.

After the Second World War the Marche region could be considered an agricultural area, with sixty percent of the active population employed in agriculture and around twenty percent in industrial activities. Thirty years later, in 1981, those employed in agriculture represented 11.5% of the active population, while those employed in industrial activities 44.7% (see Table 1). During the eighties the employees in the service sector outweighed those in the industrial sector that for the first time showed a slight decline; in spite of this, the share of employees in the manufacturing sector is higher than the Italian average, making the Marche region one of the most industrialised regions in Italy.

Table 1. Active population in the Marche region % by sector of activity

	1951	*1961*	*1971*	*1981*	*1991*
Agriculture	60.2	45.6	25.3	11.5	7.8
Industry	21.9	30.7	40.8	44.7	41.7
Services	17.8	23.7	33.8	43.8	50.5
Absolute value ('000)	641	581	514	564	563

Source: ISTAT – Population Census, 1951, 1961, 1971, 1981, 1991

The industry structure of the Marche region presents remarkable similarities with that of the other NEC regions: it is based on small firms organised in local systems and operating principally in 'traditional' sectors (footwear, clothing, furniture, machinery). Table 2 shows the share of employees by class of employees of the firms in 1996.

Table 2. Firms and employees in the manufacturing sector of the Marche region by industry and class of employees – 1996

Industry	*Absolute value*		*Employees by class of employees %*				
	Firms	Emplo-yees	1-9	10-19	20-49	50-249	>250
Food, beverage and tobacco	2,064	11,429	49.2	17.0	12.4	12.4	9.0
Textile and clothing	2,754	24,329	21.1	23.1	29.9	23.7	2.2
Leather and leather products	4,988	46,007	26.7	22.3	22.0	24.3	4.8
Wooden products	1,602	7,172	48.6	20.8	16.4	14.1	
Paper and printing	884	6,841	31.1	21.1	19.1	13.2	15.4
Coke, petroleum and other energy products	15	717	5.6		10.5	31.6	52.3
Chemical products and synthetic fibres	110	1,671	16.8	10.8	15.3	13.5	43.5
Rubber and plastics	538	6,934	18.2	2.0	27.3	25.7	8.9
Building products	661	5,175	29.1	22.0	21.4	21.4	6.1
Iron, steel and other metal products	2,387	20,244	28.4	20.6	26.8	18.5	4.7
Machinery and equipment	1,155	22,115	10.5	9.8	10.8	18.9	51.1
Electrical and electronic machinery	1,629	10,485	30.0	14.0	19.7	28.6	7.6
Transport equipment	188	2,808	12.8	17.9	14.9	35.4	18.8
Other manufacturing products (including furniture)	2,841	22,132	28.5	21.3	22.2	12.8	5.4
Total	21,816	188,059	26.3	19.4	21.2	21.6	11.5

Source: ISTAT, Census of Industry and Service, 1996

Two aspects deserve attention: the first is that more than fifty percent of employees are in firms with less than fifty employees (the EU definition of small firms); the second is that while the average size of firms has slowly increased during recent decades, there is still a concentration of employees in the class 10-49.

Table 3 shows the share of employees by sectors. During recent decades the region has strengthened its specialisation in the fashion industry (clothing and footwear), and at the same time furniture and mechanics have also become more important. Within mechanics the most important production is that of household appliances. In the last decade there has also been a rapid growth of activities in the electronics sector, mostly originating as a diversification from the traditional production of musical instruments. Indeed both the mechanical and electronics sectors showed an increase in employees between 1981 and 1991, while the other main sectors showed a slight decline.

Table 3. Employees in the manufacturing sector of the Marche region by industry, and related specialisation indexes.

	% on total			*Specialisation Index**		
Industry	1981	1991	1996	1981	1991	1996
Food, beverage and tobacco	5.3	6.8	6.1	0.65	0.73	0.65
Textile and clothing	16.4	16.5	12.9	1.01	1.05	0.91
Leather and leather products	28.4	25.5	24.5	5.99	5.50	5.16
Wooden products	6.0	3.7	3.8	1.47	1.06	1.09
Paper and printing	3.0	3.5	3.6	0.62	0.64	0.68
Coke, petroleum and other energy products	0.1	0.4	0.4	0.16	0.72	0.68
Chemical products and synthetic fibres	0.6	1.0	0.9	0.11	0.21	0.20
Rubber and plastics	2.8	3.0	3.7	0.75	0.87	0.91
Building products	4.0	2.9	2.8	0.69	0.55	0.54
Iron, steel and other metal products	8.3	9.4	10.8	0.57	0.64	0.70
Machinery and equipment	5.8	9.6	11.8	0.58	0.93	1.04
Electrical and electronic machinery	3.7	4.6	5.6	0.41	0.47	0.60
Transport equipment	1.7	1.3	1.5	0.24	0.20	0.25
Other manufacturing products (including furniture)	13.8	11.7	11.8	2.60	1.96	1.81
Total	100.0	100.0	100.0			

* $SI = (E_{j,M} / E_M)/(E_{j,I} / E_I)$; where E is the number of employees, j is the industry, M stands for Marche and I for Italy.
Source: ISTAT, Census of Industry and Service, 1981, 1991, 1996

Most of the manufacturing activities of the Marche region (principally footwear, furniture, clothing and mechanics) are organised in industrial districts. Any analysis of industrial districts, whatever its scope and aim, should rely on a clear identification of the subject being analysed. Although the term industrial district is widely used to identify the manufacturing activities of the NEC regions, the identification of local systems showing the features of an industrial district is not a trivial question for two reasons. Firstly the theoretical definition of the industrial district is too vague to give clear guidance in empirical work, and secondly some of the factors used to define it are difficult to quantify and measure[5].

The main elements identified by the theoretical literature that should characterise an industrial district are the following: i) the spatial proximity of firms; ii) the presence of a dense network of small, specialised firms exchanging goods along a production chain; iii) the fact that the firms are embedded in a specific local frame of institutions that support (or enhance) a combination of cooperation and competition among the firms.

Regarding point i) the definition does not specify how large the area should be and how to delimit it. As far as point ii) is concerned the main problem is that of deciding what a small firm is: in the Italian manufacturing system there are a large number of very small firms (less than 20 or even less than 10 employees) but, as discussed in the previous section, it is unclear to what extent this is the result of the district organisation rather than of the general Italian institutional setting. The elements in point iii) are the most difficult to identify and measure. This is why they are seldom used for empirical identification of industrial districts, although theoretical literature considers them among the most important factors in explaining the competitiveness of industrial districts.

An empirical methodology that is widely used in Italy to identify industrial districts is that elaborated by ISTAT and Sforzi, based on the definition of travel-to-work areas called local labour systems or LLS (ISTAT, 1997). Using this methodology, Italy is divided in 784 LLS, defined as neighbouring municipalities that are almost self-contained in terms of daily commuting (the latest data used to identify the LLS are those of the 1991 population census). An LLS is qualified as an industrial district according to three indicators: 1) the share of the manufacturing employees; 2) the specialisation in one particular sector; 3) the concentration of manufacturing employees in small firms. All three indicators are calculated in comparison with the national average.

There are several drawbacks in this procedure (Brusco and Paba, 1997; Cannari and Signorini, 1999): a) The number of industrial districts identified depends on the specific value of the indicators (for example the definition of the small firm)[6]; b) It has a 'one to zero' outcome, so that an LLS is not considered as an industrial district if one of the indicators is even slightly outside the fixed threshold; c) It overlooks some of the social variables (institutions, attitudes, network relationships) which are important aspects for the definition of an industrial district.

Notwithstanding these limitations, the ISTAT-Sforzi methodology is largely accepted in the empirical identification of industrial districts and is the basis on which the relevance of such districts has been recently measured (Brusco and Paba, 1998; Signorini, 2000). For this reason we will start by following it.

---

[5] This is one of the reasons why the term industrial district has been used to identify very different situations, from business clusters made up of a few firms to large agglomerations of small firms (Amin and Robins, 1990; Paniccia, 1998; Staber and Morrison, 1999).

[6] Urban agglomerations are normally excluded, even when they show the typical features of industrial districts, because of the high share of employees in the service sector.

According to the ISTAT-Sforzi procedure, the Marche region is divided in 44 LLS. Based on the structural characteristics previously mentioned, 7 of these LLS are classified as non-industrialised areas (large towns and mountain areas) and 2 as dominated by large firms. The remaining ones show the structural features of industrial districts, being characterised by the presence of manufacturing SMEs specialised in one sector (Table 4).

Table 4. Local Labour Systems of the Marche region by sector of specialisation and size of firms

	Large firms	*Small Firms*	*Small and medium sized firms*	*Non industrialised areas*	*Total*
Food products			1		1
Leather and footwear	1	11	4		16
Furniture		6			6
Textile and clothing		7	2		9
Mechanics	1				1
Other industries		2	2		4
Non manufacturing				7	7
Total	2	26	9	7	44

Source: ISTAT (1997)

The territorial distribution of manufacturing activities shows great differences among sectors: footwear and mechanics are the most concentrated, while textile and clothing are widespread across the region (Table 5).

Table 5. Percentage of employees in the LLS on the total of the Marche region for the same sector

Sector of specialisation	%
Leather and footwear	85.8
Mechanics	66.1
Furniture	53.3
Other industries	33.5
Textile and clothing	31.3
Food products	6.6

Source: ISTAT

Taken together the 9 textile and clothing LLS represent less than one third of the total employees in the sector. On the contrary, the employees in the footwear LLS account for more than 80% of the total; those in the mechanics LLS account for two thirds; and those in the furniture LLS for more than half. The LLS identified through the ISTAT-Sforzi methodology are rather heterogeneous in terms of absolute size, degree of specialisation and weight of the sector in the region (see Table 6).

Table 6. Some structural features of the manufacturing LLS of the Marche region (1996)

LLS	*Sector*	*Type of LLS (1)*	*Manufacturing Employees*	*Employees in the sector of specialisation*	*Degree of specialisation (2)*	*% of employees in the district on the total (3)*
Morciano di Romagna	Furniture	SF	245	59	24.1	0.3
Cagli	Textile and clothing	SF	1707	604	35.4	3.2
Fano	Furniture	SF	7256	1397	19.3	6.4
Fossombrone	Furniture	SF	2018	686	34.0	3.1
Mondolfo	Textile and clothing	SF	2772	647	23.3	3.5
Pergola	Textile and clothing	SF	1284	513	40.0	2.8
Pesaro	Furniture	SF	16856	7215	42.8	33.0
Piandimeleto	Textile and clothing	SF	805	392	48.7	1.8
Sant'angelo in Vado	Furniture	SF	684	295	43.1	1.6
Sassocorvaro	Textile and clothing	SF	1063	369	34.7	2.0
Urbania	Textile and clothing	SMF	1295	785	60.6	4.2
Urbino	Furniture	SF	3533	1069	30.3	4.9
Fabriano	Mechanics	LF	20448	14614	71.5	66.1
Jesi	Food products	SMF	9373	2692	28.7	6.6
Osimo	Other manufacturing	SMF	12610	2329	18.5	12.4
Ostra	Textile and clothing	SF	1805	552	30.6	3.0
Sassoferrato	Leather and footwear	LF	979	484	49.4	1.1
Senigallia	Textile and clothing	SF	4396	1208	27.5	6.5
Serra de' Conti	Leather and footwear	SF	2600	668	25.7	1.5
Cingoli	Textile and clothing	SMF	1056	442	41.9	2.4
Civitanova Marche	Leather and footwear	SMF	10037	6475	64.5	14.1
Monte San Giusto	Leather and footwear	SF	5051	4420	87.5	9.6
Recanati	Other manufacturing	SF	7609	1702	22.4	7.8
Sarnano	Leather and footwear	SF	855	332	38.8	0.7
Tolentino	Leather and footwear	SF	4846	1646	34.0	3.6
Treia	Furniture	SMF	2397	884	36.9	4.0
Urbisaglia	Textile and clothing	SMF	869	367	42.2	2.0
Ascoli Piceno	Other manufacturing	SMF	5498	1041	18.9	5.6
Comunanza	Leather and footwear	SF	2024	811	40.1	1.8
Fermo	Leather and footwear	SF	8840	6296	71.2	13.7
Montefiore dell'Aso	Leather and footwear	SF	1109	425	38.3	0.9
Montegiorgio	Leather and footwear	SF	5109	2631	51.5	5.7
Montegranaro	Leather and footwear	SF	4895	4586	93.7	10.0
Monte San Pietrangeli	Leather and footwear	SF	1066	962	90.2	2.1
Offida	Leather and footwear	SMF	1672	711	42.5	1.5
Porto Sant'Elpidio	Leather and footwear	SF	9949	9032	90.8	19.6
San Benedetto del Tronto	Other industries	SF	7502	1438	19.2	7.7
Average			4652	2183	43.6	7.5
Min			245	59	18.5	0.3
Max			20448	14614	93.7	66.1

(1) SF= small firms; SMF=small and medium sized firms; LF=large firms
(2) Employees in the sector of specialisation on total manufacturing employees
(3) Employees of the sector of specialisation in the district on total employees of the same sector in the Marche region

Even if we refer only to the LLS characterised by the presence of small firms, we find a great variety of situations. The use of the LLS as a territorial unit in order to identify industrial districts presents another important shortcoming. Focusing on the daily commuting by the employees it overlooks one of the most important aspects of the district organisation: i.e. the supply relationships along the production chain. Ignoring this aspect has two important consequences.

The first is that in some cases the methodology classifies as a district an LLS made up of one or a few large firms (operating in the final stage of a production chain) and a large number of small firms supplying the former. Although there is a predominance of small firms in the area, the system is better defined as a firm network rather than as an industrial district. There are at least three cases of such LLS in the Marche region: the Treia LLS, specialised in furniture products and dominated by the presence of the Lube Group (a kitchen maker); the Urbisaglia LLS, specialised in clothing products, dominated by the presence of the Urbis group; the Tolentino LLS, specialised in leather products in which there are two dominant firms. On the contrary, the Fabriano LLS is not classified as a district area because of the presence of a few large firms (the Merloni groups). Nevertheless, the area has one of the largest European concentrations of household appliance producers with more than ten large and medium sized groups (operating in the final stages of the production chain) and hundreds of independent companies operating as their suppliers. Except for the average size of its firms, it shows all the features that characterise a localised industry.

The second consequence of ignoring the supply relationships among firms when identifying industrial districts is that these relationships do not necessarily come within the same boundaries as those of daily commuting. This is evident if we observe the territorial distribution of LLS by sector of specialisation (Figure 1).

Figure 1. Local Labour Systems of the Marche region by sector of specialisation

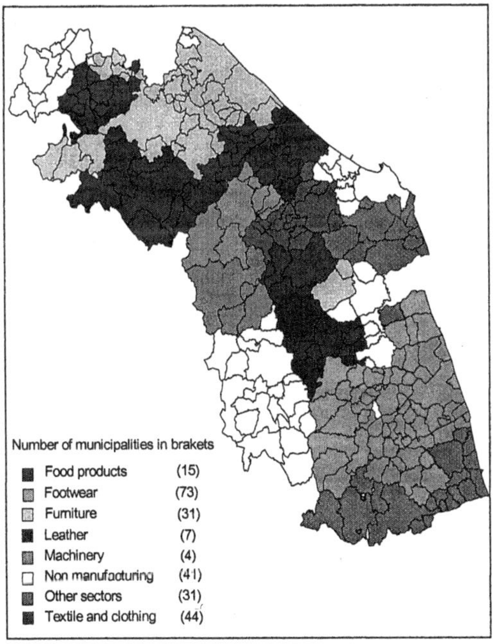

LLS specialised in the same sector are mostly contiguous because they are part of the same 'system'. For example, of the 15 footwear LLS, 13 are contiguous and form what is widely known as the 'footwear district' in the southern part of the region[7]. Starting from the original nucleus around a few small towns on the Adriatic coast, the development of the district during the sixties and seventies spread the manufacturing activities to the neighbouring municipalities[8].

The same considerations could be made for the 6 furniture LLS, 5 of which are concentrated in the northern part of the Region (around the town of Pesaro).

Another important shortcoming of the LLS methodology in identifying industrial districts is that it does not take into account the absolute size of the local system. As can be seen from Table 6, the methodology identifies LLS ranging from a few hundred employees to more than ten thousands. This is an aspect that has been almost completely overlooked by the empirical literature on industrial districts, although it is unreasonable that the advantages of the spatial agglomeration of specialised firms can be achieved in a local system with just a few hundred employees. From the original formulation by Marshall to the recent theorising by Porter it seems evident that the advantages of the spatial agglomeration of firms (see Figure 2) are associated to the concentration in an area of a relevant portion of a specific production. It is not by chance that Marshall, in analysing the advantages of the spatial agglomeration of firms, uses the term 'localised industry' and explicitly mentions the 'large aggregate production of the same kind'. However, most of the empirical literature on industrial districts has failed to take into consideration the absolute size of the local system and the relative importance of district production on the industry as a whole.

This seems a major drawback as it is increasing acknowledged that the superior performance of firms in industrial districts is mainly associated with dynamic efficiency (i.e. the ability to generate innovation through the accumulation of idiosyncratic knowledge) rather than with static efficiency (associated with the peculiar structure of input-output relationships between firms in the district).

At the same time, we think that some Italian literature has over-emphasised the importance of small firms for the definition of the industrial district. Marshall clearly stated that both small and large firms can benefit from localisation in a district and Porter includes, among his clusters, local systems based on large firms.

---

[7] An example of supply relationships in this system is represented by soles production. Because of the economies of scale, the production of synthetic soles is concentrated in a few medium sized firms that serve the footwear producers located in all the LLS.

[8] This has also been stimulated by the provision of public subsidies for the less developed areas of the internal part of the region. For this reason, the largest footwear company of the district (Tod's) has its head-quarters in one of the peripheral municipalities (Comunanza) although some of its plants and most of its suppliers are located in the original municipalities of the district.

Figure 2. Advantages of spatial agglomeration of specialised firms

Advantages of localised industries according to Marshall: Marshall (1920), book IV, ch. X	*Advantages of clusters according to Porter: Porter (1998)*
Better circulation of information regarding the industry: "The mysteries of the trade become no mysteries"	Access to specialised information. Extensive market, technical, and competitive information accumulates within a cluster
"...the economic use of expensive machinery can sometimes be attained in a very high degree in a district in which there is a large aggregate production of the same kind"	Complementarities among firms
"... a localised industry gains a great advantage from the fact that it offers a constant market for skill"	Better access to specialised employees and suppliers
"... if one man starts a new idea, it is taken up by others and combined with suggestions of their own; and thus it becomes the source of further new ideas"	Better ability to generate knowledge ("learning by interacting") and to innovate
	Better motivation and measurement. Local rivalry is highly motivating. Peer pressure amplifies competitive pressure within a cluster.
	Access to specialised institutions and public good

Taking into account these considerations we have identified 9 areas that can be considered as industrial districts. In some cases they correspond to an LLS; in others they are formed by contiguous LLS specialised in the same sector (Figure 3).

We excluded some of the textile and clothing LLS and the other industries LLS. The latter were excluded because they do not show any clear specialisation. Some textile and clothing districts were excluded for two reasons: the first was that, as previously mentioned, they represent a non significant quota of the sector within the region (textile and clothing activities are so widespread in the region that is difficult to single out industrial districts); the second reason was that the majority of firms in this sector are sub-suppliers of larger companies located outside the region (Brusco and Bigarelli, 1995).

Figure 3. Industrial districts considered in the analysis

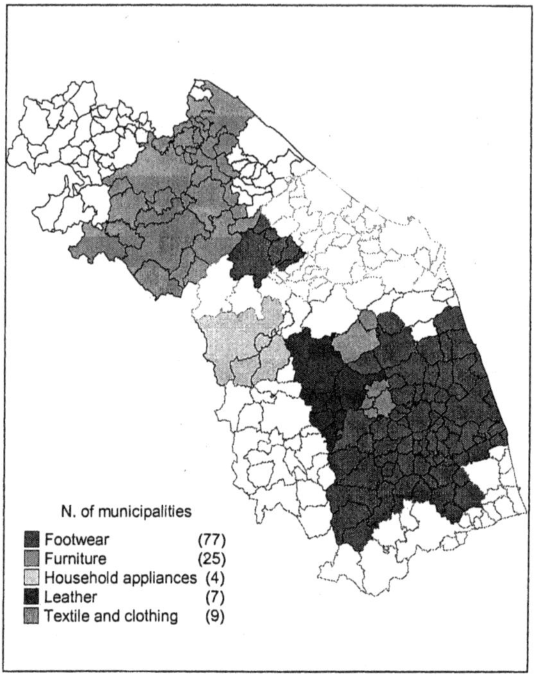

4 The role of large and medium-sized firms in the industrial districts of the Marche region

Some structural characteristics of the industrial districts considered in the analysis are shown in Table 7.

Of the districts analysed, three show a significant absolute size: the footwear, the household appliance and the furniture districts[9]. These three districts concentrate a significant share of the regional employees in the sector of specialisation. The footwear district is the most important in terms of employees and geographical extension. It is also one of the most specialised: two thirds of the manufacturing employees in the district belong to the footwear sector[10].

---

[9] The main features of these three districts and the characteristics of their main companies have been analyzed by Balloni and Iacobucci (2000).

[10] The footwear district of the Marche region has the largest territorial concentration of footwear production in Italy and probably the largest in Europe.

In order to analyse the role of the largest firms within these industrial districts we have taken advantage of a data set regarding the main manufacturing companies of the Marche region collected by the Merloni Foundation. This data set contains information about 200 manufacturing companies, including all those with a turnover exceeding 15 million Euro, as well as a representative sample of those having a turnover of 5 to 15 million Euro. In terms of value added, in 1996 these firms represented around 1/4 of the manufacturing industry of the region. The Merloni Foundation data set contains balance sheet data as well as other information about the companies: employees, exports, ownership, etc. Using this data set we isolated for each industrial district the largest companies in terms of employees in order to assess their relevance in the district they belong to (Table 7).

Only the footwear district of Civitanova-Fermo and the furniture district of Pesaro show the typical 'fragmentation' required by the theoretical definition of industrial districts. In all the other districts the weight of the largest firms within the district is rather high. In the case of the household appliance district of Fabriano this is due to the presence of a few large firms (the Merloni groups). The other smaller districts show a dichotomic structure with a few medium sized firms and a large number of very small firms, mainly operating as suppliers to the former. This is the case of the furniture district of Treia, the clothing district of Urbisaglia and the leather district of Tolentino.

For the furniture district of Pesaro and the footwear district of Civitanova-Fermo, we must take into account that the largest firms show a high level of decentralisation of activities[11]. For this reason the weight of the largest companies can be best understood if we consider their quota in the final output of the district rather than in terms of employees.

Assessing the weight of the largest companies in the output of the district is not an easy task because of the absence of data on output (sales or value added) at the municipality level. We have made a rough estimate of this weight using data on exports that are available, in Italy, at a provincial level. Although this is a higher level than needed in order to identify the exports of the districts, it is nevertheless possible to establish a relationship between the districts considered in the analysis and the province they belong to (see Table 6)[12].

---

[11] The largest firms in the furniture district of Pesaro and Treia have a degree of value added on sales ranging from 10% to 15%; the average in the Italian manufacturing sector is around 25%. In the case of the footwear district some of the largest companies perform only a very small part of the manufacturing activities, relying on an extended network of sub-suppliers.

[12] While the numerator refers to the exports of the companies located in the industrial districts, the denominator refers to the exports of the entire province; for this reason the quota of the largest firms on the exports of the districts is slightly underestimated.

Table 7. Structural characteristics of the main industrial districts of the Marche region (1996)

Sector of specialisation	*Furniture*	*Textiles and clothing*	*Footwear*	*Household appliances*	*Furniture*	*Textiles and clothing*	*Leather*	*Footwear*
Classification codes (ATECO)	361	17, 18	193	29	361	17, 18	191, 192	193
Most important town(s)	Pesaro	Urbania	Serra de' Conti	Fabriano	Treia	Urbisaglia	Tolentino	Civitanova, Fermo
Province	Pesaro	Pesaro	Ancona	Ancona	Macerata	Macerata	Macerata	Macerata, Ascoli Piceno
Number of LLS	3	3	1	2	1	1	1	12
Employees in the manufacturing sector (A)	22,407	5,982	2,600	17,541	2,397	869	4,846	57,228
Employees in the sector of specialisation (B)	8,780	2,234	666	14,614	882	387	1,286	37,309
Firms in the sector of specialisation (C)	974	159	63	144	58	18	89	4,108
Average size of firms (C/B)	9.0	14.1	10.6	1,01.5	15.2	21.5	14.4	9.1
Employees in the sector in the Region (D)	17,671	18,612	42,632	22,115	17,671	18,612	3,375	42,632
Degree of specialisation (B / A) %	39.2	37.3	25.6	83.3	36.8	44.5	26.5	65.2
Relevance in the Region (B / D) %	49.7	12.0	1.6	66.1	5.0	2.1	38.1	87.5
Employees of the 4 largest companies (E)	859	265	236	10,040	400	295*	432**	1191
Employees of the 8 largest companies (F)	1,167			10,875				1,856
Weight of the 4 largest companies (E / B) %	9.8	11.9	35.4	68.7	45.4	76.2	33.6	3.2
Weight of the 8 largest companies (F / B) %	13.3			74.4				4.8

* One company

** Two companies

Sources: ISTAT, Censimento intermedio dell'Industria e dei servizi, 1996; Fondazione Aristide Merloni

Assessing the weight of the largest firms on exports is particularly interesting because most Italian districts are highly export oriented and the international competitiveness of such productive organisations has often been underlined (Conti and Menghinello, 1996; Becattini and Menghinello, 1998; Bagella and Becchetti, 2000). The consideration of exports also has some limitations: not all the districts are export oriented and not all the largest firms concentrate their sales on foreign markets. Regarding the former aspect it should be taken into account that the furniture district of Pesaro is the least export oriented. Moreover, the largest firms in this district, that are all kitchen makers, export a very small share of their sales.

Notwithstanding these limitations we have calculated the share of exports of the four largest firms on the total at the provincial level (see Table 8)[13].

Table 8. Percentage of the 4 largest companies on the exports by province

	Pesaro		*Ancona*		*Macerata*		*Ascoli Piceno*	
	1996	1999	1996	1999	1996	1999	1996	1999
Textile and clothing	19.6	23.5			14.6*	20.8*		
Leather					41.5**	41.2**		
Footwear					12.0	13.0	14.0	20.6
Mechanics			40.5	40.5				
Furniture	4.9	8.5			66.4	50.8		
Total	17.6	23.9	37.0	35.6	20.4	22.5	16.1	20.7

* One company
** Two companies
Source: ISTAT for export data; Merloni Foundation for company data

Except for the furniture district of Pesaro, in all the other districts the weight of the largest firms on exports is higher than that of employees and is increasing over time. In the case of the footwear district the weight of the four largest firms is more than five times that of employees. In the smaller districts the weight of the largest firms is substantial.

[13] Data on exports at the provincial level are collected by ISTAT according to the location of the manufacturing activities. In determining the exports of the largest firms we have taken into account the fact that in some cases they have more than one plant, not always located within the district. In these cases we have corrected the exports of the company according to the percentage of the employees in the district. This correction concerns only some of the household appliance producers in the Fabriano district. In all other cases the companies considered have all their plants within the district.

Besides the quantitative importance of the largest firms in terms of employees or exports, we must consider the role played by these firms in controlling some of the key factors affecting the competitiveness of the district. All the largest firms operate in the final stages of the production chain and are increasingly concentrating their investment in the strategic phases of product development, marketing and distribution. In all the sectors considered (furniture, clothing, footwear) these phases are particularly important for the competitiveness of products and for the possibility of the firm to appropriate value along the production chain.

Two aspects deserve attention here. The first is that official statistics about firms systematically underestimate the true size of the largest firms as they are all organised as groups: i.e. sets of companies legally distinct but owned by the same entrepreneur(s). In some cases they are located within the same district, although most of them are located abroad. In the case of the footwear district the two most important companies (Tod's and Fornari), the only two companies analysed that produce a consolidated balance sheet, had 1,033 employees at the end of 2000; this figure increased to 1,550 if all the employees in the controlled companies were considered. Overall the two groups were made up of 19 companies, most of them located abroad. In the case of the furniture district the largest company (Scavolini) had 388 employees or 500 if the group is considered. In this case the group is composed of two manufacturing companies, both located within the district. In the case of the clothing district of Urbisaglia the largest company (Urbis) had 295 employees that increased to 521 when the group was considered. The group is made up of 21 companies: 6 of them are production companies, all located within the district and the rest are distribution companies, 5 of which located abroad[14].

The structure of these groups clearly shows that their strategy is mainly addressed at controlling the distribution phase, with specific reference to foreign markets. Recent empirical investigations have demonstrated that while export activity is achievable even by smaller firms, size becomes a critical factor for internationalisation, i.e. the setting up of production and distribution units abroad (Mariotti and Mutinelli, 2001; Balloni and Iacobucci, 2001).

The second aspect that in our opinion deserves attention in order to assess the role of the largest firms in industrial districts is the rapid change in the factors affecting their competitiveness. Until the beginning of the nineties the competitiveness of the district products had been boosted by the periodical devaluation of the Italian currency[15]; since 1999 this possibility has been ruled out by the advent of the Euro. In this new scenario the products of industrial districts have undergone increasing pressure from foreign competitors, especially those located in countries with lower labour costs. The main response of the

---

[14] Overall the Urbis group has more employees than the LLS where it is located. In spite of this, according to the ISTAT-Sforzi methodology the LLS is classified as a district of small and medium sized firms.

[15] Between the end of 1992 and the middle of 1993 the Italian lira was devaluated by an average of 25 percent against the major currencies.

district firms to this competitive pressure has followed two main patterns, sometimes developed as complementary and sometimes as alternative strategies. The first is the systematic improvement in the quality of products. Although remaining within the same sector, firms systematically move towards the higher segments of the market. To achieve this, companies are devoting more and more resources to R&D, marketing and the control of the distribution phase (Balloni and Iacobucci, 2001). The improvement in the quality of the products requires a more strict control of the production phases. This has been obtained through a reduction in the number of sub-suppliers and the setting up of more collaborative relationships with them. Although the companies within these networks remain formally independent, the relationships among them are more and more hierarchical.

The second major strategy followed by the largest groups is the internationalisation of production activities. This process is taking two main forms: the decentralisation of some phases of the production process (the labour intense ones) in order to reduce production costs; the decentralisation of the entire production process in order to offer the foreign markets products which have the quality and costs they can afford.

These changes will increase the disparities between small and large firms within industrial districts. Operating in the highest segments of the market requires greater concentration of investments in R&D and marketing as well as attention to financial management. The investments in these activities show economies of scale and threshold effects so that they cannot be efficiently made by small firms.

5 Conclusions

The main aim of this work was to examine some of the structural features of the industrial districts of the Marche region, with specific reference to the role of medium sized and large firms.

By concentrating on the structural characteristics of industrial districts and specifically on the role of small firms, literature has underestimated two important elements for the interpretation of this form of production organisation: the first is the historical evolution of the internal and external conditions affecting the competitiveness of industrial districts; the second is the heterogeneity of firms within industrial districts.

We have examined these two aspects focusing on the industrial districts of one region of the so-called 'Third Italy': the Marche region.

Although theoretical literature on industrial districts has done a lot to identify the characteristic features of this production organisation and the factors responsible for its competitiveness, it offers less guidance in the empirical identification of the districts. As a first approximation we followed the ISTAT-Sforzi methodology based on the identification of local labour systems (LLS). This methodology has been widely used in recent empirical studies in order to ascertain the weight of employees in industrial districts (Brusco and Paba, 1997; Cannari and Signorini, 1999) or to compare the performance of district versus non-district firms (Fabiani *et al.*, 2001).

In the case of the Marche region we showed that this procedure identifies very different situations in terms of absolute size and structural characteristics. Furthermore, in some cases, notwithstanding a large presence of small firms, the local system is dominated by the presence of one or a few large firms for which the former operate as sub-suppliers.

We also believe that the LLS is not always the appropriate unit of analysis in order to identify industrial districts. Concentrating on the daily commuting by workers, the procedure overlooks one of the important features of industrial districts, i.e. the supply relationships among firms. Some empirical studies regarding the main industrial districts of the Marche region have shown that the supply networks of firms exceed the area of an LLS (see also Iuzzolino, 2000). This is also evident from the fact that the majority of LLS specialised in the same sector are contiguous. This is why in identifying the industrial districts of the Marche region in some cases we have put together several contiguous LLS specialised in the same sector.

In two of the main industrial districts of the region (the furniture and the footwear district) the role of small firms is pre-eminent. Indeed these districts show the characteristic 'fragmentation' of Italian districts with a large share of employees in very small firms (those with less than 10 employees). Regarding this aspect we have tried to demonstrate that heterogeneity among firms within industrial districts is higher that it appears to be from census statistics. This is the result of two elements. On one hand we must consider that the industrialisation process of the Marche region (as well as the other NEC regions) is a recent phenomenon and that it takes time to develop the necessary conditions for the development of large organisations. Most of the largest firms operating in Marche districts were set up during the sixties and seventies and a lot of them are still run by the founders. Until a few years ago even the largest firms were owned and managed by members of the founder's family and this is still the case for most of the medium sized firms. We expect that the largest firms will continue to gain importance within the system. In 1990 only one firm in the region was listed on the Italian stock exchange. At the end of 2001 there were 5 companies listed and we expect this number will probably grow substantially in next few years.

On the other hand because of the changes in the external conditions affecting the competitiveness of industrial districts, the importance of the manufacturing activities (that are the 'core' activity of small firms) will be reduced, in favour of other activities along the production chain, such as product design and innovation; marketing and distribution; financial management. The largest firms have already started concentrating their investments in these activities.

In the case of the most important districts (like the furniture and the footwear districts) the growth process of the largest firms will not eliminate the characteristic features of this productive organisation and the advantages attached to it. Indeed, most of the advantages of industrial districts can also be observed in local systems characterised by the presence of medium sized and large firms. A case in point is the household appliance district of Fabriano. Being characterised by the presence of a few large firms it would not be classified as an industrial district, if we refer to the 'ideal type' district based on small firms co-operating within a production chain. However, the absolute size of the local system, the large number of final producers, the presence of a complex network of sub-suppliers, the cultural homogeneity of the area and the presence of specialised institutions should induce us to recognise the characteristic features of a 'localised industry'.

For this reason we do not believe that the increasing importance of medium sized firms within the industrial districts of the Marche region will undermine their ability to foster innovation and efficiency. On the contrary, we believe that the reorganisation of these systems around a few leading firms is a way to preserve their competitive advantage, given the new internal and external conditions.

Whether we should continue to call them industrial districts or not will depend on whether we are more interested in comparing their structure with that of the 'ideal type' or whether we are more interested in the factors affecting the competitiveness of the firms within these systems.

References

ACCORNERO A. (1999), ""Poter" crescere e "voler" crescere: i piccoli imprenditori ex dipendenti", in: TRAÙ F. (a cura di), *La "questione dimensionale" nell'industria italiana*, Il Mulino, Bologna.

AMIN A., ROBINS K. (1990), "The Re-Emergence of Regional Economies", *Environment and Planning D*, 8.

BAGELLA M., BECCHETTI L. (Eds.) (2000), *The Competitive Advantage of Industrial Districts. Theoretical and Empirical Analysis*, Physica-Verlag, Heidelberg.

BALLONI V., IACOBUCCI D. (2001), "I "nuovi protagonisti" dell'industria italiana", *L'Industria*, 22 (4), 633-675.

BECATTINI G. (1991), "The Industrial District As a Creative Milieu", in: BENKO G., DUNFORD M. (Eds.), *Industrial Change and Regional Development: The Trasformation of New Industrial Spaces*, Belhaven Press/Pinter, London.

BECATTINI G. (1997), "Totalità e cambiamento: il paradigma dei distretti industriali, *Sviluppo Locale*, 4 , 75-94.

BECATTINI G. (2000), *Il distretto industriale*, Rosenberg & Sellier, Torino.

BECATTINI G., MENGHINELLO S. (1998), "Contributo e ruolo del made in Italy "distrettuale" nelle esportazioni nazionali di manufatti", *Sviluppo Locale*, 5.

BRUSCO S., CAINELLI G., FORNI M., FRANCHI M., MALUSARDI A., RIGHETTI R. (1996), "The Evolution of Industrial Districts in Emilia-Romagna", in: COSSENTINO F., PYKE F., SENGENBERGER W. (EDS.), *Local and Regional Response to Global Pressure: The Case of Italy and Its Industrial Districts*, International Institute for Labour Studies, Geneva.

BRUSCO S., BIGARELLI D. (1995), "Struttura industriale e fabbisogni formativi nei settori della maglieria e delle confezioni in Italia. Un'analisi per regione 1993", *Rivista Italiana di Economia*, 1, 7-47.

BRUSCO S., PABA S. (1997), "Per una storia dei distretti industriali italiani dal secondo dopoguerra agli anni novanta", in: BARCA F. (a cura di), *Storia del Capitalismo Italiano dal Dopoguerra ad Oggi*, Donzelli Editore, Roma.

CANNARI L., SIGNORINI L.F. (2000), "Nuovi strumenti per la classificazione dei sistemi locali, in: SIGNORINI L.F. (a cura di), *Lo Sviluppo locale. Un'indagine della Banca d'Italia sui distretti industriali*, Meridiana Libri, Corigliano Calabro.

CONTI G., MENGHINELLO S. (1996), "Territorio e competitività: l'importanza dei sistemi locali per le esportazioni italiane di manufatti. Un'analisi per province (1985-94)", in: *ICE, Rapporto sul Commercio Estero 1995*, Roma.

FABIANI S., PELLEGRINI F., ROMAGNANO E., SIGNORINI L.F. (2000), *"Efficiensy and localisation: the case of italian discricts"*, in: BAGELLA M., BECCHETTI L. (EDS), *The competitive advantage of industrial discricts. Theoretical and Empirical Analysis*, Physica Verlag, Heidelberg.

FORTIS M. (1996), *Crescita economica e specializzazioni produttive. Sistemi locali e imprese del "made in Italy"*, Vita e pensiero, Milano.

FUÀ G. (1983), "L'industrializzazione del Nord Est e del Centro", in: FUÀ G., ZACCHIA C. (a cura di), in: *Industrializzazione senza fratture*, Il Mulino, Bologna.

FUÀ G. (1988), Small-Scale Industry in Rural Areas: the Italian Experience, in: ARROW K.J. (Ed.), *The Balance between Industry and Agricolture in Economic Development (Proceedings of the Eight World Congress of the International Economic Association) Vol. 1*, MacMillan, London.

GUELPA F. (1999), "Specializzazione per area funzionale e performance d'Impresa, in: TRAÙ F. (a cura di), *La "Questione dimensionale" nell'industria italiana*, Il Mulino, Bologna.

HARRISON B. (1992), Industrial Districts: Old Wine in New Bottles?, *Regional Studies*, 26(5) , 469-483.

ISTAT (1997), *I Sistemi locali del lavoro 1991*, ISTAT, Roma.

ISTAT (1999), *L'Occupazione non regolare nelle nuove stime di contabilità nazionale. Anni 1992-1997, Statistiche in Breve* ISTAT, Roma.

ISTAT (2001), *Struttura e competitività del sistema delle imprese industriali e dei servizi nel 1998, Statistiche in breve* ISTAT, Roma.

IUZZOLINO G. (2000), "I distretti industriali nei dati censuari", in: SIGNORINI L.F. (a cura di), *Lo Sviluppo locale. Un'indagine della Banca d'Italia sui distretti industriali*, Meridiana Libri, Corigliano Calabro.

MARSHALL A. (1920), *Principles of Economics*, MacMillan, London.

PANICCIA I. (1998), "One, a Hundred, Thousands of Industrial Districts: Organizational Variety in Local Networks of Small and Medium-Sized Enterprises", *Organization Studies*, 19, 667-699.

PIORE M.J., SABEL C.F. (1984), *The Second Industrial Divide*, Basic Books, New York.

PORTER M.E. (1998), "Clusters and the New Economics of Competition", *Harvard Business Review*, november-december.

RABELLOTTI R. (1997), *External Economies and Cooperation in Industrial Districts*, MacMillan, St. Martin's Press, Basingstoke, New York.

RABELLOTTI R., SCHMITZ H. (1999), "The Internal Heterogeneity of Industrial Districts in Italy, Brazil and Mexico", *Regional Studies*, vol. 33, n. 2, pp. 97-108.

SIGNORINI L.F. (2000), "L'"effetto distretto": motivazioni e risultati di un progetto di Ricerca, in: SIGNORINI L.F. (a cura di), *Lo sviluppo locale. Un'indagine della Banca d'Italia sui distretti industriali*, Meridiana Libri, Corigliano Calabro.

STABER U., MORRISON C. (1999), "The Empirical Foundations of Industrial District Theory", Paper presented at the *ISRN Workshop on Globalization and Regional Innovation Systems,* Toronto, May 17-19, 1999.

WHITFORD, J. (2001), "The Decline of a Model? Challenge and Response in the Italian Industrial Districts, *Economy and Society*, vol. 30, n. 1, pp. 38-65.

Entrepreneurship, labour organisation and labour mobility in industrial districts*

Massimo Omiccioli[1] and Fabio Quintiliani[2]

[1]Bank of Italy, Research Department, Via Nazionale 91, 00184 Rome, Italy
[2]Bank of Italy, Piazza Cavour 6, 40124 Bologna, Italy

1 Introduction

In this work we present new evidence on the governance structure of firms and the functioning of the labour market in industrial districts. The evidence is derived from a comprehensive field survey carried out by the Bank of Italy in 1998.[1] The survey was intended to cover many aspects of industrial districts in Italy, ranging from the organisation of production (with special emphasis to subcontracting) to financial relationships. Using a very detailed questionnaire, a sample of 700 firms operating in 15 industrial districts was interviewed.[2] The second section of the questionnaire, in particular, dealt with: *a)* ownership and management structure of firms; *b)* labour organisation and mobility. This section provided the basis for a parallel survey carried out on 200 firms belonging to non-district areas, chosen in the same regions where the industrial districts examined were located. The availability of this 'control sample' enabled us to highlight the differences in structure and behaviour between isolated small firms and firms operating within industrial districts.

Literature on the subject has underlined the peculiarities of the functioning of the labour market in industrial districts: as compared to other production systems, a higher degree of labour mobility and flexibility would constitute a fundamental element of competitive advantage.[3] From Marshall on, it has been emphasised that labour mobility among firms is made easier and more efficient – both for firms and workers – by the network of community relations inside industrial districts, that allows to share and reduce risks arising from a higher instability of work relationships.[4] Mobility among firms, in addition, is considered to be the privileged channel for the circulation of 'contextual' or 'tacit' knowledge (knowledge that is

* The authors wish to thank G. Solinas and R. Torrini for their helpful comments on previous versions of this paper. The views expressed in this article are those of the authors and do not involve the responsibility of the Bank of Italy.

[1] Signorini (2000), pp. 289-383, describes the survey in detail.

[2] For more information on the sample, see Iuzzolino (2000).

[3] For a recent survey on the subject, see Birindelli (1999); see also Casavola, Pellegrini and Romagnano (2000).

[4] See Becattini (1991), pages 56-57.

hard to be codified and transferred to others), which represents the technological basis of industrial districts.[5]

In the case of industrial districts, moreover, the concepts of labour market and labour mobility are to be intended in a wider sense, not limited to payroll jobs. A crucial aspect of labour mobility in industrial districts is supposed to be the frequent change from the condition of payroll employee to that of entrepreneur/self-employed. The prospective of a possible change, moreover, seems to act as a stimulus towards higher (*internal* and *external*) flexibility also as concerns payroll employees, at least for those workers whose careers – due to the increased possibility of accumulating human capital in industrial districts – can more easily end up into autonomous entrepreneurial initiatives (Solinas, 1982 and 1991; Brusco and Paba, 1997). It should also be added that in industrial districts the closer participation of employees to the destiny of the firm is supposed to act in the sense of favouring labour flexibility within firms. Finally, the social climate of the district would also favour a greater mobility of self-employed (or entrepreneurial) work, which is also supported by a higher rate of firm entry and exit. All these theoretical suggestions supported the decision to deal jointly with ownership and management structure issues, on one hand, and issues of labour organisation and mobility, on the other.

2 Ownership and management structure

When comparing firms from district and non-districts areas, we found a higher percentage of companies controlled by local entities in industrial districts, as compared to other areas: 92% of the firms in industrial districts resulted to be controlled by residents in the Local Labour System (LLS), as against 86% in non-district local systems (Table A.1). Conversely, 12% of non-district firms resulted to be controlled by non-local national entities and 2% by foreign entities, as compared to 7% and 1%, respectively, for district firms. On the whole, although the differences were not dramatic, the 'local' or 'community' character of industrial districts was confirmed.

The share of local controllers was greater in the districts of Central Italy (96%) than in the South (91.5%) and in the North (about 90%). Local controllers accounted for a considerably high percentage of firms in the size class below 50 employees (almost 95% of the firms) and diminished sensibly for those with a number of employees equal or superior to 50 (82.3% of the firms).

The widespread presence in Italy of business groups formed by small and medium size enterprises is a phenomenon that has aroused wide interest over the last few years, in relation to both ownership structures of Italian firms and the trans-

---

[5] See Becattini (1998), pages 83-88.

formations under way in industrial districts.[6] In our sample, district firms belonging to business groups accounted for 15.4% of the total, a homogeneous percentage between the North, the Center and the South of the country (Table A.2). The share of firms belonging to business groups increases progressively as firm size increases: it ranges from 6 per cent for firms with less than 10 employees to 10% for firms with 10 to 19 employees, 13% for firms with 20 to 49 employees, to finally reach 40% for firms with at least 50 employees. In non-district areas the diffusion of firms' groups seemed more limited (involving only 10 per cent of the firms under examination), because of their extremely reduced presence in the South and in middle size classes (between 10 e 49 employees). District firms belonging to groups showed a relatively lower frequency of local ownership as against the other district firms (77% as against 95%); however, also in this case the 'closeness' of districts was still much higher and clearly superior to that of non-district areas (which was slightly over 50%).

The management constituted a further element of difference between district and non-district firms. District firms were directly managed by the owners or by their relatives in 94% of the cases, against a percentage of about 90% in non-district firms (Table A.3). While the difference did not appear to be quantitatively very impressive, a closer identification between control and management of the firm (at individual or familiar level) seemed, however, to be confirmed.

Table 1. Previous activity carried out by persons in charge of the firm's management *(percentages)*

Area	*Education*	*Payroll employees*	*Self-employed*	*Manager in a different firm*	*Other activity*
Districts	19.3	36.9	23.5	16.4	3.9
Non-districts	20.1	37.7	25.2	12.3	4.7

Literature on industrial districts has identified higher social mobility and higher entrepreneurship rate as further traits of industrial districts. By looking at the managers' background, Table 1 shows that no strong differences emerged between district and non-district firms. The only relevant difference concerned the higher frequency, in districts, of firms' managers with an analogous experience in another firm. This depended on the fact that, within industrial districts, there is a higher horizontal mobility between one firm and the other both for managers and – as we will see below – for general employees. For 14% of district firms the managers had had a similar experience in other firms in the district; for non-

[6] For general reference, see Barca *et al.* (1994), Barbetta, *et al.* (1996), Cannari and Gola (1996), Sassatelli (1999), Carone and Iacobucci (1999); on the case of the *Marche* see Balloni and Iacobucci (1997) and on *Emilia Romagna* Bianchi, *et al.* (1999). On the importance of business groups in the recent evolution of industrial districts, see Brusco *et al.* (1997), Dei Ottati (1997) and Crestanello (1997).

district firms; on the contrary, the percentage is 4 percentage points lower (Table A.4).

Table 2. Managers' background and firm's ownership when the manager previously had a payroll job *(percentages)*

Manager	*Percentage shares*		*The manager is also owner of the firm*	
Background	Districts	Non Districts	Districts	Non Districts
Same firm	26.3	38.7	91.0	81.8
Other firm in the same LLS	64.5	49.9	94.8	79.1
Other firm in a different LLS	9.2	11.4	89.3	72.7

A more in-depth analysis, however, highlighted further and more marked differences between district and non-districts. In particular, work careers leading to a management position were significantly different in industrial districts as compared to what happened in other local systems. Data collected in Table 2 show that in industrial districts the typical career starts with a payroll job in a different firm in the district; after that, the worker decides to 'start up a new business'. In other local systems, on the contrary, different social mobility mechanisms can be observed: ownership of a firm is a less frequent conclusion; the decision to start new entrepreneurial initiatives is less frequent; careers within the same firm are more common.

The diversity in the mechanics of social mobility was also supported by other elements which emerged from the analysis (Table A.4). In industrial districts the smaller the size of the firm, the more likely it was that a firm's manager previously had a payroll job in other firms of the district; this relationship was not evidenced for non-district firms. In non-district systems, on the contrary, the likelihood that the managers of a firm had a previous job experience in the same firm depended on the size of the firm; also in this case the same relationship cannot be traced in district firms.

A further aspect that emerged during the survey concerned the starting up of new entrepreneurial initiatives by way of attributing capital shares to employees. Over the preceding five years, only 4% of the firms in industrial districts had financed the start up of new activities, a percentage substantially similar to that observed in other Local Labour Systems (Table A.5). In both samples, it should be noted that instances of start up of new entrepreneurial activities were more frequent in the mechanical engineering industry and in northern areas, whereas the phenomenon was much more rare in the South.

3 Labour organisation and mobility

Literature has underlined that one of the main reasons for the success of industrial districts is the high degree of productive flexibility, due to the specialisation of firms by stage of production, in particular through the widespread recourse to outsourcing and subcontracting. In industrial districts, as a consequence, flexibility is a feature of the production system as a whole (especially in the relations among its different components), rather than of the specific forms of labour employment within individual firms.

The results of the survey showed that in industrial districts utilisation of job relations different from the traditional full-time permanent job contract was not more widespread than in non-district firms (Table A.6); as concerned fixed term contracts, on the contrary, they were significantly less widespread in industrial districts (6% of the employed) as compared to other Labour Local Systems (12.5%). A different form of flexibility in job relations (related to employees' remuneration) consisted in employees' participation in the firms returns and also in this instance district firms made less use of this instrument (11.5% of the cases) as compared to non-district firms (17%; Table A.7).[7]

It cannot be excluded, however, that there may exist in industrial districts different forms of 'implicit contracts' providing for employees' participation in firms' revenues, which do not necessarily take the form of formal or informal agreements, but are nevertheless apt to provide incentives to the employees' commitment in the production process. The larger utilisation of over-time work in industrial districts could be one of these indirect mechanisms: in the case of temporary increases in the demand, over-time work was utilised in industrial districts by 60% of the firms in the sample, as against a percentage of 51% in non-district firms (Table A.8).[8] A similar function could be performed by employment conditions that would guarantee – at least to a 'central' group of workers – the preservation of the job. This could also provide an interpretation for the limited use of fixed term contracts in industrial districts, together with the flexibility offered by the possibility of non reconfirming such contracts in the presence of a slowdown in the demand (this possibility was used by 23% of district firms, as against 29% of the others).

The instruments of flexibility that typically characterise industrial districts consist of the variations in the recourse to home-based work and subcontracting for supplies. In the event of temporary increases in production, district firms intensified on average the recourse to home-based work and subcontracting in 41% of

[7] Both in districts and in non-district areas, forms of profit-sharing by employees are more frequently used by firms located in the North, bigger in size and belonging to the mechanical engineering industry.

[8] There was no evidence of differentiated attitudes in relation to the recourse to forms of time flexibility that do not affect the annual amount of work hours; such forms are more frequently used in conditions of demand reduction (49% of firms) rather than in the opposite case (35%).

the cases, as against 27% in other local systems (Table A.8). In the event of a reduction in production, district firms reduced orders to subcontractors or home-workers in 44% of the instances, as against 27% circa in the non-district areas under study (Table A.9).

As to the analysis of employment turnover, 70% of interviewed firms in industrial districts searched new personnel in 1997 (Table A.10). The total number of vacant job positions amounted to 19.6% of the average employment rate in the year. In non-district local systems, firms that have searched new personnel accounted for 67%, to cover an overall total of 711 positions, equal to – also in this instance – 19.6% of the average employment rate.

Specialised workers are the professional figures that firms operating in industrial district areas predominantly tend to look for, as compared to non-district areas (22 and 18%, respectively). In non-district local systems, 22.3% of the firms searched apprentices, as against 19.6% in industrial districts. The percentage of firms that declared to have searched non-specialised workers was slightly below 60% in both areas, whereas the hiring of home-workers was limited both in districts and in non-district areas (0.9 e 1.5%).

From Marshall on, it has been highlighted that one of the main competitive advantages of industrial districts is the opportunity to *pool* specialised labour resources, characterised by a high degree of mobility between firms; labour mobility is facilitated and made more efficient – both for firms and employees – by the network of community relations characterising the district. The results of the survey supplied clear empirical evidence of such features. If we look at the geographical background of workers hired in 1997, the degree of 'self-containment' of industrial districts appeared to be much higher than that of other local labour systems, especially as concerned high-skilled workers (Table A.11). In industrial districts 80% of specialised workers hired in 1997 was from the same local labour system, as against a percentage of 50% in non-district firms. The difference was significant also in the case of low-skilled workers, where the percentage of local newly hired workers was over 85% in districts, as against 71% for firms in other Local Labour Systems. Only as concerned apprentices the two percentage rates seemed to be fundamentally similar.

Similar evidence emerged from the examination of the percentage of new employees coming from other firms operating in the same sector (Table A.12). For high-skilled workers the percentage rate amounted to 56% for industrial districts, as against 45% for other firms; in the case of low-skilled workers the percentage is 44% for district firms, as against 25% in the other firms. In industrial districts, moreover, mobility among firms of the same sector also concerned apprentices. This is a further evidence of the fact that in industrial districts workers' careers develop through the passage from one firm to another, starting at the very beginning of work experience. In the case of apprentices the percentage of employees with previous experience in firms of the same sectors was slightly below 15% in district firms, whereas it was close to nil in the other firms.

Finally, there is evidence of a wider recourse to informal channels (relatives, friends, acquaintances) for the recruitment of workers in industrial districts, due to the important role played by community networks (Table A.13).[9] As concerned search for personnel carried out in 1997, informal channels were activated by district firms in 73% of cases in the search for apprentices (as against 50% by non-district firms), 64% in the search for non specialised workers (as against 58% of other firms), in 68% of cases in the search for specialised workers (as against 37% by non-district firms).

4 Conclusions

The field survey highlighted some peculiar traits of industrial districts:

- although to a limited extent, ownership and management structure of district firms is marked by a higher degree of localism and stronger identification between ownership and management; in industrial districts, moreover, there is some evidence of a more widespread presence of business groups;
- differences in workers' careers are more marked: in districts such careers are more likely to produce autonomous entrepreneurial initiatives, whereas in the other local systems careers within the same firm are more common;
- in industrial districts the flexibility in the use of the workforce is the result of the interaction between the different components of the system (in particular through the instrument of subcontracting for supplies) rather than the recourse – within individual firms – to flexible contractual forms (which are more often adopted, on the contrary, in non-district firms);
- as concerns recruitment of workers (in particular specialised workers), industrial districts show a high degree of self-containment and a high degree of internal mobility, which is evident even in the early stages of apprenticeship;
- finally, recruitment of workers in industrial districts shows a clear prevalence of informal channels; this is further evidence of the role played by community relationships in reducing information and search costs.

---

[9] Becattini wrote (1990, p. 42): "An evaluation of a worker's personal and professional qualities, for his most fruitful employment, is much easier for a firm deeply immersed in the community's relations than for an isolated firm or for the firm lost in urban anonymity. The information on a prospective employee available to a firm in the district at a very low cost and with hardly any effort, is much more 'revealing' than that which can be gathered elsewhere by means of formal structures. It is reasonable to assume that even the large firm, in spite of all its sophisticated information structures, psychological tests, etc., may be, in some cases, at a disadvantage with respect to the small firm which operates in the district".

References

BALLONI V., D. IACOBUCCI (1997), "Cambiamenti in atto nell'organizzazione dell'industria marchigiana", *Economia Marche*, vol. 16, n. 1, pp.29-66.

BARBETTA, G. P., C. PIGA, M. VIVARELLI (1996), *Il fenomeno dei gruppi di imprese in Italia*, Mediocredito Centrale, Osservatorio sulle piccole e medie imprese, Quaderni di politica industriale.

BARCA, F., M. BIANCO, L. CANNARI, R. CESARI, C. GOLA, G. MANITTA, G. SALVO, L.F. SIGNORINI (1994), *Assetti proprietari e mercato delle imprese,* vol. I: *Proprietà, modelli di controllo e riallocazione nelle imprese industriali italiane*, Il Mulino, Bologna.

BECATTINI, G. (1990), "The Marshallian Industrial District as a Socio-Economic Notion", in: F. PYKE, G. BECATTINI, W. SENGERBERGER (Eds.), *Industrial Districts and InterFfirm Co-Operation in Italy*, International Labor Office, Geneva.

BECATTINI, G. (1998), *Distretti industriali e made in Italy*, Bollati Boringhieri, Torino.

BIANCHI, R., M. S. BRIOSCHI, G. CAINELLI (1999), "Struttura e localizzazione dei gruppi di imprese in Emilia Romagna", in: IDSE-CNR, *Trasformazioni strutturali e competitività dei sistemi locali di produzione. Rapporto sul cambiamento strutturale dell'economia italiana*, Franco Angeli, Milano.

BIRINDELLI, L. (a cura di) (1999), *Mobilità del lavoro nei distretti: elementi teorici, metodologie ed evidenze fattuali*, Mediocredito Centrale, Osservatorio sulle piccole e medie imprese, "Quaderni di politica industriale", n. 29.

BRUSCO, S., G. CAINELLI, F. FORNI, M. FRANCHI, A. MALUSARDI, R. RIGHETTI (1997), "L'evoluzione dei sistemi produttivi locali in Emilia-Romagna", in: COSSENTINO, F. PYKE E W. SENGERBERGER (a cura di), *Le risposte locali e regionali alla pressione globale: il caso dell'Italia e dei suoi distretti industriali*, Il Mulino, Bologna.

BRUSCO, S., PABA S. (1997), "Per una storia dei distretti industriali italiani dal secondo dopoguerra agli anni novanta", in: BARCA F. (a cura di), *Storia del capitalismo italiano dal dopoguerra a oggi,* Donzelli, Roma.

CANNARI, L. e C. GOLA (1996), *La diffusione dei gruppi industriali in Italia*, in AA.VV., *I gruppi di società*, Giuffrè, Milano.

CARONE, A., IACOBUCCI D. (1999), "I gruppi di piccole e medie imprese nell'industria italiana", in: TRAÙ F. (a cura di), *La "questione dimensionale" nell'industria italiana*, Il Mulino, Bologna.

CASAVOLA, P., PELLEGRINI G., ROMAGNANO E. (2000), "Imprese e mercato del lavoro nei distretti industriali italiani", in: SIGNORINI L.F. (a cura di), *Lo sviluppo locale. Un'indagine della Banca d'Italia sui distretti industriali*, Meridiana Libri, Roma.

CRESTANELLO, P. (1997), "I distretti industriali in Veneto: cambiamenti e tendenze", in: COSSENTINO F., PYKE F., SENGERBERGER W. (a cura di), *Le risposte locali e regionali alla pressione globale: il caso dell'Italia e dei suoi distretti industriali*, Il Mulino, Bologna.

DEI OTTATI, G. (1997), "Cambiamento e vitalità dei distretti industriali della Toscana", in: COSSENTINO F., PYKE F., SENGERBERGER W.(a cura di), *Le risposte locali e regionali alla pressione globale: il caso dell'Italia e dei suoi distretti industriali*, Il Mulino, Bologna.

SASSATELLI, M. (a cura di) (1999), *Solidità organizzativa e posizionamento competitivo dei gruppi industriali di piccole e medie imprese*, Mediocredito Centrale, Osservatorio sulle piccole e medie imprese, "Quaderni di politica industriale", n. 31.

SIGNORINI, F. L., (a cura di) (2000), *Lo sviluppo locale. Un'indagine della Banca d'Italia sui distretti industriali*, Meridiana Libri, Roma.

SOLINAS, G. (1982), "Labour Market Segmentation and Worker's Careers: the Case of Italian Knitwear Industry" , *Cambridge Journal of Economics*, n. 6, pp.

SOLINAS, G. (1991), "L'anzianità aziendale. Alcune considerazioni sul caso emiliano", *Politica Economica,* n. 3.

Appendix

Table A.1. Firms' ownership structure (%)

Items	1991			1997		
	Local	National	Foreign	Local	National	Foreign
	(a) District firms					
Total	91.9	7.2	0.9	92.1	6.6	1.3
Geographical areas						
North	88.8	9.6	1.6	89.6	7.6	2.8
Center	96.4	3.1	0.5	96.0	3.5	0.5
South	91.6	8.0	0.4	91.5	8.1	0.4
No of employees						
1-9 employees	94.4	4.2	1.4	94.5	4.1	1.4
10-19 employees	94.5	5.0	0.5	93.6	5.3	1.1
20-49 employees	94.0	6.0	0.0	94.6	5.4	0.0
50 employees and over	81.5	16.1	2.4	82.3	13.7	4.0
Sectors						
Textiles and clothing	86.4	12.4	1.3	86.0	12.7	1.3
Leather and footwear	98.2	1.1	0.6	98.3	1.1	0.6
Furniture	93.9	5.0	1.1	93.0	5.4	1.6
Mechanical engineering	90.8	9.2	0.0	93.4	4.7	1.9
	(b) Non-district firms					
Total	85.0	13.4	1.5	85.7	12.4	2.0
Geographical areas						
North	88.0	9.8	2.2	88.1	9.7	2.2
Center	78.8	19.7	1.5	79.2	18.1	2.7
South	88.9	11.1	0.0	92.1	7.9	0.0
No of employees						
1-9 employees	92.1	7.9	0.0	92.3	7.7	0.0
10-19 employees	92.0	8.0	0.0	90.6	9.4	0.0
20-49 employees	87.5	10.9	1.5	88.3	10.3	1.5
50 employees and over	66.7	28.6	4.8	69.8	23.3	7.0
Sectors						
Textiles and clothing	76.0	22.0	2.0	77.4	18.9	3.7
Leather and footwear	86.9	13.1	0.0	88.1	11.9	0.0
Furniture	94.1	5.9	0.0	94.2	5.8	0.0
Mechanical engineering	82.6	13.0	4.4	84.8	10.9	4.4

Table A.2. Groups of firms: diffusion and ownership structure (%)

Items	*% of firms*	*Ownership structure*		
	Belonging to groups	Local	National	Foreign
		(a) District firms		
Total	15.4	76.8	20.4	2.8
Geographical areas				
North	15.7	66.7	25.6	7.7
Center	15.0	87.1	12.9	0.0
South	15.4	78.9	21.1	0.0
No of employees				
1-9 employees	6.1	77.8	22.2	0.0
10-19 employees	9.6	73.7	21.0	5.3
20-49 employees	13.2	90.3	9.7	0.0
50 employees and over	39.5	69.4	26.5	4.1
Sectors				
Textiles and clothing	13.8	54.8	45.2	0.0
Leather and footwear	14.8	100.0	0.0	0.0
Furniture	17.2	78.8	18.2	3.0
Mechanical engineering	16.8	77.8	11.1	11.1
		(b) Non-district firms		
Total	10.0	52.4	42.9	4.8
Geographical areas				
North	10.9	70.0	30.0	0.0
Center	13.0	40.0	50.0	10.0
South	2.6	0.0	100.0	0.0
No of employees				
1-9 employees	7.3	66.7	33.3	0.0
10-19 employees	3.8	50.0	50.0	0.0
20-49 employees	3.0	50.0	50.0	0.0
50 employees and over	32.5	50.0	42.9	7.1
Sectors				
Textiles and clothing	7.7	0.0	75.0	25.0
Leather and footwear	14.3	57.2	42.8	0.0
Furniture	0.0	0.0	0.0	0.0
Mechanical engineering	11.9	60.0	40.0	0.0

Table A.3. Management structure of firms (%)

Items	*Firms managed by the owners or by their relatives*	*Firms managed by persons not linked to the owners by family ties*	*Firms managed by the owners and by persons not linked to the owners by family ties*
		(a) District firms	
Total	94.0	5.1	0.9
Geographical areas			
North	92.2	7.0	0.8
Center	93.5	4.5	2.0
South	96.3	3.7	0.0
No of employees			
1-9 employees	96.6	2.7	0.7
10-19 employees	95.7	3.2	1.1
20-49 employees	96.6	3.4	0.0
50 employees and over	83.7	13.9	2.4
Sectors			
Textiles and clothing	92.5	7.1	0.4
Leather and footwear	97.7	2.3	0.0
Furniture	93.4	3.8	2.7
Mechanical engineering	92.5	7.5	0.0
		(b) Non-district firms	
Total	89.8	9.7	0.5
Geographical areas			
North	92.1	7.9	0.0
Center	82.1	16.4	1.5
South	97.4	2.6	0.0
No of employees			
1-9 employees	95.0	5.0	0.0
10-19 employees	94.1	5.9	0.0
20-49 employees	87.7	10.8	1.5
50 employees and over	82.1	17.9	0.0
Sectors			
Textiles and clothing	78.4	19.6	2.0
Leather and footwear	92.7	7.3	0.0
Furniture	97.2	2.8	0.0
Mechanical engineering	92.9	7.1	0.0

Table A.4. Activity previously performed by the firm's managers *(%)*

Activity	*No of employees* 1-9	10-19	20-49	50-over	*Total*
	(a) District firms				
Education	14.7	18.8	20.5	23.6	19.3
Payroll job at the same firm	5.9	12.1	10.1	9.6	9.7
Payroll job at another firm of the LLS	33.0	24.7	22.2	14.0	23.8
Payroll job at a firm outside the LLS	3.5	2.7	3.9	3.7	3.4
Self employment connected to the firm activity	17.6	16.6	18.6	16.9	17.6
Self employment not connected to the firm activity	4.7	6.2	6.8	5.1	5.9
Management of a different firm of the LLS	16.4	10.8	13.6	14.7	13.6
Management of a different firm outside the LLS	0.6	2.7	1.4	8.9	2.8
Other	3.5	5.4	2.9	3.7	3.9
	(b) Non-district firms				
Education	23.1	17.9	21.4	16.7	20.1
Payroll job at the same firm	7.7	14.3	13.1	26.2	14.6
Payroll job at another firm of the LLS	15.4	16.0	22.6	19.1	18.8
Payroll job at a firm outside the SLL	3.9	3.6	6.0	2.4	4.3
Self employment connected to the firm activity	21.1	25.0	15.5	16.7	19.2
Self employment not connected to the firm activity	3.9	12.5	3.6	4.7	6.0
Management of a different firm of the LLS	13.5	5.4	13.1	4.7	9.8
Management of a different firm outside the LLS	3.9	1.8	1.2	4.7	2.5
Other	7.7	3.6	3.6	4.7	4.7

Table A.5. Start up of new entrepreneurial activities with attribution of capital shares to employees *(%)*

Items	*No*	*Yes*
	(a) District firms	
Total	95.8	4.2
Geographical areas		
North	94.3	5.7
Center	96.5	3.5
South	96.8	3.2
No of employees		
1-9 employees	97.3	2.7
10-19 employees	96.8	3.2
20-49 employees	97.9	2.1
50 employees and over	88.6	11.4
Sectors		
Textiles and clothing	98.2	1.8
Leather and footwear	96.6	3.4
Furniture	94.6	5.4
Mechanical engineering	91.6	8.4
	(b) Non-district firms	
Total	95.0	5.0
Geographical areas		
North	91.3	8.7
Center	97.1	2.9
South	100.0	0.0
No of employees		
1-9 employees	92.7	7.3
10-19 employees	92.3	7.7
20-49 employees	97.1	2.9
50 employees and over	97.5	2.5
Sectors		
Textiles and clothing	98.1	1.9
Leather and footwear	100.0	0.0
Furniture	97.3	2.7
Mechanical engineering	88.1	11.9

Table A.6. Average employment and atypical work contracts (*units and percentages*)

Items	*Average employment in 1997*	*Fixed-term workers*	*Part-time workers*	*Non EC immigrant workers*	*Home-workers*	*Appren-tices*
			(a) District firms			
Total	37	5.9	3.0	2.6	1.0	3.6
Geographical areas						
North	39	5.8	5.2	3.7	0.7	2.7
Center	31	6.6	2.4	4.2	2.1	3.3
South	41	5.4	1.3	0.2	0.4	4.8
No of employees						
1-9 employees	5	4.4	4.2	2.4	1.0	2.6
10-19 employees	14	5.4	3.4	2.3	0.6	4.8
20-49 employees	30	6.0	2.4	2.6	1.0	3.6
50 employees and over	125	8.2	2.4	3.0	1.6	3.1
Sectors						
Textiles and clothing	36	4.0	3.3	1.3	1.1	3.8
Leather and footwear	36	7.1	1.3	2.5	1.6	2.4
Furniture	38	8.2	3.0	3.4	0.8	4.9
Mechanical engineering	40	4.0	5.2	3.8	0.3	2.9
			(b) Non-district firms			
Total	43	12.5	3.9	1.2	2.5	3.3
Geographical areas						
North	40	8.8	6.1	2.2	2.9	3.0
Center	54	16.1	1.5	0.2	2.0	3.3
South	30	14.4	2.4	0.4	2.3	4.2
No of employees						
1-9 employees	6	12.8	4.2	1.2	0.0	0.9
10-19 employees	13	7.3	3.2	0.8	3.2	6.1
20-49 employees	31	14.2	1.0	1.5	0.8	3.2
50 employees and over	140	15.6	9.3	1.0	8.0	2.1
Sectors						
Textiles and clothing	48	13.8	2.9	0.1	2.3	2.8
Leather and footwear	50	17.4	2.2	0.6	4.2	5.4
Furniture	25	10.5	2.1	1.1	0.1	1.7
Mechanical engineering	47	6.2	4.0	2.9	0.8	2.5

Table A.7. Incentive and profit-sharing forms *(%)*

Items	*No*	*Yes. Established in contractual clauses*	*Yes. Through informal agreements*
		(a) District firms	
Total	88.5	5.7	5.8
Geographical areas			
North	81.0	9.1	9.9
Center	89.9	6.6	3.5
South	95.2	1.3	3.5
No of employees			
1-9 employees	97.2	0.7	2.1
10-19 employees	93.4	2.7	3.8
20-49 employees	88.5	4.4	7.1
50 employees and over	70.8	18.3	10.8
Sectors			
Textiles and clothing	92.3	3.6	4.1
Leather and footwear	90.8	6.1	3.1
Furniture	85.6	6.6	7.7
Mechanical engineering	82.1	7.5	10.4
		(b) Non-district firms	
Total	82.9	7.0	10.1
Geographical areas			
North	76.9	8.8	14.3
Center	84.3	5.7	10.0
South	94.7	5.3	0.0
No of employees			
1-9 employees	95.0	2.5	2.5
10-19 employees	80.8	3.8	15.4
20-49 employees	85.3	7.4	7.4
50 employees and over	69.2	15.4	15.4
Sectors			
Textiles and clothing	86.8	3.8	9.4
Leather and footwear	83.3	9.5	7.1
Furniture	91.7	5.6	2.8
Mechanical engineering	65.9	12.2	22.0

Table A.8. Recourse to flexibility forms of in case of temporary production increases *(%)*

Items	*Work-time flexibility*	*Overtime work*	*Part-time workers*	*Establishment of week-days new turns*	*Establishment of holiday or night turns*	*Greater recourse to sub-contractors or home-workers*
			(a) District firms			
Total	34.8	59.5	30.8	8.1	7.8	40.9
Geographical areas						
North	26.6	78.3	30.3	7.0	9.4	62.3
Center	47.2	60.3	30.2	6.5	4.0	29.6
South	32.6	39.7	31.8	10.5	9.2	28.5
No of employees						
1-9 employees	30.6	43.1	11.1	2.1	5.6	45.1
10-19 employees	31.7	63.4	26.8	5.5	2.7	39.9
20-49 employees	35.3	58.6	39.2	9.5	9.5	38.8
50 employees and over	43.1	74.8	43.9	16.3	14.6	41.5
Sectors						
Textiles and clothing	39.8	44.3	24.4	7.2	6.8	43.0
Leather and footwear	33.5	52.4	41.8	10.6	8.8	27.1
Furniture	40.8	70.1	22.3	6.0	4.9	38.0
Mechanical engineering	15.9	84.1	41.1	9.3	13.1	63.6
			(b) Non-district firms			
Total	36.0	50.8	30.5	8.6	6.6	27.4
Geographical areas						
North	23.9	62.5	30.7	10.2	9.1	25.0
Center	48.6	40.0	35.7	5.7	5.7	32.9
South	41.0	43.6	20.5	10.3	2.6	23.1
No of employees						
1-9 employees	30.8	25.6	20.5	0.0	7.7	30.8
10-19 employees	42.3	44.2	19.2	9.6	0.0	28.8
20-49 employees	38.2	61.8	29.4	5.9	2.9	26.5
50 employees and over	28.9	65.8	57.9	21.1	21.1	23.7
Sectors						
Textiles and clothing	41.5	37.7	34.0	7.5	3.8	28.3
Leather and footwear	47.6	40.5	26.2	11.9	7.1	40.5
Furniture	35.3	50.0	32.4	2.9	5.9	26.5
Mechanical engineering	17.1	75.6	24.4	9.8	0.0	31.7

Table A.9. Recourse to flexibility forms of in case of temporary production decline (%)

Items	*Work-time flexibility*	*Wage Supplementation Fund*	*Reduction of fixed-term employment*	*Firing*	*Local social shock absorbers*	*Lesser recourse to subcontractors or homeworkers*
	(a) District firms					
Total	48.7	26.8	22.5	4.2	3.9	43.6
Geographical areas						
North	38.8	26.0	27.7	1.7	4.5	66.9
Center	61.9	30.9	21.6	3.1	4.6	32.5
South	48.0	24.2	17.6	7.9	2.6	28.2
No of employees						
1-9 employees	46.0	11.5	7.9	5.0	2.2	46.8
10-19 employees	46.9	26.0	16.9	4.0	2.8	40.7
20-49 employees	51.3	32.0	26.5	5.3	4.0	41.2
50 employees and over	49.6	33.0	39.7	1.7	7.4	48.8
Sectors						
Textiles and clothing	47.7	20.6	17.8	4.2	1.4	44.4
Leather and footwear	53.9	41.8	26.7	6.1	6.1	26.1
Furniture	58.7	20.7	17.3	3.9	5.0	42.5
Mechanical engineering	25.7	26.7	34.3	1.9	3.8	71.4
	(b) Non-district firms					
Total	50.5	25.3	28.9	4.7	5.8	26.8
Geographical areas						
North	47.7	18.6	30.2	3.5	4.7	26.7
Center	55.2	32.8	32.8	4.5	9.0	28.4
South	48.6	27.0	18.9	8.1	2.7	24.3
No of employees						
1-9 employees	56.8	24.3	13.5	5.4	2.7	27.0
10-19 employees	59.6	14.9	23.4	4.3	4.3	31.9
20-49 employees	46.3	28.4	32.8	7.5	9.0	28.4
50 employees and over	41.0	33.3	43.6	0.0	5.1	17.9
Sectors						
Textiles and clothing	48.0	30.0	30.0	6.0	4.0	28.0
Leather and footwear	50.0	47.5	20.0	2.5	12.5	32.5
Furniture	47.1	14.7	35.3	11.8	5.9	23.5
Mechanical engineering	46.2	12.8	25.6	2.6	2.6	38.5

Table A.10. Personnel recruitment in 1997 (%)

Items	*% of firms that searched for personnel*					*Positions to be covered in % of average empl.*
	Overall	*Home-workers*	*Appren-tices*	*Low-skilled workers*	*High-skilled workers*	
			(a) District firms			
Total	70.1	0.9	19.6	59.1	21.7	19.6
Geographical areas						
North	74.5	0.4	20.2	63.8	30.8	18.3
Center	74.2	1.6	19.6	66.0	16.5	20.8
South	62.6	0.8	19.0	49.0	16.9	20.0
No of employees						
1-9 employees	44.0	0.7	5.0	35.5	7.8	34.6
10-19 employees	68.1	0.0	21.3	52.2	15.9	21.0
20-49 employees	77.2	1.3	26.3	68.2	22.0	17.2
50 employees and over	90.8	1.7	21.0	80.7	47.1	13.3
Sectors						
Textiles and clothing	57.7	1.3	11.3	46.4	15.8	18.0
Leather and footwear	78.7	1.8	22.5	67.4	16.0	20.2
Furniture	73.9	0.0	23.9	63.6	28.8	20.1
Mechanical engineering	76.2	0.0	24.8	64.8	31.5	20.2
			(b) Non-district firms			
Total	67.0	1.5	22.3	56.8	17.8	19.6
Geographical areas						
North	76.7	1.1	26.7	60.1	25.5	18.7
Center	57.4	1.5	14.7	51.5	13.3	20.1
South	61.5	2.6	25.6	58.9	7.7	21.5
No of employees						
1-9 employees	43.9	0.0	9.7	34.2	7.3	53.6
10-19 employees	67.3	0.0	28.9	48.1	11.5	18.8
20-49 employees	66.7	1.5	24.3	63.7	9.1	13.6
50 employees and over	92.1	5.2	23.7	81.6	52.6	10.6
Sectors						
Textiles and clothing	53.8	0.0	19.2	51.9	9.6	19.1
Leather and footwear	70.7	7.3	29.3	58.5	9.8	13.4
Furniture	62.2	0.0	16.2	54.1	21.6	29.7
Mechanical engineering	82.9	0.0	19.5	73.1	26.9	18.2

Table A.11. Geographical background of recently hired employees (*percentages*)

Channels	*Apprentices*	*Low-skilled workers*	*High-skilled workers*
		(a) District firms	
Local Labour System	85.1	85.2	80.3
Region	10.4	15.1	24.3
Other Regions in the Center-North	0.0	2.7	6.6
Other Regions in the South	6.0	9.9	5.9
European Union	0.7	0.2	0.7
Rest of the World	4.5	11.9	2.6
		(b) Non-district firms	
Local Labour System	84.1	71.3	50.0
Region	18.2	30.4	36.8
Other Regions in the Center-North	0.0	7.0	23.7
Other Regions in the South	2.3	3.5	5.3
European Union	0.0	0.0	5.3
Rest of the World	0.0	5.2	0.0

Table A.12. Previous activity of recently hired employees (%)

Channels	*Apprentices*	*Low-skilled Workers*	*High-skilled workers*
		(a) District firms	
School or professional training	61.9	19.8	16.4
Firms in the same sector	14.9	44.4	55.9
Other firms	13.4	43.7	34.9
Other activities	17.9	11.4	5.9
		(b) Non-district firms	
School or professional training	61.4	27.8	26.3
Firms in the same sector	2.3	25.2	44.7
Other firms	9.1	41.7	42.1
Other activities	29.5	27.8	10.5

Table A.13. Channels used for the recruitment of workers (%)

Channels	*Apprentices*	*Low-skilled workers*	*High-skilled workers*
		(a) District firms	
Public employment agencies	13.4	25.4	12.5
Private associations	3.7	5.9	8.6
Private employment agencies	2.2	5.2	16.4
Schools	5.2	8.9	15.1
Relatives, friends or acquaintances cono-	73.1	63.7	68.4
Other	19.4	26.2	23.7
		(b) Non-district firms	
Public employment agencies	25.0	25.2	15.8
Private associations	2.3	5.2	13.2
Private employment agencies	0.0	3.5	15.8
Schools	13.6	13.9	23.7
Relatives, friends or acquaintances cono-	50.0	58.3	36.8
Other	22.7	14.8	28.9

Integration and convergence in the internal dynamics of industrial districts. A case study[•]

Mario Nosvelli

CERIS-DSE, National Research Council, Via Bassini 15, 20133, Milan, Italy

1 Introduction

Italian industrial districts have captured the attention of many economists as a specific model of local development[1]. Analyses that focus on studying specific industrial districts contribute, on the one hand, towards explaining the ongoing success of 'made in Italy' abroad (Maggioni and Nosvelli, 2000; Becattini, 1998) and, on the other hand, towards justifying the particular structure of production centred around small and medium-sized firms (Becattini, 1997). This work aims to contribute to the analysis of the interactions among different sectors that constitute the production structure of the district, and in particular what happens when a district includes a sub-district as defined by the criteria for industrial districts definition established by ISTAT in Italy.

First of all, I will verify the possibility that an area characterised by a specialised production structure, may also contain areas in which the structure of production reveals high levels of specialisation in sectors that differ from the dominant one. This would signify the presence of sub-districts, and, as a consequence, sustain the concept of a district as a composite economic reality in which the typology of sector specialisation does not hinder the full development of other sectors. The presence within the district of Asse Sempione in Lombardy, specialised in the textile and clothing sector, of an area around the municipality of Parabiago traditionally specialised also in footwear has stimulated this analysis which, in fact, deals with the characteristics of industrial districts in this geographical area.

The study of the above-mentioned interactions between sectors and geographical areas requires appropriate statistical and economic tools. Besides those indices that make it possible to compare the levels of specialisation, shift share analysis allows us to produce a breakdown of the components in the growth of local systems. These components make it possible to identify the contribution made by the sectoral structure of the economy to the evolution of the areas studied.

[•] A preliminary version of this work, written in collaboration with M. Castelli, was published in Italian in *Piccola Impresa – Small Business,* n. 1 – 2002.

[1] On the basis of ISTAT data for 1991, 199 industrial districts have been individuated (ISTAT,1996). For a theory of the development of local economic systems see Bramanti and Maggioni (1997).

The work is organised as follows. In section 2, I illustrate the peculiarities of the Parabiago area, namely the existence of an industrial specialisation in the production of footwear within the textile district of Asse Sempione. I show that Parabiago meets all the criteria, defined by theory and codified in national and regional laws, set as requirements to be considered an industrial district. In section 3, I examine, by using *location quotient* and the shift share analysis, the evolution of the sub-district by comparing it with the Lombardy region and the district of Asse Sempione. In the last section, I offer some considerations on the complex relationship existing between district and sub-district over the last few years then suggesting some possible industrial policy interventions to improve the competitiveness of both.

2 The area of Parabiago within the textile district of Asse Sempione

After the Italian national Law 317 of 1991 on SMEs and industrial districts, the Regional Government of Lombardy laid down the criteria and parameters to identify industrial districts in 1994. The application of these criteria and parameters have made it possible to identify the presence of 21 industrial districts spread over nearly all the provinces[2] and in various manufacturing sectors.[3]

The industrial district of Asse Sempione which is situated in the North West of Lombardy. It is made up of 50 municipalities, 27 of which are located in the province of Varese and 23 in the province of Milan. The district covers a surface area of 505.4 square kilometers and has a population of 556,518 inhabitants (ISTAT, 1998) with an average density of 1,101 inhabitants per square kilometre.

This area qualifies as a district due to its specialisation in textile and clothing sectors whose the number of employees was about 32,000[4] in 1998. This figure is equal to almost 18% of employees in the two above-mentioned sectors for the whole of Lombardy. This points out the importance of the district at a regional and also at a national level.

Within the district, a smaller area is to be found which can also be clearly distinguished from the rest of the district due to significant internal homogeneity. In fact, the municipalities of Parabiago, San Vittore Olona, and Cerro Maggiore appear to be a district area specialising in footwear and with recognised, quality production on the national market.

Historically, in fact, alongside agriculture and stock farming, this area was characterised by a significant presence of textile companies[5] and fustian factories

[2] The only province that do not have industrial districts is Sondrio. For an analysis of the territorial homogeneity within the districts in Lombardy see Nosvelli (1999).

[3] Fortis (1999).

[4] Unioncamere Lombardy (1998).

[5] Pagetti (1996).

in particular, with consolidated knowledge base in these productions[6]. Manufacturing grew steadily until the end of the 19th century when it reached its peak. The following years saw fluctuations in the textile sector: years of rapid growth alternated with periods of sharp decline. It was also due to these periods of decline that the conditions were created for the emergence of the footwear sector in the area around Parabiago. In fact, many local entrepreneurs believed it necessary to diversify their investments by moving, at least in part, to the footwear sector. This was made possible by a series of advantages linked to the new business and which contributed to its diffusion: (i) limited amount of investment needed to start up; (ii) proximity of dyeing factories; (iii) availability of raw materials; (iv) availability of labour; (v) growing demand.

The first report on a craftsman's workshop for the production of shoes in Parabiago dates back to Attilio Fumagalli in 1899[7]. Since then the footwear sector has grown both in Parabiago and in the neighbouring municipalities of San Vittore Olona and Cerro Maggiore. These municipalities have become an important production area in the provinces of Milan and Varese thanks to the consolidated tradition in this sector.

Although the local footwear companies have had negative results[8] and has seen its relative importance reduced within the economic reality, it continues to be the dominant business in the area. In fact, in the above-mentioned municipalities, the work force employed in the production of shoes amounts to 1,685 which corresponds to 7.3% of the entire footwear sector in Lombardy (Unioncamere Lombardy, 1998)[9]. From a demographic point of view the population of Parabiago, San Vittore Olona and Cerro Maggiore is over 45,000, corresponding to about 8% of the Asse Sempione district. These three municipalities, therefore, represent a significant share of the whole district's economy, even though the latter covers a vast area comprising, as we have already mentioned, 50 municipalities[10].

Footwear, the sector in which the sub-district of Parabiago is specialised, has seen a decline in the number of employees accompanied by a slight increase in the number of local units from 151 in 1992 to 165 in 1998 (Unioncamere Lombardy, 1992, 1998). On the one hand there is a clear reduction in the average size of the companies, due to the increase in the automation of production processes. On the other hand the growing fragmentation of the production system reveals a fundamental weakness of the companies in this sector. In fact, they are not able to improve their competitiveness since they do not have the size required for innovating and competing on the national and international markets.

---

[6] Vianello (1934)

[7] Ceriani (1948)

[8] It needs to be mentioned that foot wear is going through a negative period at a national level due, above all to strong competition from some foreign countries (ANCI, 2000).

[9] In Lombardy, besides Parabiago, there are other footwear districts. The one that is recognised by laws on industrial districts, and which is also the most prestigious, is that of Vigevano, which specialises in the production of both shoes and shoe making machines

[10] The average density of the three municipalities is 1,621 inhabitants per square kilometer.

In order to give a complete picture of the evolution of the average structure of companies, Table 1 (based on Unioncamere data referring to 1998) shows that there was only one firm with more than 100 employees while there are 85 with less than 5 employees. It also clearly shows that the majority of workers are concentrated in Parabiago, the municipality around which the footwear sector under investigation revolves.

Table 1. Footwear sector establishments and employees in the municipalities of the district and sub-district

No. employees	*Cerro Maggiore*		*Parabiago*		*San Vittore Olona*		*Total Sub-District*	
	Est.	Empl.	Est.	Empl.	Est.	Empl.	Est.	Empl.
1-2	10	12	39	51	8	12	57	75
3-5	4	18	18	67	6	23	28	108
6-9	2	16	21	154	2	12	25	182
10-19	6	88	21	289	3	32	30	409
20-49	0	0	18	525	2	60	20	585
50-99	0	0	2	116	0	0	2	116
100-199	0	0	1	157	0	0	1	157
200 and over	0	0	0	0	0	0	0	0
Total	22	134	120	1,359	21	139	163	1,632

Source: Unioncamere (1998)

'Production of footwear', corresponding to the group 19.30 of the ATECO classification is the dominant activity in this sector, while the group 19.10 ('preparation and tanning of leather') is less widespread (Unioncamere, 1994). One may reasonably suppose that, in the three municipalities forming the footwear sub-district, companies are supplied with their raw material (in particular leather) that has already been worked and they, therefore, mainly carry out the process of assembling of the final product's various components.

In conclusion, the firms of footwear area of Parabiago distinguishes themselves by being highly fragmented, concentrating in processes of product assembly, often working as sub-contractors.

2.1 On the existence of the footwear district of Parabiago

The main way to identify an industrial district is the application of criteria stated by national legislation (Law No. 317, 1991). However, it is necessary to point out that in economic literature there is more than one criterion for the identification of industrial districts since empirical analyses have been adapted to theoretical arguments that have, at different times, been considered relevant. For this reason,

Lombardy has recently passed a regional law[11] which partly modifies the criteria for defining industrial districts. However, this paper makes reference to article 36 – law 317 of 1991 as implemented by the Ministerial decree of April 21st 1993 and adopted by the regional government of Lombardy in 1994. The reason for this choice is that the new law of 2001 is less restrictive than the former one and, in a certain way, less aimed at the identification of districts. In fact, the new regional law of 2001 makes no reference to the number of establishments in the manufacturing sector which is, however, significant for an evaluation of the importance of the production structure in an area.

In one of his works (see ISTAT – IRPET, 1986), Sforzi defined an algorithm with which it would be possible to identify 'the local labour market' and 'local systems of a district type nature'. Notwithstanding the reliability of that methodology, in this paper the procedure used for the identification of districts differentiates itself from that by Sforzi in two respects. The first concerns the fact that in this case data is referred to municipalities and not to 'local labour systems'. The latter are based essentially on worker commuters which is not, however, considered in the methodology of the reference law on industrial districts. The second concerns the threshold value for size below which companies are considered as small and medium sized companies. In this paper the threshold figure of 200 employees has been chosen, rather than 250 employees, as it seems to be a more effective representation of the districts of small and very small companies in Lombardy[12].

The criteria utilised here, that have been taken from the above-mentioned Lombard legislation of 1994, in spite of the limits linked to its rigidity, as pointed out by Brusco and Paba (1997), seem on one hand to be more thorough and detailed, and on the other hand more effective in the analysis of industrial districts according to Becattini's definition (1987, 1991). Two further advantages of the criteria utilised seem to derive from the use of a fixed measure of evaluation and from the identification of districts on the basis of five different indices, which makes it possible to evaluate geographical areas from different perspectives[13].

For an area to be considered a district, the indicators listed below must exceed some threshold values, defined on the basis of their proportionality to indicators derived from national data[14]. The following are the five indices established by the decree of April 21st 1993:

---

[11] This is a regional law of January 5th 2000 No.1 passed on March 16th 2001 by the Regional government of Lombardy

[12] The threshold level of 250 employees is used by the European Commission (see Commission des Communautes Europeennes, 1996)

[13] Of course it is necessary to remember that the preliminary condition for the identification of an area as a district lies in the capacity to link specific local features to the needs to open up to the outside or rather "harmoniously match world culture with the genius loci" (Brusco, 1994). For a more in-depth analysis see also Becattini and Rullani (1993).

[14] The first index is expressed in 100's and the other four indices in units.

1) Index of manufacturing industrialisation. This is given by the share of employees in the manufacturing sector over the total employees in the area. The fixed threshold value is that this index must be 30% higher than the corresponding national average.
2) Index of entrepreneurial density. This is given by the ratio between establishments and population. This figure must be higher than the threshold value as expressed by the corresponding national ratio.
3) Index of product specialisation. This indicator is given by the ratio of the number of employees in a specific sector to the total number of employees in manufacturing. The resulting figure must exceed the relative national average by 30%.
4) Index of concentration. This indicator is given by the ratio between the number of employees in a given sector and 30% of the total number of those employed in the manufacturing sector.
5) Index of dimensional concentration. This is given by the ratio between the number of those employed in small and medium sized companies and 50% of employees in the total of local units in relation to the sector specialisation.

In the specific case of the district of Asse Sempione in Lombardy the application of the indices reveals the presence of a sub-district which specialises in foot wear.

The result (Table 2) shows that the three municipalities under consideration constitute a sub-district in so far as they satisfy all the requirements stated by the law and the established threshold values are exceeded.

Table 2. Application of the indices stated by the law to the municipalities in the sub-district of Parabiago, 1992 data

Municipalities	*Industrialisation*	*Entre. density*	*Prod. special*	*Concentration*	*Dimen. concen*
Parabiago	116.12	1.52	7.04	1.43	2
Cerro Maggiore	113.56	1.27	2.41	10.49	2
SanVittoreOlona	106.67	1.65	5.22	1.06	2
Total area	113.90	1.46	5.58	1.13	2
Threshold	100.00	1.00	1.00	1.00	1

Source: our elaboration of ISTAT data (1992)

It is possible to apply the same indices for the year 1996 as well in order to verify that this sub-district also possesses the same characteristics over time. Table 3, which refers to 1996, also confirm the existence of a sub-district.

Table 3. Application of the indices stated by the law to the municipalities in the sub-district of Parabiago, 1996 data

Municipalities	*Industrialisation*	*Entre. density*	*Prod. special.*	*Concentration*	*Dimens. concen*
Parabiago	125.49	1.36	6.70	1.38	2
Cerro Maggiore	116.38	1.11	1.73	0.36	2
SanVittoreOlona	109.14	1.59	3.84	0.79	2
Total area	120.01	1.32	4.87	1.01	2
Threshold	100	1	1	1	1

Source: our elaboration of ISTAT data (1997)

A comparison of Tables 2 and Table 3 show some not particularly positive trends relating to the footwear sector for the first half of the 90's. On the one hand, the increase in the index of industrialisation suggests that the growth of the service sector in the local economy proceeds at a slower rate as compared with the rest of Italy. On the other hand, the decrease in the other indices [15] shows a decline in the footwear sector of Parabiago. The concentration index, in particular, highlights the most negative aspects: in 1996 the overall figure is above threshold value but this underlines the fact that the fall in employment recorded by the footwear is more significant than that recorded by the manufacturing sector. Such growth differentials will be dealt with later in this paper.

Notwithstanding the fact that the footwear sector is present also in other municipalities in the area of Parabiago, the municipalities under consideration are the only ones representing a district within the district of Asse Sempione. In fact, with the help of the same indices it is possible to show that the municipalities neighbouring with Parabiago, Cerro Maggiore, and San Vittore Olona do not meet the minimum requirements to belong to this sub-district.

[15] Only the index relating to dimensional concentration underwent no variation as compared to the previous period. This indicates that local shoe making business is concentrated in small and medium sized companies.

Table 4. Application of indices to neighbouring municipalities of the sub-district

Municipalities	*Industrialisation*	*Entre. Dens.*	*Prod. special.*	*Concentration*	*Dimens. Concen.*
Arluno	98.70	1.44	0	0	0
Busto Garolfo	117.19	1.37	0.01	0.01	2
Canegrato	114.00	1.15	0.94	0.19	2
Casogrezzo	156.33	1.24	0	0	0
Legnano	100.71	1.08	0.05	0.01	2
Rescaldina	153.67	1.43	0	0	0
Nerviano	107.98	1.37	0.80	0.16	2
Threshold	100	1	1	1	1

Source: our elaboration of ISTAT data (1997)

As can be clearly seen from Table 4, almost all the municipalities exceed threshold level for industrialisation and entrepreneurial density, but none reach a figure higher than 1 for production specialisation or concentration. None of the above-mentioned municipalities meet the requirements for a districts as do Parabiago, San Vittore Olona and Cerro Maggiore.

Lastly, it is necessary to mention another area within the district of Asse Sempione, from now on to be called the 'area of Turbigo' after the most important municipality therein, where the footwear sector is particularly developed. In this case as well, however, the prerequisites are not met for the area to be identified as a district, as the index of concentration does not meet the required level (Table 5).

Table 5. Indices for the foot wear sector in the area of Turbigo

Municipalities	*Industrialisation*	*Entre. dens.*	*Prod. special.*	*Concentration*	*Dimens. concen*
Buscate	141.25	2.32	3.29	0.68	2
Castano Primo	131.55	2.61	2.18	0.45	2
Nosate	139.36	2.63	4.15	0.85	2
Robe.Co.Induno	162.93	2.81	4.00	0.83	2
Turbigo	116.52	2.60	4.62	0.95	2
Threshold	100	1	1	1	1

Source: our elaboration on ISTAT data (1997)

It is possible to conclude, therefore, that although the shoe making sector is widespread in the district of Asse Sempione, only the municipalities being taken into consideration, i.e. Parabiago, Cerro Maggiore, and San Vittore Olona meet all the requirements to be classified as a district.

This confirms that within an area with a predominant product specialisation, there can also be different specialisations constituting sub-districts in which two predominant sectors overlap. Such pluri-specialisation derives from the area's history, where a set of skills have been exploited and applied to different but connected sectors, each of which has developed the technical know how with originality[16].

3 The evolution of the district and sub-district

3.1 The location quotient and shift-share analysis

The previous descriptive analysis has shown the existence of a sub-district in footwear within the bigger textile district of Asse Sempione. Now, the objective is to look at the characteristics of the two production structures and compare them by analysing two phases of twenty year length. Since we aim to study the interaction between the sub-district and district, their different specialisations will be examined. For this aim, two instruments of economic statistics appear to be particularly effective.

The first instrument is the *location quotient (LQ)* which may be expressed – in relation to sector *i* - by the following ratio:

$$LQ = \frac{\frac{E_{ir}}{E_{it}}}{\frac{E_{ibr}}{E_{ibt}}}$$

where *E* is employment, *r* is the study region (area), *br* is the benchmarking region (area) [17] and *t* refers to the total employment in the sector. This measures the degree of concentration of a sector in an area compared to the concentration found for a greater area of reference[18].

It is evident that the *location quotient* gives a 'relative' indicator in that it compares the ratio between the number of employees in a specific sector to the total number of employees in two different areas and indicates in which of these areas the phenomenon is more significant in terms of concentration. If the value of the *location quotient* is more than one, it means that the phenomenon being analysed is 'over-represented' in the area being studied or, in other words, that it is of greater importance than in the macro area of reference. Whenever this figure is less than one, the opposite situation is the case. It is easy to see that the choice of the area to make the comparison with is of crucial importance, since a

[16] Brusco and Paba (1997) confirmed this hypothesis by showing how in many traditional districts there are specialisations running alongside the predominant one which are in some way connected to the former more typical specialisation. Internal diversification seems to be an integral part of districts' evolution.

[17] Larger than the study region.

[18] Further details can be found in Hayter (1998) and Maggioni (2001)

benchmark of little significance can reduce the efficacy of measurement carried out with the *location quotient.*

In the rest of this Section the *location quotient* will be used to calculate the degree of sector specialisation in the area of the sub-district of Parabiago both in comparison with Lombardy and also with the district of Asse Sempione, the two most significant areas of reference for the sub-district[19].

This index appears to be particularly effective in this analysis in so far as it gives immediate findings on specialisation and location.

The second tool used is *shift-share* analysis, a statistical method whose first applications date back to the forties, within an analysis of the distribution of industrial concentration in England (Jones, 1940). In economic analysis, it has been extensively used to study growth at the local level thanks to the simplicity of its premises and the efficacy of its findings[20]. The purpose of shift share applied to regional/local growth analysis is to break down a set of data in terms of the factors concerning structure and location.

The use of this method make it possible to distinguish between the advantages gained from competitiveness relating to all sectors in an area and those deriving from greater specialisation in those sectors with higher levels of growth. The use of this tool to make comparisons makes it possible, therefore, to study the evolution of economic phenomena alongside with the area adopted as the *benchmark*[21].

It is necessary to make the premise that different methods of application for shift share analysis exist[22], which have adapted this tool according to the subject and purpose of various studies in different cases. The shift share method used in this work is based on the traditional one (see Biffignandi, 1993), widely used in the literature and particularly effective in a descriptive analysis of territorial phenomena. In particular, the chosen approach (see Smith, 2001) aims to make a comparison between the growth in employment in one region as compared to the nation, distinguishing the role played by three elements: national, structural (or sector mix) and local (or regional *shift*).

The national element shows the increase or decrease in employment due to economic conditions and general trends[23]. If the region had a sector mix and growth rate equal to the national one then no difference would be found. The national element can be found by multiplying the growth rate of the national employment by the regional employment (in absolute values) at a point of time t[24]. The structural element, or 'sector-mix effect', evaluates whether in the

---

19 Comparison with the province – the intermediate level among district and region – is not relevant since Milan is by far the largest city in its province and its size dwarfs all other municipalities. Furthermore part of the Parabiago sub-district lies in Varese province.

20 For examples and applications see: Dunn (1960), Biffignandi (1992), Esteban (2000).

21 For territorial evaluation methodology see Antonelli and Nosvelli (2002).

22 In Biffignandi (1993) four different versions of this method are taken into consideration.

23 The national value derives from the fact that the largest territory of reference in many analyses is often that of the nation.

24 See column M in tables A and B.

regional economic structure high growth sectors or conversely reduced growth sectors are relatively more represented. [25] This is obtained from the difference between the levels of growth of the sector and the entire economy at a national level and then by multiplying this difference by the actual value of regional employment at a point in time [26]. Regional *shift*, finally, estimates the growth rate of single sectors by comparing the performance of each sector at a national level with the same sector at a regional level. Regional *shift* is the result of the difference between the growth rate of the two territorial levels multiplied by the value of regional employment at the point of time t[27].

Regional growth is obtained from the sum of three elements, as the following equation illustrates:

Regional growth = National growth + Sector Mix + Regional Shift [1]

In order to focus more on the differences between regional and national growth, equation [1] can be re-written in the following way:

Regional growth – National growth = Sector Mix + Regional Shift [2]

It needs to be pointed out that variants different to those in the equations [1] and [2] are used here since reference is made to the geographical areas being studied. Consequently, the element corresponding to 'region' in equations [1] and [2] will, in the following Sections, be the sub-district of Parabiago, while the element expressing 'nation' in equations [1] and [2] is substituted by two different geographical references, first Lombardy and then the district of Asse Sempione.

3.2 The ASPO data bank

The data bank used for the application is ASPO, a district statistical archive for employment, which has been used also as a source of some of the preceding tables. ASPO was created by Unioncamere and has been elaborating data since 1981 with the aim of setting up an archive of companies and employees for different geographical levels and ensure periodic up-dates.

There are two peculiarities of this data bank which make it particularly useful for the following analysis.

The data used here refer to 1981 and 1998, the years relating to the first and last data retrieval[28]. 1981 was chosen as a starting point[29] as this made it possible to

[25] Hayter (1998)

[26] See column P in tables A and B.

[27] See column R in tables A and B.

[28] ASPO also presents data for 1992, 1994, and 1996.

[29] Other possible starting points were 1992, 1994, 1996.

analyse a long stretch of time. Taking into account the fact that the analysis focuses on the evolution of the structure, it seemed a reasonable choice in so far as this length of time offers greater opportunities for understanding the structural transformations in progress.

Secondly, the ASPO data are elaborated on the basis of the district, and this makes this data base more reliable and easier to use for the study of the district of Asse Sempione. In addition, data are available for municipalities and therefore it is easier to retrieve the data relating to the district of Parabiago.

3.3 Empirical results

Before presenting the empirical results, it is useful to point out some background data that need to be taken into account. Over the period of time under consideration, manufacturing sector declined in terms of employees in all the areas referred to: in Lombardy by 22.8%, in the Asse Sempione district by 30.3% and in the sub-district by 44.7%. As far as the service sector is concerned, the sub-district and district have recorded an increase which is higher than that of the whole region, namely 37% and 32% compared to 28%.

These data show how both district and sub-district are going through a most critical phase in which the changes in the production structure are affecting the traditional economic characteristics of these areas. Such changes, within the district, are even more relevant in so far as they make significant quantitative changes to the production specialisations with strong roots in the local culture and traditions.

The *location quotient* (*LQ*) results are obtained from the ratio between the data for the sub-district of Parabiago and both those for Lombardy and the Asse Sempione district. Attention has focused on the typical sectors of the Asse Sempione district, i.e. textile and clothing, and the typical sector of the sub-district Parabiago, namely leather and the leather goods sector.

Table 6 shows that according to the data for 1998, the footwear sector is clearly 'over-represented' in the sub-district. The figures for textiles, the main characteristic of the Asse Sempione district definitely does not indicate the same level of specialisation as foot wear in the sub-district even if the levels exceed those for the whole region. As far as textile and clothing are concerned, as compared to the district, it can be noted that concentration figure is less than one: this indicates that the relative density is not significant in the comparison of the two areas.

Table 6. Location quotient: comparison between 1981 and 1998[30]

	Sub-district/region		*Sub-district/district*	
Sector	LQ81	LQ98	LQ81	LQ98
Textile	2.34	2.47	0.58	0.72
Clothing	0.18	0.45	0.20	0.21
Leather/leather goods sector	53.11	19.74	5.98	6.71

Source: Unioncamere (1999)

Furthermore, it can be noted that between 1981 and 1998 the level of specialisation in the sub-district has diminished as far as footwear is concerned while it has increased for the textile and clothing sector. It seems therefore that the sub-district has been more attracted to and integrated into the predominant activity of the district it belongs to. In other words, the level of assimilation between the sub-district and the district has increased along with the increase in importance of textile and clothing compared to other manufacturing activities, which nevertheless remains important for the districts.

Moving on to shift share analysis, general results in terms of both percentage and absolute value are presented for an in depth examination of the sub-district performance. For the sub-district/region comparison equation [2] the following results emerge:

Table 7. Sub-district vs region: shift share results

Sub-district Growth - Regional Growth	=	Sector Mix + Sub-district Shift
-19% - (-1%)	=	+14% + (-32%)
2.628 - (-229)	=	+ 1.926 + (-4.325)

These values clearly show that the principal cause of the troubles of the sub-district lies in greater reduction of the number of employees recorded in the sub-district than in the region (-4,325). This means that although the sector mix brings essentially positive results for the sub-district, a poor performance results for companies in the sub-district as compared to the regional ones. This indicates that the sub-district's poor performance is caused by structural factors connected to the economy of the territory. The sub-district, therefore, is going through a crisis, which, if considered in relation to regional performance, cannot be traced back to the typology of the production specialisation.

If shift share analysis is applied to the comparison between sub-district and district, from the application of equation [2] totally different results are obtained from those previously shown.

---

[30] The *location quotient* is applied to the total for the sectors and therefore the concentration is lower than the indicator relating to the district in Section 2 that are related only to manufacturing.

Table 8. Sub-district vs district: shift share results

Sub-district Growth - District Growth	=	Sector Mix + Sub-district Shift
-19% - (-11%)	=	-4% + (-4%)
-2628 - (-1434)	=	-546 + (-647)

In this case it can be noted that the industrial mix and *shift* of the sub-district take on negative values and are not very different in terms of absolute values. This means that the worst performance recorded in the sub-district is attributable both to sector mix and (to a greater extent) to the fact that the companies of the sub-district have recorded worse results than those in the district. Therefore, the sub-district shares at least in part the difficulties recorded by the district, in particular in the sectors related to the specialisation they have in common, as explained later in further detail. However, unlike the district, the sub-district pays for the fact that the sectors in which it is specialised tend to worsen employment figures further. These findings show that the sub-district has more negative results than the district it belongs to and it seems that over the years between 1981 and 1998 the trend towards a divergence prevails between the two. This trend is only partly explained by the dynamics in progress in the entire area.

In the comparison between the sectors, we will dwell on those relevant for the specialisations of the district and sub-district. The appendix shows the analytic tables related to the application of the *share shift* method. More precisely, the following columns are analysed: M for the region and district growth, N and O for sector mix, P and Q for sub-district shift.

With regards to the comparison with Lombardy, Table A shows that in the clothing and textile sector the negative results are due to the concentration of companies belonging to sectors with negative growth rates in the sub-district. As already revealed by the LQ, this concentration is a determining factor with regards to the negative results, since the companies of the sector located in the sub-district of Parabiago, while hardly being affected by the negative phase in progress in the region, end up being penalised by the fact that they are very concentrated in the sub-district[31].

The footwear sector, which the sub-district of Parabiago specialises in, manifests clear differences from the textile and clothing sector. The negative trend of the sector is entirely due to territorial factors, given that the sector-mix effect is clearly positive. In other words, the negative trend in comparison with that of the region is explained by the companies' performance and not by their relative importance in the local production structure. It is, therefore, a question of a sector that must seek internal causes – in the management of production, innovation, etc., as discussed in Section 4 - for the negative performance recorded.

[31] In order to evaluate the diffusion of this sector it is necessary to consider both the LQ and the indicators of manufacturing concentration calculated in Section 2.

Table B compares the district of Asse Sempione with the shoe making sub-district of Parabiago. A detailed analysis of Table B shows that in two of the three industrial sectors (textile and shoes), the importance of the sector effect is a determining factor for their growth rate.

The footwear specialisation, as indicated by the analysis of the totals, plays a decisive role in determining two diverging trends. Footwear, in particular, illustrates how concentration has been important in determining the differentials in the growth rates, notwithstanding the positive performance of the shift effect in the sub-district. Textiles, at different levels, follows the same trend showing problems deriving from the sector mix.

These sectors show that, notwithstanding the current difficult phase, they have managed to obtain positive results compared to companies belonging to the district. This highlights the competitiveness and the excellence of the companies which, in spite of the downtrend, have sustained the local economy.

The clothing sector differentiates itself in so far as it presents a positive result, albeit weak, in terms of sector mix and a negative result in terms of sub-district shift. In this case the companies in the district appear to be more competitive than those of the sub-district even though the values are decidedly low and less significant than those relating to textiles and footwear.

Previous analyses (Castelli and Nosvelli, 2002) underlined the move towards the tertiary, which probably represents a reaction to the negative trend in manufacturing. This trend confirms the previously mentioned divergence – at least in terms of performance – from the district of which the sub-district continues to be an integral part.

4 The Asse Sempione district and the Parabiago sub-district: sectors and territory, integration and autonomy

The relationship between the district of Asse Sempione and the sub-district of Parabiago has a history both in terms of integration and divergence with regard to both economic structure and development paths. This is an apparent contradiction which may be explained by the different sector specialisation, the fact of sharing the same geographical area, and having the clothing sector as a common base. As the application of the shift share method highlighted, the territorial effect is demonstrated in the comparison made with Lombardy, while the specialisation effect emerges clearly from the comparison with the district. Often such effects have acted in different ways, sometimes acting as connectors and sometimes as divaricators along the path of development of these two economic and territorial realities[32].

At the beginning of the Nineteenth century the footwear sector arose out of a crisis in textiles which gave rise to a search for a different product and resulted in the emergence of the footwear companies concentrated in the sub-district of

[32] For the specialisation effects see Fortis (1996, 1999).

Parabiago, where, probably, that crisis had been felt more acutely. The emergence of a sector such as footwear is not a matter of coincidence but was favoured by the presence of some phases of production that could be carried out both for the production of textiles and also for shoes. This is especially true for the dyeing factories that are necessary for dyeing both textiles and leather and thus represented and still represent a fundamental stage in the working of both lines of production. Today, there is a collaboration in marketing which could lead to the creation of production lines for co-ordinating cloths, clothing and shoes. The human resources to be exploited, production styles, and, above all, local culture and traditions have always represented the greatest forces of cohesion of this area.

Alongside these clear cohesive links, it is, however, necessary to investigate the economic causes for the divergences in the two sectors and two areas.

The highly specialised sub-district has been reinforced over time by the entrepreneurial traditions of the footwear sector, with the development of the skills and links between entrepreneurs. This has led to the permanence and diffusion of footwear in the sub-district. The sector, however, is structured in a way that has not mirrored development patterns in other districts – such as Brenta, Montebelluna or San Mauro Pascoli – which, in recent years, have grown significantly thanks to product innovations and investments in training (ANCI, 2000)[33].

In the sub-district of Parabiago the crisis has been sharpened by some features that generally hinder the districts' growth and which in this case have been more penalising due to the lack of alternative strategies. In the Parabiago's sub-district the average firm size is small and business is prevalently based on sub-contractors. Furthermore, the generational change-over is particularly slow. These phenomena can explain the low innovation efforts, the limited development in marketing and product promotion as well as the limited penetration of international markets. Although these elements are common to other footwear industrial districts, they have brought about a profound crisis in the sector whose companies, unlike those of other districts did not find remedies to regain competitiveness. Some of the factors of success such as niche products, workers' skills, and quality brands are not or cannot be exploited.

One objective that could be set is to establish different interactions with the district to which the sub-district belongs by weaving stronger links both between sectors and also institutions and the territory.

With this aim a careful evaluation must be made of the findings from the shift share analysis, i.e. that the sub-district of Parabiago has to deal with the difficulties faced by the shoe making sector and the Parabiago micro-area, along with those linked to textile sector and the Asse Sempione macro-area. Because of this overlapping, the multiplicative effects should not be disregarded in the evaluation of the sub-district's performance as they may have amplified the negative outcome for single sectors.

---

[33] For shoe making districts and the reasons for their performance see Mo (1999a, 1999b), Ordandini (1997), ERVET (1995).

The district/sub-district relationship, however, may be seen as an opportunity rather than an obstacle, since it can lead to product diversification as it has occurred in many traditional industrial districts (see Brusco and Paba, 1997). Two specific opportunities may be arise in this regard. The first one is the possibility of focussing on inter-related quality production, the so-called co-ordinating fashion lines, could help to penetrate the market with the district's high quality brands. The second one is the opportunity to set up inter-sectorial associations – i.e. a new and efficient institutional framework – able to sustain and coordinate the development of small companies, innovation and create commercial channels.

To sum up, autonomy in the process of development that has marked the two territories seems to have left a deep mark on the structural evolution of the sub-district which has found itself having to deal with crisis on two fronts. The possible way out is that of exploiting existing integration potential both in terms of sectors and of institutions. This change in the pattern of development hopefully can represent the path towards a re-birth for both the Asse Sempione district and the shoe making sub-district of Parabiago.

5 Conclusions

On the basis of the criteria used in Italy for industrial districts identification, this paper has identified the existence of a sub-district in the area around the municipality of Parabiago, a set of neighbouring municipalities that, although they belong to a district recognised by the regional law, form by themselves a homogeneous economic reality that has the features of a district. This sub-district distinguishes itself from the district it belongs to – i.e. Asse Sempione – due to the fact that it includes the footwear sector as well as the specialisation in textiles and clothing that it shares with the district. In this sector, the sub-district has consolidated a century-long tradition and market prestige thanks to high quality production.

This first result demonstrates that also within economically homogeneous areas with a high level of specialisation and constituting districts, further areas and different specialisations may be identified which themselves constitute districts in accordance with the classifications formulated in the literature. This analysis paves the way to the study of the internal dynamics of districts, which can indicate how the economy of a district represents a complex set of companies with interlinking production potential. A more in-depth investigation into the internal structure and evolution of a district can be of great help towards understanding the performance and interactions with the 'outside', namely the companies and institutions not belonging to the district.

The use of the shift share technique has turned out to be particularly effective for investigating the relation between economic specialisation and territorial dimension. The use of benchmarking in shift share makes it possible to compare the internal districts dynamics so as to the get a better picture of the specific features of the sub-district. In particular, the existence of a relationship between district and sub-district emerges, which is, on the one hand, integrative and, on the other hand, divergent.

Integration is a result of the sub-district sharing with the district not just the sector of specialisation but also some production processes and services apart from knowledge/competences that may be shared by both sectors. This leads to the two areas achieving results that are basically similar as shown in the comparison with the regional results.

Divergence emerges from the evidence showing that Parabiago suffers due to the fragile structure of footwear industry, i.e. the production specialisation that has been carried out for much time alongside textile and clothing. Its location within the district has not only sharpened the sub-district's crisis but has also heightened its effects by adding difficulty to difficulty.

The differentiation of production specialisation and the ever-increasing recourse to tertiary activities have been escape routes taken by the sub-district to overcome a critical phase. Even in this case, it has played an autonomous role, at least in part, showing a divergence as compared to the district.

We believe, however, that a different degree of integration between district and sub-district, in terms of production organisation and institutions, could contribute towards a reinforcement of those interactions that can be the key to success.

References

AGRATI G., BEZZA B., MORONI A., RE R., VIGNATI M. (1985), *Cerro Maggiore pagine di storia*, Edizioni Landoni, Legnano.

ANCI (2000), *L'industria calzaturiera italiana – pre-consuntivo 2000*, Edizioni ANCI, Milano.

ANTONELLI G., NOSVELLI M. (a cura di) (2002), *Monitoraggio e valutazione delle politiche del lavoro per una 'nuova economia'*, Il Mulino, Bologna.

BECATTINI G. (1987), (a cura di), *Mercato e forze locali: il distretto industriale*, Il Mulino, Bologna.

BECATTINI G. (1991), "Il distretto industriale marshalliano come concetto socio-economico", in: PYKE F., BECATTINI G., SENGENBERGER W. (a cura di), *Quaderni della Banca Toscana*, n. 34, Firenze.

BECATTINI G. (1998), *Distretti industriali e made in Italy,* Bollati e Boringhieri, Torino.

BECATTINI G., RULLANI E. (1993), "Sistema locale e mercato globale", *Economia e politica industriale,* n. 80.

BIFFIGNANDI S. (1992), "L'analisi *shift share* in un approccio statistico integrato: discussione e proposte nell'ambito dello studio dell'evoluzione del commercio internazionale", *Quaderni del Dipartimento di Matematica, Statistica, Informatica e Applicazioni*, Università degli Studi di Bergamo, n.8.

BIFFIGNANDI S. (1993), *Aspetti metodologici e interpretativi della tecnica shift share*, Cedam, Padova

BRAMANTI A., MAGGIONI A.M. (a cura di) (1997), *La dinamica dei sistemi produttivi territoriali: teorie, tecniche, politiche*, Franco Angeli, Milano.

BRUSCO S. (1994), "Sistemi globali e sistemi locali", *Economia e politica industriale*, n. 84.

BRUSCO S., PABA S. (1997), "Per una storia dei distretti italiani dal secondo dopoguerra agli anni novanta", in: BARCA F. (a cura di), *Storia del capitalismo italiano*, Donzelli Editore, Roma.

CAMERA DI COMMERCIO, INDUSTRIA, AGRICOLTURA, ARTIGIANATO (1999), *I distretti industriali in Lombardia*, www.lom.camcom.it/distretti

CASTELLI M., NOSVELLI M. (2002), "Dinamiche intradistrettuali tra divergenza e integrazione. Il caso del distretto di Parabiago nel distretto Asse Sempione", *Piccola Impresa – Small Business*, n.1, pp. 99-129.

CERIANI M. (1948) *Storia di Parabiago: vicende e sviluppi dalle origini ad oggi*, Unione Tipografica, Milano.

COMMISSION DES COMMUNAUTES EUROPEEENNES (1996), *Recommandation de la Commission concernant la dèfinition des petites et moyennes enterprises*, 03/IV/96.

DUNN E.S. (1960), "A statistical and analytical technique for regional analysis", *Regional Science Association Papers and Proceedings*, n.6.

ERVET (1995), *I calzaturifici emiliano romagnoli nel mutato quadro della concorrenza internazionale*, Marsilio Editori, Venice.

ESTEBAN J. (2000), "Regional convergence in Europe and the sector mix: shift-share analysis", *Regional Science and Urban Economics*, n.30, pp.353-364.

FORTIS M. (1996), *Crescita economica e specializzazioni produttive*, Vita e Pensiero, Milano.

FORTIS M. (1999), *Aree distrettuali prealpine*, Vita e Pensiero, Milano.

GIUNTA REGIONALE LOMBARDA (2001), *Deliberazione n. VII/3839, seduta del 16 marzo 2001*, Milano.

HAYTER R. (1998), *The Dynamics of Industrial Location*, Wiley, Chichester.

ISTAT – IRPET (1986), "I mercati locali del lavoro in Italia", *Quaderni di discussione ISTAT*, Roma.

ISTAT (1992), *Censimento generale dell'industria e dei servizi*, Roma.

ISTAT (1996), *Rapporto Annuale*, Roma.

ISTAT (1997), *Censimento intermedio dell'industria e dei servizi*, Roma.

JONES J.H. (1940), A memorandum on the location of sector, Appendix II, *Royal Commission on the distribution of the industrial population* (Barlow Report), HMSO Cmnd, 6153.

MAGGIONI M.A. (2001), "The Development of High-Tech Clusters: Theoretical Insights and Policy Implications", in: FELDMAN M., MASSARD N. (Eds.) *Knowledge spillovers and the geography of innovation*, Kluwer, Dodrecht.

MAGGIONI M.A., NOSVELLI M. (2000), "Il 'made in Italy' nella struttura del commercio internazionale: l'analisi reticolare dei paesi OCSE", in: QUADRIO CURZIO A. (a cura di) , *Il "Made in Italy oltre il 2000*, Il Mulino, Bologna.

MO F. (1999a) "Il distretto calzaturiero resta leader", *Il Sole 24 Ore*, 28 giugno.

MO F. (1999b), E le calzature ripartono, *Il Sole 24 Ore*, 13 marzo.

NOSVELLI M. (1999), "I mercati locali del lavoro: un'analisi dei distretti lombardi", *Economia e Lavoro*, vol. 33, n.1.

ORDANDINI A. (1997), "I principali distretti calzaturieri italiani", in: *Scritti di Economia Aziendale*, n.45, EGEA, Milano.

PAGETTI F. (1996), *Dinamiche socio-economiche e trasformazioni territoriali di una regione complessa: la Lombardia,* ISU Università Cattolica, Milano.

SMITH G.W. (2001), *Shift share Analysis on Employment Growth* 1969-99, Washington State University, http://niip.wsu.edu/washington/ssharewa.htm

UNIONCAMERE LOMBARDA (1992, 1998), *Aspo statistico: addetti e unità locali per le provincie di Varese e Milano*, Unioncamere, Milano.

UNIONCAMERE LOMBARDA (1998), *Aspo statistico: addetti, unità locali e classi dimensionali per le provincie di Milano e Varese*, Milano.

VIANELLO C. A. (1934), *Il settecento milanese*, Edizioni Baldini e Castoldi, Milano.

Appendix

Table A. *Shift-share* analysis: Parabiago sub-district Vs Lombardy

	A	*B*	*C*	*D*	*E*	*F*	*G*	*H*
	Sub-district employee				(B-A)/A	D-B	Lombardy data	G*A
Sectors	Employee 1981	%	Employee 1998	%	Var.% '81-'98	Net Var.	Var % Lombardy	Net standard variation (1)
Textile	1876	14	1177	11	-37	-699	-27	-515,8
Clothing	108	1	95	1	-12	-13	-56	-60,0
Foot wear and leather	3841	28	1685	15	-56	-2156	44	1682,6
Total	13609	100	10981	100	-19	-2628	-2	1376,6

(1) Sub-district employment variation if each sector had grown at the same rate as its national corresponding sector

	I	*L*	*M*	*N*	*O*	*P*	*Q*
	A+H	Var. Lombardy Tot. (2)	L*A	G-L	O*A	E-G	Q*A
Sectors	Empl. Standard	Lombardy growth %	Lombardy growth a.v. (3)	Sector mix %	Sector mix a.v.	Shift sub-dist %	Shift sub-dist. a.v.
Textile	1360	-1	-31,6	-26	-484	-10	-183
Clothing	48	-1	-1,8	-54	-58	44	47
Foot wear and leather	5524	-1	-64,7	45	1747	-100	-3839
Total	14986	-1	-229	14	1926	-32	-4425

(2) Overall sector growth at regional level. Expressed in %.

(3) a.v. absolute values

Source: elaboration on Unioncamere (1981, 1998)

Table B. *Shift-share* analysis: Parabiago sub-district Vs Asse Sempione district

	A	*B*	*C*	*D*	*E*	*F*	*G*	*H*
	Sub-district employee				(B-A)/A	D-B	Lombardy data	G*A
Sectors	Employee 1981	%	Employee 1998	%	Var. % '81-'98	Net Var.	Var % Lombardy	Net standard variation (1)
Textile	1876	14	1177	11	-37	-699	-0,44	-823,9
Clothing	108	1	95	1	-12	-13	-0,05	-5,7
Foot wear and leather	3841	28	1685	15	-56	-2156	-0,57	-2174,8
Total	13609	100	10981	100	-19	-2628	-0,11	-1980,5

(1) Sub-district employment variation if each sector had grown at the same rate as its national corresponding sector

	I	*L*	*M*	*N*	*O*	*P*	*Q*
	A+H	Var. Lombardy Tot. (2)	L*A	G-L	O*A	E-G	Q*A
Sectors	Empl. Standard	Lombardy growth %	Lombardy growth a.v. (3)	Sector mix %	Sector mix v.a.	Shift sub-dist %	Shift sub-dist a.v.
Textile	1052	-11	-197,72	-33	-626	7	125
Clothing	102	-11	-11,38	5	6	-7	-7
Foot wear and leather	1666	-11	-404,82	-46	-1770	0	19
Total	11628	-11	-1434	-4	-546	-4	-647

(2) Overall sector growth at regional level. Expressed in %.
(3) a.v. absolute values
Source: elaboration on Unioncamere (1981, 1998)

Part III. Innovation

Can a Marshallian industrial district be innovative? The case of Italy

Giulio Cainelli[1] and Nicola De Liso[2]

[1]University of Bari and CERIS-DSE, National Research Council, Via Bassini 15, 20133 Milan, Italy
[2] University of Lecce, Via per Monteroni snc, 73100, Lecce, Italy

1 Introduction

The Italian manufacturing industry is characterised by a widely known positive 'anomaly': namely, a system made up of a large number of small-sized firms, many of which are concentrated in bounded geographical areas. These local systems of small-sized firms have been named Marshallian industrial districts (Brusco, 1982; Becattini, 1989). Scholars such as Becattini, Brusco and others have highlighted the basic features of this form of industrial organisation: (i) district firms are often specialised in traditional sectors such as textile, leather, footwear, wood, and so on (Brusco *et al.*, 1996); (ii) district firms tend to be rather small; (iii) district firms are localised in a bounded geographical area – generally composed of five or six municipalities; (iv) the physical proximity between firms engenders positive spillovers which concern the diffusion of information, knowledge and ideas – also because of skilled workers moving from one firm to another; (v) the role played by intentional innovative effort is very limited. The latter feature has led to the idea that industrial districts are technologically laggard and suffer disadvantages in generating technological innovation. In other words, according to this view, district firms are not able to generate a sufficient stream of innovative activities.

The aim of this contribution is to show that Italian manufacturing firms belonging to Marshallian industrial districts carry out a higher intentional effort than is usually acknowledged. This intentional innovative effort, together with the positive 'Marshallian' externalities, makes it possible to reach a high performance in terms of productivity, which in turn makes it possible to compete in international markets in which low-labour-cost countries tend to specialise in mature products.

Our view is supported by both a descriptive and econometric analysis which makes use of a new and original data-set. In fact, this data-set refers to information at the firm level built up by integrating three different statistical sources, namely the Community Innovation Survey (CIS), the Italian Structural Business Statistics (SCI) and the Italian Business Register (ASIA).

The empirical analysis uses a panel of 1,218 district and non-district firms belonging to traditional sectors. Data refers to the years 1992 and 1995.

In order to test our basic hypothesis we have estimated an augmented Cobb-Douglas production function. The estimation of this equation makes it possible to empirically identify three different determinants of firms'productivity: (i) the intentional innovative activity; (ii) the 'district effect', that is the (positive) effect due to belonging to a district; and finally (iii) the joint district and technological innovation effect which may be interpreted – according to our theoretical setting – as a sort of complementarity between codified and tacit knowledge.

This chapter is organised as follows. Section 2 analyses some theoretical aspects concerning problems of division of labour, codified and tacit knowledge in industrial districts; section 3 discusses the characteristics of the data-set and presents some performance indicators; in section 4 the econometric investigation and its results are illustrated; finally, section 5 contains the conclusions.

2 Division of labour, codified and tacit knowledge within industrial districts

2.1 Theoretical issues

One of the main features of industrial districts concerns the role played by tacit knowledge in defining the competitive advantage of the firms belonging to this form of industrial organisation. One can try to understand what tacit knowledge is by comparing it with codified knowledge. Codified knowledge refers to any form of knowledge which can be accessed, transmitted and improved by anyone possessing certain abilities (e.g. engineering principles understood and improved by engineers). Tacit knowledge is embodied in either individuals, organisations and systems, but those taking part in using and building it are not necessarily aware of the knowledge creation and learning processes taking place. Both tacit and codified knowledge affect, and are affected by, learning mechanisms as well as the division of labour.

As is widely known in the industrial district literature[1], a good deal of the competitive advantage of industrial districts can be attributed to (tacit) forms of knowledge spillover between firms, while at the same time, a minor role is attributed to intentional efforts.

While this perspective may explain part of the phenomena under analysis, we believe that other aspects have to be investigated. Indeed, identifying the tacit unintentional form as the main source of knowledge spillover is a rather narrow view of this situation. Also, we want to stress the fact that there exists a relationship between the (district) division of labour, forms of learning and the diffusion of knowledge. In particular, there exists a strong link between Smithian and Arrowian – see below – forms of learning, which in the district production systems come to be highly complementary.

---

[1] See Becattini (1989), Langlois and Robertson (1995) and Best (1990).

The division of labour arises as a peculiar endogenous feature of the industrial district. The overall production process may be broken down into several phases and each phase may be subdivided into different operations. This gives rise to what may be defined an *organic* district – an adjective that reminds the Marxian one as referred to manufactures[2] – in which the actual process of division of labour within and between firms is different in different districts. An example may clarify what we mean. If one considers two districts producing the same good, one will observe different ways of organising the process of production inside the firm as well as within the district – the latter seen as a system. This is what one observes in the case of the footwear districts of San Mauro Pascoli and Fusignano (Cainelli and Nuti, 1996; Brioschi *et al.*, 2002).

Given that the process of production can be organised according to different ways of dividing labour – inside the firm and within the district – we will necessarily observe different paths of development of both individual firms and the system as a whole. Different forms of division of labour create different environments in which learning mechanisms take place.

The two primary forms of learning we want to refer to are the pure Smithian mechanism of learning-by-doing and learning mechanisms *à la* Arrow. The former refers to 'simple' learning through repetition: the more one performs an operation the faster he/she becomes; learning *à la* Arrow is different, as it considers not only the Smithian mechanism, but also the stimuli which (new) capital goods engender within the process of production: "... each new machine produced and put into use is capable of changing the environment in which production takes place, so that learning is taking place with continually new stimuli" (Arrow, 1962, p. 157).

Both forms of learning occur in any firm – whether belonging to an industrial district or not.

However, two points have to be stressed when we consider firms belonging to a district. First, the learning processes which take place within a single firm have a direct effect on the other firms that work on the same stage(s) of the production process; this occurs through the mechanisms of imitation-competition; an example is the acquisition of new textile machinery in the Italian textile district of Carpi during the 1990s. Secondly, firms' learning processes engender spillovers among firms that belong to different stages of the process of production; an example is that of the tile district of Sassuolo, in which some firms have specialized in producing dedicated machinery (Brusco *et al.*, 1996).

Learning and the division of labour are also affected by final demand and the broader strategic context faced by firms and the district. One of the strengths of the industrial district lies in the capability to adapt quickly and efficiently to changes in demand – typically product customisation – and more generally to changes in the overall competition context.

Thus, as a rule, we will observe district-specific forms of division of labour and learning. This means that there exists variety, which is characterised in at least

[2] Marx (1867 [1976], chapter 14) speaks of *heterogeneous* and *organic* manufactures; the latter is the most advanced form of manufacture, typical of the Industrial Revolution.

two ways: variety in firms' internal organisational characteristics and variety in the organisation of the relationships between firms within the district. One feature of the district firm is that its internal organisation may be affected by the other district firms.

Organisational variety together with the mechanisms which make the district as a whole stronger than its individual components give rise to the higher productivity of district firms, as well as to the higher survival capability with respect to 'scattered' firms.

Learning, evolution within the processes of division of labour, responsiveness to changes in demand and market conditions can give rise to technological innovations; put it another way, innovativeness is a structural feature of district firms.

This conclusion allows us to better qualify the traditional industrial district theory; in fact, the latter has paid more attention to the role of Marshallian externalities as determinants of better performance of district firms rather than focusing on the intentional nature of innovative activities. Our analysis, on the other hand, focuses on the complementarity between intentional and (Marshallian) unintentional innovativeness.

2.2 From theory to empirical analysis: some clarifications

It is worthwhile to clarify the relationships which exist between the concepts referred to up to now.

(i) First of all, let us recall the fact that the industrial districts literature refers to some mechanisms which make the districts themselves stronger than their individual components. These mechanisms – Marshallian externalities, agglomeration economies, local knowledge spillovers, Smithian learning-by-doing – may have a strong positive effect on innovation and thus on firms' performance. We have to stress the fact that these mechanisms are largely unintentional. In this sense we will speak of unintentional innovation.

(ii) Secondly, within firms and within districts one can observe the existence of intentional activities aimed at producing new products or at acquiring new capital goods. Moreover, these forms of intentional activities tend to generate formalised and codified knowledge. A clarifying example is the development of new products and consequently the necessary new skills/capabilities which must be created in order to produce these new goods. To this intentional stage one can add the Arrowian process of learning, which affects the firm's performance, and which is likely to lead to proper innovation. Here we have the coexistence of both intentional and unintentional learning and innovation.

(iii) The econometric analysis which follows hereafter will concentrate on the analysis of the three main factors behind firms'productivity, namely, the Marshallian externalities, the intentional innovative activity and the joint district and innovation effect. This means that we will not consider the division of labour[3].

(iv) The phenomena under scrutiny are quite difficult to be empirically measured. For this reason, as usual in these analyses, it is necessary to use proxies. As we will further clarify in section 4.1, we will measure the unintentional and Marshallian mechanisms by means of a 'district dummy'; the intentional innovative activity of firms' is captured by 'innovation dummies' which also allow us to distinguish the type of technological innovation introduced by the firm: namely, product or process innovations; finally, the complementarity between unintentional and intentional mechanisms is synthesised by a joint 'district and innovation dummy'.

To sum up, in what follows we try to measure some of the key forces which characterise industrial districts, paying attention also to the reinforcing mechanism given by the circumstance of being innovative *and* member of a district at the same time. This joint effect is very important both for the individual firm and the district as a whole: in fact, as there exist close relationships between district firms, the transmission of productivity spurts is much more likely to occur within a district environment rather than in an environment in which firms have a 'simple' market relationship. Put it another way, within a district the economic relationships between firms include – besides the 'ordinary' exchange of goods and services at a given price – a reciprocal influence in terms of processes of production and a bi-directional stream of information.

3 The empirical analysis

3.1 The data set

Our empirical analysis is based on a new firm-level data set built up by matching three different statistical sources: namely, the Community Innovation Survey (CIS), the Italian Structural Business Statistics (SCI – *Sistema dei Conti delle Imprese*) and the Italian Business Register (ASIA – *Archivio Statistico delle Imprese Attive*).

Data from CIS allows us to measure intentional (codified) innovative activities carried out by the firms of our sample; in fact, the CIS survey gathers explicit information on process and product innovations. In particular, with regards to this statistical source, we have taken into account information referred to the period 1990-1992.

Data from SCI and ASIA provide us with microeconomic structural information at the firm level such as value added, number of employees, physical capital, and so on.

[3] For a theoretical model of the division of labour and learning-by-doing see De Liso *et al.* (2001).

The resulting sample of the integration of these three sources consists of 4,783 manufacturing firms with 20 or more employees. Within this set of firms we have selected a panel of 1,218 firms which operate in traditional sectors in two different years: 1992 and 1995. This choice is motivated by the fact that Italian industrial districts specialise in traditional sectors. We must make it clear that not all of these 1,218 firms belong to a district; put it another way, in our sample we are able to distinguish between district and non-district firms. This distinction is based on the statistical procedure used by the Italian National Statistical Institute according to which the Italian districts are identified (ISTAT, 1997).

3.2 Some performance indicators: a descriptive analysis

Table 1 shows some performance indicators (labour productivity, working hours per worker, ROI and ROE) as well as factor costs (in particular labour cost) for firms operating in traditional sectors. Indicators are disaggregated for firms both belonging and not belonging to an industrial district, and for both innovative and non-innovative firms.

As one can see, innovative industrial district firms perform better than those in an industrial district that do not innovate as well as than those not belonging to an industrial district. This is true for all size classes. For example, within the class 20 to 49 employees the ROI is 13.1% for district, innovative firms whereas the value for firms that innovate but are not inside an industrial district is 11.4%. It is interesting to note that indicators do not significantly differ between non-innovative firms – be they inside or outside the industrial district. The only exception is for firms in the class 50-249 employees which seem to benefit from an industrial district effect.

As for labour productivity, there is a marked difference between innovative district firms and innovative non-district firms in the '20-49' and '250 or more employees' classes.

Innovative firms belonging to industrial districts, while having good performance, do not show differences in the labour cost structure when compared to firms not belonging to an industrial district. Actually, in the most common size class – i.e. the smallest – the hourly wage per employee is higher for district firms than for non-district firms. This result contrasts with the common view according to which one of the competitive factors characterising industrial districts is the low cost of labour.

Table 1. Performance indicators for district and non-district firms, 1995

	Non-district firms		District firms	
	Innovative	Non Innovative	Innovative	Non Innovative
20-49 employees				
Labour productivity (million liras)	70.9	64.0	77.5	65.3
Labour costs	42.5	40.2	44.1	40.9
Working hours per employee	1,739	1,713	1,748	1,701
Hourly wages (000 liras)	16.7	16.1	17.0	16.2
ROI (%)	11.4	10.6	13.1	11.7
ROE (%)	13.9	10.2	16.6	14.0
50-249 employees				
Labour productivity (million liras)	88.9	73.2	89.6	80.8
Labour costs	51.3	47.6	50.3	47.5
Working hours per employee	1,713	1,703	1,712	1,705
Hourly wages (000 liras)	20.7	19.2	19.7	18.7
ROI (%)	10.5	7.6	11.1	10.0
ROE (%)	10.3	5.1	12.5	11.0
250 or more employees				
Labour productivity (million liras)	99.6	80.0	119.6	81.6
Labour costs	59.7	48.2	60.1	49.8
Working hours per employee	1,669	1,665	1,726	1,680
Hourly wages (000 liras)	25.3	20.0	23.2	19.7
ROI (%)	8.7	7.7	8.0	9.5
ROE (%)	4.5	4.2	7.9	9.6

4 The econometric investigation

4.1 The modelling strategy

In this section we empirically analyse the role played by intentional innovative activities and Marshallian agglomeration forces on productivity differentials between firms operating in traditional sectors. Furthermore, we test the joint effect of these two forces.

In order to perform this exercise we use an augmented Cobb-Douglas production function[4]. The equation to be estimated is:

[4] We have tested whether the Cobb-Douglas production function is the most appropriate functional form and we have found that this is true for our data. These results are available upon request from the authors.

$$\ln Y_{i,t} = a_0 \cdot \ln A_{i,t} + a_1 \cdot \ln L_{i,t} + a_2 \cdot \ln K_{i,t} + v_i + u_{i,t} \quad [1]$$

where $Y_{i,t}$ denotes the value added of firm i in period t, $L_{i,t}$ the number of employees, $K_{i,t}$ is a proxy for the capital stock and, finally, $A_{i,t}$ represents the state of firm's technology (to which we will come back later). v_i denotes the firm's unobserved time-invariant fixed effect, while $u_{i,t}$ is the error term with the usual statistical properties.

The term v_i included in equation [1] is justified by the possible presence of omitted variables – such as the ability of the firm's management – that can be correlated with the other explanatory variables. Had we not considered v_i we would have experienced a major econometric problem as we would have obtained biased and inconsistent estimates of the coefficients. However, as v_i cannot be observed, we have to find a way to eliminate it. We can do it thanks to the panel structure of our data set. In fact this term can be removed taking the long differences of equation [1]; by means of this operation we can remove the firms' unobserved time-invariant fixed effects, obtaining the following long-differenced production function:

$$\Delta_3 \ln Y_{i,t} = \lambda_0 \cdot \Delta_3 \ln A_{i,t} + \lambda_1 \cdot \Delta_3 \ln L_{i,t} + \lambda_2 \cdot \Delta_3 \ln K_{i,t} + \Delta_3 u_{i,t} \quad [2]$$

Δ_3 denotes the long-difference operator – where t is the year 1995 and t-3 is the year 1992; it is worthwhile to stress the fact that $\Delta_3 A_{i,t}$ now measures the rate of technological progress of firm i.

As far as the rate of firms' technological progress is concerned we assume that it can be explained, according to our theoretical setting, on the basis of three different factors: such as (i) the 'district effect' – which captures the Marshallian externalities; (ii) the intentional innovative activity – which we will call 'innovation effect'; (iii) the joint district and innovation effect. In econometric terms, these effects can be measured by using three dummies:

$$\Delta_3 \ln A_{i,t} = b_0 + b_1 \cdot DIS_i + b_2 \cdot INN_{i,t-3,j} + b_3 \cdot \left(DIS_i \cdot INN_{i,t-3,j}\right) \quad [3]$$

The first one is DIS_i which takes value 1 or 0 according to whether firm i belongs to an industrial district (1) or not (0). The second dummy is $INN_{i,t-3,j}$ takes a value equal to 1 if firm i has introduced a technological innovation of type j in the last three years ending in t, and 0 otherwise; we have to point out that the innovation dummy is taken with a three-year time lag. This lag is necessary because innovation requires time to be implemented and to become effective. The third is an interactive dummy obtained by multiplying the first two dummies; the variable thus constructed captures the joint district and innovation effect; in this sense we can speak of complementarity between codified knowledge deriving from the intentional innovative activity and Marshallian tacit knowledge.

4.2 The results

Before presenting our empirical results we want to make three points.

(i) First, the value added and the physical capital at 1995 prices have been deflated and expressed according to 1992 prices by using appropriate price indices. In particular, value added (*Y*) has been deflated with the appropriate two digit production price index provided by the National Statistical Institute (ISTAT), while the physical capital (*K*) has been deflated using the aggregate business investment deflator. In this way, we have made comparable the value of these variables between 1992 and 1995.

(ii) Second, regional and size dummies and a vector of firms' characteristics – such as a proxy for the organisational structure of firms (the proportion of white-collar workers to total workforce) and for the belonging of a firm to a business group – have been inserted in all the specifications used in the econometric analysis. This has been done in order to take into account firms' observed heterogeneity.

(iii) Third, all the estimates, which also include a constant term, have been carried out by using a 'robust' Instrumental Variable (IV) estimator to take into account potential measurement error problems. In fact as far as the capital variable is concerned, a measurement error is likely to be present that can produce biases and inconsistent estimates of the unknown coefficients of the Cobb-Douglas production function. To take into account these econometric problems, we apply an IV estimator to equation [2], using as instrument a ranking scheme of the capital variable in question. This kind of instruments are considered especially appropriate when right-hand side variables are measured with error (Bowden and Turkington, 1984). We rank the change in $\ln(K_{i,t})$ and use this rank as an instrument. Intuitively this rank should be highly correlated with the properly measured variable, but not with the measurement error, thus providing a good instrument.

Now, let us go back to the econometric results. Table 2 shows the findings of the estimate of different versions of the augmented Cobb-Douglas production function.

In the specifications shown in Table 2, all of which use the control variables previously mentioned (see point (ii) above), the sign of the two coefficients – labor and capital – is statistically significant and positive as expected. The values of the adjusted R^2 are also particularly high. This implies that our regression equations explain a significant share of the variation in the dependent variable: that is the firms' productivity. Let us now examine the single specifications.

The first column shows that the district dummy is positive and statistically significant in explaining firms' productivity differentials in traditional sectors. This confirms the industrial districts literature idea according to which being part of a district has positive effects on the firm's performance. In other words, the mechanisms connected to co-operation and competition between firms, to local knowledge spillover, and more in general to the Marshallian industrial atmosphere, generate positive externalities. This district effect is supported by both theoretical, empirical and econometric evidence (Becattini *et al.*, 2003; Brioschi *et al.*, 2002; Signorini, 1994).

Table 2. Long-differenced, firm-level, production function (1992-1995)

	[1]	[2]	[3]	[4]	[5]
Dep. variable: Change in (real) value added 1992-1995					
ESTIMATION METHOD	IV	IV	IV	IV	IV
Δln(Labour)	0.568**	0.570**	0.569**	0.567**	0.567**
	[0.029]	[0.029]	[0.029]	[0.029]	[0.029]
Δln(Capital)	0.325**	0.326**	0.326**	0.325**	0.326**
	[0.010]	[0.010]	[0.010]	[0.010]	[0.010]
Size dummies (2)	Yes	Yes	Yes	Yes	Yes
Regional dummies (3)	Yes	Yes	Yes	Yes	Yes
Firms' characteristics (2)	Yes	Yes	Yes	Yes	Yes
District	0.016*	...	0.016*	0.017*	...
	[0.009]		[0.009]	[0.009]	
Innovation (1992)	...	-0.009	-0.009	...	...
		[0.009]	[0.009]		
Product innovation (1992)	...	...	...	0.025**	...
				[0.011]	
Process innovation (1992)	...	...	...	-0.021**	...
				[0.010]	
District×product innovation (1992)	...	...	...	...	0.032**
					[0.012]
District×process innovation (1992)	...	...	...	...	-0.012
					[0.012]
N. Obs.	1,218	1,218	1,218	1,218	1,218
Adjusted R^2	0.722	0.721	0.722	0.723	0.722

Note: Standard errors in parenthesis. ** denotes significant at 5% and * denotes significant at 10%. The estimates also include a constant term.

The second column shows that the dummy which captures intentional innovative activity is negative and is not statistically significant. This should not come as a surprise. In fact this result refers to traditional sectors, which are usually believed to be technologically laggard. Put it another way, this result seems to confirm the common view which identifies traditional sectors in this way. However, as we will show later, when we distinguish between *process* and *product* innovation the picture we have changes.

Column three considers simultaneously the district and the innovation dummy, confirming the previous results. In fact the district dummy is still positive and statistically significant, while the innovative dummy is still negative and not statistically significant. This means that firms operating in traditional sectors benefit from unintentional forces and not from intentional innovative activities as generally suggested by the traditional industrial district theory. In other words, these findings are consistent with the general perception about the functioning mechanisms operating within a Marshallian industrial district.

An element of advantage of our data-set is that we can discriminate the type of technological innovation introduced by the firms: i.e. product versus process innovation. In this way, we can empirically test the role played by each of these two forms of technological innovation on firms' productivity.

In column four two different technological innovation dummies have been introduced into the Cobb-Douglas production function. The first refers to product innovation carried out by the firms, while the second one refers to process innovation. As one can see from the analysis of Table 2, the coefficient of the product innovation dummy is positive and statistically significant, while that referred to process innovation is also statistically significant, but it is negative. In other words, the productivity of firms operating in traditional sectors seems to depend on forms of technological innovation such as the development of new products or the improvement of the existing ones.

Put it another way, product innovation in firms belonging to traditional sectors may be considered one of the key factors behind the competitive advantage, instead of the usually highlighted factors such as low labour costs. Product innovation was indicated by Schumpeter as one of the main forms of innovation and maybe too often the emphasis is laid on process innovation.

The above result is reinforced by the findings shown in column five. In fact, this column shows that the interactive dummy capturing the joint district and product innovation effect is positive and statistically significant. We should note that the value of this coefficient is higher than any of the other values considered. Put it another way, district firms which introduce product innovations are characterised by better performances. This result differs from what is generally emphasised about the nature of industrial districts – i.e., according to traditional theory, intentional innovative activity plays a minor role. A clear distinction has thus to be drawn: while district firms do not perform well in terms of process innovation, the perform very well in terms of product innovation – and this latter form is often an intentional form of innovation. Product innovation plays a central role in defining the competitive advantage and, therefore, the economic performance of firms operating in these sectors. This advantage is amplified by being a member of a Marshallian industrial district.

5 Conclusions

In Italy more than 47% of manufacturing firms are concentrated in traditional sectors such as food, textiles, leather and shoe, wood, furniture, etc. In terms of value added, the share is 38.1%. Among the European countries, only in Spain we find similar patterns – 49,5% in terms of employment and 37,6% in terms of value added (data refer to 1995). Most of these firms are localised in Marshallian industrial districts.

As is well known, the district literature has emphasised the role played by co-operative and competitive relationships between firms, local knowledge spillover and, more generally, the industrial atmosphere. All of these factors engender positive externalities, thus positively affecting district firms' performance. However, these positive factors are usually associated with a low level of innovativeness: the fact that within industrial districts, firms carry out little innovative activities is largely acknowledged.

Our work on the one hand confirms the existence of the typical district effect, while on the other hand it identifies a much higher innovativeness than usually acknowledged. In fact, our work shows that traditional sectors, at least in Italy, perform more formalized innovative activities than they were thought to, and this component has to be 'added' to the spillover effects which have been always recognised as important, even though difficult to quantify. This formalised innovative activity tend to take the form of product innovation. In fact, this type of formal innovative activity seems to find into production structures such as those of the Italian industrial district a favourable environment which exerts a positive and significant influence on firms' economic performance. Another key result consists of the empirical identification of the joint district and innovation effect which suggests a more complex interactive mechanism between Marshallian externalities and intentional innovative activities. In fact, what emerges from our econometric investigation is that there exists complementarity between formal and informal (district) innovative activities. In other words, the transmission of ideas, imitation and, broadly speaking, knowledge spillovers, seem to stimulate for some groups of firms and for some industrial districts more formal innovation activity as well.

These findings have interesting policy implications as the latter have been largely based on the idea that the innovativeness of industrial districts had to be attributed to informal phenomena such as 'unintentional' knowledge spillovers and 'simple' imitation. Our results indicate that policies aimed at stimulating local systems should also take into account the role played by intentional innovation.

References

ARROW K. (1962), "The Economic Implications of Learning-by-Doing, *Review of Economic Studies*, vol. 29, pp. 155-173.

BECATTINI G. (1989), "Sectors and/or Districts: Some Remarks on the Conceptual Foundations of Industrial Economics", in: GOODMAN E., BAMFORD J. (Eds.), *Small Firms and Industrial Districts in Italy*, Routledge, London.

BECATTINI G. (1990), "The Marshallian Industrial District as a Socioeconomic Notion", in: PYKE F., BECATTINI G., SENGENBERGER W. (Eds.), *Industrial Districts and Inter-firm Co-operation in Italy*, International Institute for Labour Studies, Geneva.

BECATTINI G., BELLANDI M., DEI OTTATI G., SFORZI F. (2003), *From Industrial Districts to Local Development*, Elgar, Cheltenham.

BEST M. (1990), *The New Competition*, Polity Press, Cambridge.

BLACK S., LYNCH L. M. (2000), "What's Driving the New Economy: the Benefits of Workplace Innovation", *NBER Working Papers*, n. 7479.

BLACK S., LYNCH L.M. (2001), "How to Compete: the Impact of Workplace Practises and Information Technology on Productivity", *Review of Economics and Statistics*, vol. 83, pp. 434-445.

BOWDEN R J., TURKINGTON D. A. (1984), *Instrumental Variables*, Cambridge University Press, Cambridge and New York.

BRIOSCHI F., BRIOSCHI M. S., CAINELLI G. (2002), "From the Industrial District to the District Group. An Insight into the Evolution of Local Capitalism in Italy", *Regional Studies*, vol. 36, n. 9, pp. 1037-1052.

BRUSCO S., SABEL C. F. (1981), "Artisan Production and Economic Growth", in: WILKINSON F. (Ed.), *The Dynamics of Labour Market Segmentation*, Academy Press.

BRUSCO S. (1982), "The Emilian Model: Productive Decentralisation and Social Integration", *Cambridge Journal of Economics*, vol. 6, pp. 167-184.

BRUSCO S. (1986), "Small Firms and Industrial Districts: the Experience of Italy", in: KEEBLE D., WEVER E. (Eds), *New Firms and Regional Development in Europe*, Croom Helm, London.

BRUSCO S., CAINELLI G., FORNI F., FRANCHI M., MALUSARDI A., RIGHETTI R. (1996), "The Evolution of Industrial Districts in Emilia-Romagna", in: COSSENTINO F., PYKE F., SENGENBERGER W. (Eds.), *Local and Regional Response to Global Pressure: the Case of Italy and Its Industrial Districts*, International Labour Office, Geneva.

CAINELLI G, NUTI F. (1996) "Directions of Change in Italy's Manufacturing Industrial Districts. The Case of the Emilian Footwear Districts of Fusignano and San Mauro Pascoli", *Journal of Industry Studies*, n. 2, pp. 105-118, 1996.

CAINELLI G., DE LISO N., MONDUCCI R., PERANI G. (2001), "Technological Innovation and Firm Performance in Italian Traditional Manufacturing Sectors", in: *Innovation and Enterprise Creation*, Innovation Papers n. 18, European Commission.

CAROLI E., VAN REENEN J. (2001), "Skill Biased Organisational Change? Evidence from a Panel of British and French Establishments, *Quarterly Journal of Economics*, vol. 116, pp. 1449-1492.

CONTINI B., REVELLI R., CUNEO S. (1992), "Productivity and Imperfect Competition", *Journal of Economic Behavior and Organization*, vol. 18, pp. 229-248.

DE LISO N., FILATRELLA G., WEAVER N. (2001), "On Endogenous Growth and Increasing Returns: Modeling Learning-by-Doing and the Division of Labour", *Journal of Economic Behavior and Organization*, vol. 46, pp. 39-55.

ISTAT (1997), *I Sistemi locali del lavoro 1991*, Argomenti n. 10, Istituto Poligrafico e Zecca dello Stato, Roma.

LANGLOIS R., ROBERTSON P.L. (1995), *Firms, Markets and Economic Change. A Dynamic Theory of Business Institutions*, Routledge, London.

MARX K. (1867, Repr. 1976), *Capital, Vol. I*, Harmondsworth, Penguin Books.
POLANYI M. (1983), *The Tacit Dimension*, Peter Smith, Gloucester, Mass.
ROSENBERG N. (1982), *Inside the Black Box: Technology and Economics*, Cambridge University Press, Cambridge.
SCHUMPETER J.A. (1939), *Business Cycles*, McGraw-Hill, New York.
SCHUMPETER J.A. (1943), *Capitalism, Socialism, Democracy*, Allen & Unwin, London.
SIGNORINI L. F. (1994), "The Price of Prato, or Measuring the Industrial District Effects", *Papers in Regional Science*, vol. 73, pp. 369-392.

Are industrial districts more conducive to innovative production? The case of Emilia Romagna

Riccardo Leoncini[1] and Francesca Lotti[2]

[1] University of Bologna and CERIS-DSE, National Research Council, Via Bassini 15, 20133 Milan, Italy
[2] Sant'Anna School, Via Carducci 40, 56100, Pisa, Italy

1 Introduction

During the last years, the debate about the characteristics of local development has undergone a substantial expansion. New interdisciplinary theoretical approaches have spurred a series of methodological and empirical instruments, which allowed for more precise and effective analyses of the contributions of local technological change to long-run economic growth. This process has been paralleled by a substantial increase in the effort of international organisations responsible for producing data, in refining and improving the coverage of the main features of innovative activity. For instance, OECD acted both as locus for theoretical and methodological debate and as data collector and organiser of new and improved sets of indicators. Also EUROSTAT acted so as to enlarge and harmonise already existing datasets (for example, the Community Innovation Survey).

The availability of Technology Watch, a survey carried out by ASTER upon a sample of firms from Emilia Romagna, allows us to extend to a regional level an analysis that has been typically carried out either at national level or by means of case studies.

The aim of the present contribution is to perform an analysis of the relationships between techno-economic performance and firms' organisation. Therefore, we will empirically test some hypothesis regarding the innovative behaviour of the firms of our sample stressing in particular the ways in which the problems arising from organisational relationships both within the firms and from the external environment are dealt with. In particular, the relationships with the external environment (both inter and intra industry) can assume different aspects, and they have a crucial impact upon the techno-economic performance of the firms. One dimension is relatively unstructured, and refers knowledge transfers in the form of spillovers. Another dimension, a more structured one, is related to the firm being part of a more or less 'codified' structure of relationships, which can be referred to the firm being part of either a business group or an industrial district. Upon this structure of relationships, diversified public forms of intervention act to sustain and incentivate the innovative activity of the system of firms, and hence its competitiveness.

The structure of this chapter is as follows. Section 2 will deal with the theoretical background. The following section (section 3) will present the dataset and the methodology adopted. Section 4 will discuss the main results of the econometric analysis. Section 5 contains the main conclusions.

2 Innovative activity, organisation and market structure

The complexity of the patterns with which technological change affects production and competitiveness makes it impossible to confine the analysis to single and uncorrelated topics, since each of them contributes to the process of long-run growth. It is thus necessary to distinguish between two separate domains within which innovation is defined: one refers to those activities directly related to science and technology as such, while the other one refers more directly to production activities. Indeed, firms conduct their activities on both these closely related levels, and the output of such interactive process may be seriously biased, depending on the way it is measured. In this sense, the term innovative activity does not necessarily refer only to the science and technology domain, but, to a larger, interactive and more varied context in which other variables play a crucial role, such as environmental (referring to both inter firms relationships and to interactions with other public and private institutions) and organisational ones (Schumpeter, 1934).

Therefore, in order to catch-up with this kind of complexity, our analysis will be articulated on several levels. The first level of analysis is concerned with the distinction between inputs and outputs of firms' innovative activity. Thus, from a purely technological point of view, the variables commonly utilised are usually R&D expenditure as an input proxy and patents as an output one.[1]

However, since patenting activity is a proxy for potential productive capability and not for realised productive capability, it is necessary to take into account further variables aimed at measuring the effective productive capability, such as production of innovative products.

Indeed, from a Schumpeterian perspective, competition among firms is based on new or improved products, and the innovative activity of a firm is aimed at appropriating the ensuing temporary monopoly profits, once the innovation is successfully marketed. Thus, we need a better understanding of how the research activity can be translated into market opportunities in terms of effective production. Thus, since not all the input/output science and technology activity is directly translated into superior economic performance, it is necessary to take into account the extent to which innovative activity originates improvements in production.

---

[1] For more detailed surveys on the pros and cons of the various indicators used as proxies of innovative activity, see, for example, Patel and Pavitt (1995) and Griliches (1995). For a survey related to Italy, see Santarelli and Sterlacchini (1996).

Moving from this background, we estimate several equations in order to determine which are the structural and organisational characteristics of a firm that might affect the likelihood that they (*i*) invest in R&D, (*ii*) are granted at least one patent, and (*iii*) are capable of marketing innovative products.

Another field of analysis is related to the role and the impact of local knowledge spillovers and, more in general, to the role that such positive externalities (Marshall, 1920) play in the process of diffusion of a technology.[2] Different kinds of positive externalities are identified, as they refer to the most favourable market structure for the transmission of knowledge amongst firms. In particular, specialisation in some production within a certain area can be thought of as generator of intra-industry spillovers. In this way proximity and specialisation contribute to determine a suitable environment[3] for innovation and hence long-run growth.[4] Jacobs (1969) argues instead in favour of variety of the industrial system as the best incentive. In this case, the set of opportunity for a firm is made up by inter-industry spillovers.

We tried to take into account these aspects also, by including into the equation of our model, two indexes measuring the degree of specialisation and the concentration of the local market. The specialisation index (*Spec*) is defined by the ratio between the share of employees of a firm (l_i) with respect to the whole industry ($l_{i,k}$) and the share of employees of that same industry ($l_{i,k}$) with respect to total employment in the region (l_{ER}):

$$Spec_i = \frac{\frac{l_i}{l_{i,k}}}{\frac{l_{i,k}}{l_{ER}}}$$

Where the presence of a share of employees above (below) average indicates that an industry is specialised (de-specialised).

The concentration of an industry k ($Conc_k$) is measured through the Herfindahl index, in terms of employees:

---

[2] On the argument, see, for instance, Glaeser *et al.* (1992), Henderson *et al.* (1995). For an analysis of the Italian experience, see Cainelli and Leoncini (1999) and Cainelli *et al.* (2001).

[3] We refer, for instance, to the mobility of skilled workers, imitation processes and so on.

[4] There is still an open debate about which market structure is the most favourable to this kind of externalities. Following Schumpeter (1943), a monopolistic setting is the market structure allowing for the appropriation of the results of R&D, and consequently it represents an incentive for innovation. Other scholars (Becattini 1999; Brusco, 1982, Porter, 1990) have pointed out how these forms of externalities can be maximised in markets where the degree of competition is very high.

$$Conc_k = \sum s_{ik}^2$$

where s is the share of firm i in industry k.

Another kind of relationship we are interested in, and thus tried to assess, is related to the impact of other organisational variables on innovation. In particular, we have controlled for firms eventually belonging either to an industrial district (in the Marshallian tradition) or to a business group. This aspect is complementary to the one concerning the spillovers. Indeed, organisational aspects are directly affected, to a considerable extent, by the existence of a certain kind of environment which determines a set of both formal and informal flows of knowledge travelling between firms. Thus, the characteristics of knowledge spillovers are different if a firm is part of a larger set where knowledge transfer are in general relatively more structured and the barriers to the diffusion of knowledge are less rigid. In both cases (as will be clearer in the following) we hypothesise that there are quite different ways in which knowledge flows are exchanged within more structured clusters of firms (both districts and groups), with respect to less structured clusters where the only way in which knowledge can travel is by informal patterns of involuntary exchanges deriving from local knowledge spillovers. Hence, our hypothesis is that there might be differences in organising both innovative activity and production depending on the different organisational models described.

Finally, a last group of empirical tests concerns the impact of public policies supporting innovation. We analysed both the direct and the indirect impacts of public policies. The former are analysed through the role and the effectiveness of the different typologies of policy from which firms can benefit, while the latter is analysed through the impact of scientific collaborations between firms and public research institutions. The public support to technological innovation opens a very controversial issues about the additionality/substitutability principle of innovative activity funding. In fact, public support can be pursued insofar as it does not crowd-out resources away from private investments.[5] By pointing out that the former is lower than the latter, they claim that there is a market failure related to the nature of public good of R&D (Arrow, 1962), to the true (strong) uncertainty of the results of innovative activity (in the sense of Knight, 1921), to asymmetries in information relative to it, to the presence of technical indivisibility, and to sunk costs (Dosi, 1988). In this case, the social planner should play a role in stimulating private investment.

[5] There is a whole stream of literature underlying the systematic difference between private and social returns of R&D. See, among the many, Griliches (1995) and Jones and Williams (1997).

3 Data and methodology

In order to empirically test our hypotheses, we used a dataset that is a statistically representative sample of the whole manufacturing firms localised in the Emilia Romagna region (for full details, see ASTER, 1998). Our sample comprises 1,026 firms, belonging to 11 sectors (Table 1), with employees ranking from 50 to 500. The direct survey was carried out in November and December 1997, using computer-aid techniques: in this way it was possible to have very detailed information in a short time.

Table 1. Sectoral disaggregation

1	Food and beverages
2	Textile and clothing, leather and footwear
3	Wood products & furniture
4	Paper and printing products
5	Chemical products
6	Rubber & plastic products
7	Non-metallic mineral products
8	Metal products
9	Electrical machinery
10	Electrical apparatus (incl. radio, tv), Office & computing
11	Transport equipment

Many of the questions included in the questionnaire were multiple choice: this affected the econometric analysis, since we could deal with qualitative and ordinal variables only. We thus estimated a set of logit models, aimed at assessing the weight of each explicative variable on the likelihood of the phenomenon under investigation. Since the logit model is not linear, the estimated parameters do not represent the marginal effect of the dependent variables. For this reason, each comment developed afterwards will be referred to the sign and to the intensity of the estimated parameters.

4 The empirical results

The estimates were carried out by grouping the observations into two sets. The first group comprises the whole sample of firms. The second set instead comprises only those industries characterised by the presence of industrial districts. This second sample was built with the aim of gaining a better understanding of the causal links between production and organisation within firms. Hence, it was possible for us to make some conjectures upon the nature of the network externalities surrounding the firm, which we described earlier in the paper.

We thus took into account the two different organisational forms (district and group) in the following way. While a direct question was included in the questionnaire, assessing whether a firm belongs or not to a business group, in order to recognise if a firm is localised inside an industrial district we used a map developed by CNEL (1997). This taxonomy is more oriented towards a workable definition, rather than a legal one (Law N. 317/91), and collects also the structural characteristics of about 20 industrial district in Emilia Romagna. In this way, by using information about the localisation, the specialisation and the leader firms, we could easily identify both the industries characterised by the presence of districts and the firms of our sample belonging to it. We then estimated a set of logit models in order to assess, *ceteris paribus*, if belonging to a group or to a district exerts an effect upon the likelihood of innovative performance (in the form of both potential and effective innovative production).

4.1 Innovative activity and R&D expenditures

The first set of estimates was aimed at determining which structural and organisational characteristics might affect the presence/absence of R&D expenditure in the firm, under the hypothesis that this represents an autonomous innovative effort.

We estimated the following model:

$$\mathrm{Prob}(\mathrm{R\&D}=1) = f(Dim, Dest, Integr, Sect, Pav, Spec, Conc)$$

where *R&D* is equal to 1 if there is some R&D expenditure, and 0 otherwise; *Dim* is the size of the firm in terms of employees, in log scale; *Dest* is the final destination of the product (1 = final good; 2 = intermediate good; 3 = investment good); *Integr* is a dummy variable identifying an integrated firm according to a definition given by ASTER itself (see ASTER, 1998); *Sect* represent some industry dummies, which identify the sectors in accordance to Pavitt's taxonomy;[6] *Spec* is the specialisation index of the firm; *Conc* is the concentration index of the industry.

Our main hypotheses are that the firm size exerts a positive impact on the probability of performing R&D expenditure. Moreover, while the existence of sectoral specificity is nowadays regarded as a sort of stylised fact, and this explains why we used sectoral dummies in our model, and the integration of the different functions within a firm can positively affect the likelihood of having R&D expenditure, since it is a signal of an efficient business management, as well as openness towards innovation. Finally, the degree of specialisation should capture the knowledge spillovers between firms belonging to the same industry, while the concentration ratio should give us some information about the market

[6] *Pav1* = science based, *Pav2* = supplier dominated, *Pav3* = scale intensive and *Pav4* = specialised suppliers respectively (Pavitt, 1984).

structure: the joint analysis of these two indexes will supply a better understanding of the kind of prevailing externality.

With the second set of estimates we further focused our attention on the organisational aspects: we used the same variables described before, plus two dummy variables: *Dist* (= 1 if the firm belongs to an industrial district, and 0 otherwise) and *Group* (= 1 if the firm is part of a business group, and 0 otherwise). The results are reported in Table 2.

Table 2. Likelihood of performing R&D expenditure

	[1]	[2]
Constant	-4.79***	-1.78***
Dim	0.59***	...
Dest	0.26*	0.25*
Integr	0.32*	0.35*
Pav 1	1.13***	not incl.
Pav2	-0.11	not incl.
Pav3	1.11***	not incl.
Pav4	Constant	not incl.
Spec	-0.07***	-0.07*
Conc	...	...
Dist		-1.99***
Group		-0.19
Test Concordant	79,3%	79,3%
N. Obs.	302	195

Significance level: *** = 1%, ** = 5%, * = 10%.

As we expected (see column 1 of Table 2), the larger the firm, the more likely it is for a firm to invest in R&D, due to the typical scale effect of the R&D activity. The more the products are investment goods, the higher the probability for a firm to have some R&D expenditure. Also the integration within the firm positively affects such probability, suggesting that a more rational handling of the information is a good basis for the innovative process. As far as the sectoral specificities are concerned, the higher propensity toward innovation of the science based and scale intensive firms is confirmed. As far as the supplier dominated and the specialised suppliers clusters are concerned, we did not find evidence of a relationship, the estimated coefficient not being statistically significant. Finally, concerning the role of externalities, the specialisation turns out to have a negative impact on the likelihood of having R&D expenditure and the coefficient of the concentration ratio is not significant. We might then conclude that the most relevant factor in terms of positive externalities for the research activity is the degree of despecialisation of a local system. In other words, in very competitive environments, where for some reason knowledge spillovers are lacking, the firms tend to undertake formal R&D activity.

The estimates concerning the set of firms belonging to industrial districts is reported in the second column of Table 2. It should be preliminarily pointed out that, with respect to the previous specification, the sectoral classifications were dropped from the analysis, being these estimates limited to the sole sectors within which industrial districts are present. The variables related to the internal organisation of the production show coefficients very similar to the ones estimated within the previous model, except for the coefficient of the dimension, which is not significant in this case. Being a member of an industrial group doesn't affect the likelihood of having some R&D expenditures, while it turns out that a firm within an industrial district is not motivated in innovation, since the estimated coefficient is negative. These results can be interpreted as follows: if a firm belongs to an industrial group it is less necessary to carry out formal R&D activity, since the firm can benefit from knowledge spillovers from the other firms belonging to the group. In interpreting the negative coefficient of the firms belonging to an industrial district, we need to distinguish whether a leader firm exists within the district or not. In the former case, we might argue that R&D activity is concentrated in few leader firms, while the other firms are poor R&D performers, since the localisation advantages make easier the appropriation of the leader firms' R&D results. In the latter case, the innovative activity follows creation and diffusion processes more related to the tacit component. Accordingly, if this is the case, the presence of formal R&D activity is less necessary, due to the high degree of specialisation and appropriability conditions within the district.

4.2 Potential and actual innovative production

The following set of estimates is aimed at quantifying the impact of some structural and competitive firm's characteristics on the likelihood that such firm has some results in terms of both potential and actual innovative production. Therefore, we looked at firm's patents as potential output, since the choice of patenting or not, although related to market strategies (which in turns depend on the kind of product, on the firm's position in the market, and so on) does not reflect effectively the presence of innovative products. For this reason, we also analysed the presence of innovative products, whether patented or not.

In particular, as far as potential production is concerned, the probability of a firm having a patent (PAT) is hypothesised to be a function of the following variables:

$$\text{Prob}(\text{Pat}=1)=f\left(R\&D, Dest, Integr, Posint, Posnaz, Spec, Conc\right)$$

where *Posint* represent the declared position in the international market (=1 if the firm declares to be a leader in the international market), *Posnaz* is the same as before, referred to the national market and the other variables are the same as in the previous set of estimates. Like before, we added the *Group* and *Dis* variables and separate results are reported in Table 3.

Table 3. Likelihood of depositing some patents

	[1]	[2]
Constant	-2.27***	-3.41***
R&D	0.10***	0.18***
Dest	0.31**	0.67***
Integr	-0.37**	-0.59***
Posint	0.50***	...
Posnaz	0.95***	0.92***
Spec	...	...
Conc	-3.74**	-4.62*
Dist		0.69**
Group		0.87***
Test Concordant	71,7%	71,7%
N. Obs.	304	197

Significance level: *** = 1%, ** = 5%, * = 10%.

From column one of Table 3, it emerges that R&D expenditure positively impacts upon the likelihood of having some patent. Moreover, if the product marketed is an investment good, the firm is again more likely to protect it with patents, while the organisational integration turns out to affect negatively the propensity to patent. This last result is due to the fact that integration itself is an attempt to reorganise the productive processes and accordingly, this does not affect the patenting activity. In other words, the negative impact of organisation on patenting activity may depend on the higher complexity of innovative processes due to organisational integration which, in turn, implies both higher appropriability and high degrees of complexity (and thus of difficulty) in filing a patent. Having a good position in the market significantly increases the likelihood of having some patents, even if it seems that the domestic market is more relevant than the international market. Finally, in the manufacturing industries as a whole, the market structure that more encourages patenting activity is competition. In this sense, patents are a necessary mean to protect innovation, since the market structure is far from a monopoly one.

Concerning the firms belonging to industrial groups or districts (Table 3, column two), it must preliminarily be noted that the signs and the values of the estimated coefficients are very similar to the ones of the same model, estimated on all industries. Moreover, in order to have some patents, no particular organisational form has advantages: both the district and the group allow the firm to benefit from the positive externalities that such organisational forms usually provide, as far as patenting activity is concerned. This result appears to be coherent with the ones we had from the analysis of R&D activity. In fact, even if within a district or a group we have little formalised innovative activity, it is very likely that informal innovative activity might end with a patent. This result confirms that different models characterises different systemic settings, and that there easily is a tendency to underestimate innovative activity within small firms, if we are to measure it with the consolidated indexes and to overlook models in which innovation result from very strong interaction between external functions rather than internal ones.

The second level of output, i.e. the realisation of innovative products (*Prodinn*), was analysed by means of the following equation:

$$\Pr ob(\Pr odinn = 1) = (Brev, Dim, Integr, Pos\,\mathrm{int}, Spec, Conc)$$

where the variables definitions are the same as before. Also in this case, an additional regression was run to take into account the variables concerning the group and the district. The results are reported in Table 4.

Table 4. Likelihood of having innovative products

	[1]	[2]
Constant	-1.31***	-0.4
Pat	2.02***	1.49***
Dim	0.28***	0.16
Integr	0.33	...
Posint	0.78***	0.68***
Spec	-0.04***	-0.05***
Conc	...	-3.42*
Dis		-0.01
Group		0.22
Test Concordant	73.2%	70,3%
N. Obs.	307	199

Significance level: *** = 1%, ** = 5%, * = 10%.

As far the whole sample is concerned (column 1), it must be pointed out that a strong positive relationship with the patenting activity does exist: the more a firm patents, the more likely it is to have innovative products. The dimension of the firm positively affects the probability of marketing innovative products. The coefficient relative to the position of the firm in the international market is rather high. Finally, like for R&D expenditures, specialisation has a negative impact and the coefficient of concentration is not significant at all. This confirms that innovation is more likely to happen in despecialised, fragmented and competitive environments.

Looking at column 2 of Table 4, we can infer that already having some patents is a fundamental determinant of the probability of having some innovative products, while size and integration do not affect this probability. Analysing the firm's position in the international market, it turns out to positively affect the likelihood of having innovative products. Looking at the coefficient relative to the concentration ratio, our previous results are confirmed: for the manufacturing firms localised within a district, the market structure that boosts the probability of having innovative products is competition. Besides, the degree of specialisation negatively affects such probability.

Concerning the organisational form, neither the group nor the industrial district is able to affect the likelihood of realising some innovative products. In this case, the production of innovative products is tightly linked to forms of externalities, for the appropriation of which it is necessary to have a strong knowledge base applied within a context of international integration (a sort of absorpting capacity) coupled to a strong fragmentation and despecialisation. Accordingly, it should not come as a surprise that such organisational forms might hamper the single firm to innovate, since they are meant to enhance that of the whole network rather than that of the single constituent firms. And this happens according to different patterns of cooperation/competition.

5 The role of the public sector

Since the questionnaire contains specific items concerning the incentives for innovation, we could separately analyse the impact of both direct and indirect policies in support of innovation on each variable used in the previous section: the R&D expenditures, the number of patents and the number of new products. In particular, firms were asked to list the financial sources for innovation. Among the possible answers we selected the ones that more likely corresponded to direct support policies, i.e. having benefited from European projects funding, non-European international projects funding, as well as having funds according to national and regional laws. In order to assess the impact of the indirect support policies, we considered the scientific collaborations with Universities, CNR (National Research Council), ENEA (National Agency for the Energy and the Environment) and other Public Research Institutions.

We used a methodology for the comparison based on the variance analysis (ANOVA). Each time, we subdivided the sample of the responding firms in two groups according to whether the firms received such incentives or not. Then we carried out a statistical test for the null hypothesis of equality of the mean value of the variables we looked at (R&D expenditures, number of patents, number of innovative products). The results are reported in Table 5.

The first thing to note is about the so-called additionality principle, according to which public intervention is aimed at integrating the amount of resources devoted to innovation, in order to complement firms innovative efforts. Looking at the results of the test, we can see that the firms that have access to public funding (with the sole exception of the international projects) show a higher level of R&D expenditures. The effectiveness of such interventions was also measured by a test carried out on the effective innovative production: the firms that received public funding exhibit a much higher number of innovative products than the others.

Only in one case we have a "crowding-out" effect, where there are scientific collaborations with other public research institutions. As we can see from Table 5, support activities have typically a wide range: from research to experimentation stage, to marketing of goods incorporating such innovative activity.

Table 5. Effectiveness of the support policies

	R&D Expenditures			Number of Patents			Number of Innovative Products		
	F Test (a)	Pres. (b)	Abs. (c)	F Test (a)	Pres. (b)	Abs. (c)	F Test (a)	Pres. (b)	Abs. (c)
Direct Support Policies									
Source of the financing:									
EU	16.70***	4.91	2.92	1.31	1.69	1.48	43.00***	4.21	3.03
Int. Projects	0.01	3.28	3.31	0.04	1.57	1.51	12.06***	4.14	3.19
Nat. Laws	11.31***	4.25	2.86	7.01***	1.80	1.38	34.71***	3.87	2.97
Reg. Laws	5.88***	4.17	3.05	0.02	1.54	1.51	16.79***	3.80	3.09
Indirect Support Policies									
Scientific Collaborations with:									
University	10.78***	2.78	1.83	33.03***	1.65	0.98	55.09***	3.10	2.85
CNR	0.73	2.46	2.10	0.79	1.33	1.18	16.88***	3.26	2.72
ENEA	0.38	2.41	2.11	7.87***	1.70	1.15	0.01	2.90	2.88
Other Inst.	7.81***	1.10	2.27	0.75	1.32	1.17	1.50	3.10	2.85

(a) F Test on the significance of the difference of the two groups' means
(b) mean value of the group in which the phenomenon took place
(c) mean value of the group in which the phenomenon did not take place
Significance level: *** = 1%, ** = 5%, * = 10%.

In particular, concerning R&D expenditures, the European financing projects have the stronger impact, since they increase the average R&D expenditure of 68.2%, while such increase is 48.4% and 36.7% in the case of national and regional financing respectively. A similar proportion is to be found in the case of innovative products, even if for lower absolute values: 38.9% for the European projects, 30.2% for the national and 23% for the regional financing. The international non European projects have an average impact of 30%.

Such differences are much stronger in the case of patenting activity. The regional initiatives for innovation turn out to be insignificant: on the contrary, the national funding projects do have some impact. This seems to depend on the specific features of patenting activity: the domestic incentives are the best tools to provide some incentives for innovative activity that is performed at least at national level. As far as international projects are concerned, it seems quite intuitive that they can affect innovative products only. In fact, since the firms of the Emilia Romagna region are typically small, they probably participate to international joint research projects not in order to obtain a patent, but in order to have some innovative products.

Accordingly, the patenting activity seems not to be affected by any of the different types of public intervention, with the sole exception of the national initiatives. But before going further into the analysis, we have to bear in mind some caveats. The patenting activity is, of the three variables considered, the one for which the typical delay between innovative effort, innovative results and patenting activity is more relevant. For this reason, since our data are relative to just one year, some observations on patenting activity can be biased. Moreover, the average size of the observed firms is small, and this does not favour patenting strategies in a systematic way. Moreover, patenting activity seems to be more related to radical innovations, while innovative products seem to be based on more incremental-types innovation. For this reason, the patenting activity depends, in a more functional way, on the material and immaterial resources flow coming from public institutions.

A little more controversial is the role of research institutions in providing an indirect support to innovative activity. In this case, only University is able to give a substantial contribution: with a wide spectrum of knowledge, ranging from pure scientific competencies to the ones with more applied technologically-oriented content. This makes University a sort of natural research partner, especially for small firms. Nevertheless, its role appears to be differentiated in terms of intensity. Indeed, we register a significant increase in average R&D expenditure (+52%) and in patenting activity (+62.4%), but a smaller one in the case of innovative products (+8.8%) which is half the impact of the scientific collaborations with the CNR (+19.2%). Concerning patents, a similar role to that of University, is played by the ENEA, although with less effectiveness (the average increase of patenting activity is 47.8%).

6 Conclusions

The aim of this work was to analyse the innovative capacity of manufacturing firms (on the basis of the specificity of the various steps of the innovative process, from input to output elements) in Emilia Romagna on the basis of a statistically significant sample of 1,026 firms collected in 1997.

The hypotheses that we empirically tested are about various aspects of the organisational forms regarding of both internal and external kind of relationships. The latter refers to two typologies: structured (referring to either industrial districts or industrial groups) and unstructured (as result of involuntary knowledge spillover) relationships between firms.

As far as the role of structured externalities is concerned, pertaining to an industrial district or to an industrial group tends to favour potential output rather then effective one. This is because of the less vulnerable environment which the firm is part of, allowing firms to better retain the results of informal innovative activity.

These results are in accordance with the picture supplied by the indicators of unstructured externalities. Indeed, they highlight a very competitive and fragmented environment, in which it is not easy to maintain monopolistic rents due to innovation, as the most conducive to knowledge spillover. Thus, the search for more structured kinds of relationships could foster innovative activity and favour the preservation and development of competitiveness.

Finally, as far as public support to innovative behaviour is concerned, we noted in general a positive role in terms of additionality of the resources made available by the public sector, to those mobilised by the firms, with a wide spectrum of options that make up for a necessarily diversified set of choices for the public policy decision making process.

References

ARROW K. (1962), "Economic Welfare and the Allocation of Resources for Invention", in: *The Rate and Direction of Inventive Activity*, Princeton University Press, Princeton.

ASTER (1998), *Techwatch 98. L'innovazione tecnologica in Emilia Romagna. Primi risultati dell'indagine campionaria,* ASTER, Bologna.

BECATTINI G. (1990), The Marshallian Industrial District as a Socio-Economic Notion, in: PIKE F.,. (Eds.), *Industrial Districts and Inter-firm Co-operation in Italy*, International Labor Office, Geneva.

BRUSCO S. (1982), "The Emilian Model: Productive Decentralisation and Social Integration", *Cambridge Journal of Economics*, vol. 6, pp. 167-184.

CAINELLI G., LEONCINI R. (1999), "Externalities and Long-Term Local Industrial Development: Some Empirical Evidence from Italy", *Revue d'Economie Industrielle*, n. 90, pp. 25-39.

CAINELLI G., LEONCINI R., MONTINI A. (2001), "Struttura produttiva e sviluppo regionale di lungo periodo in Italia", *Moneta e Credito*, n. 216, pp. 461-485.

CNEL (1997), *Innovazione, piccole imprese e distretti industriali,* 3° Rapporto Cnel/Ceris-Cnr, Documenti Cnel, n. 7, CNEL, Roma.

DOSI G. (1988), Sources, Procedures, and Microeconomic Effects of Innovation, *Journal of Economic Literature*, vol. 26, pp. 1120-1171.

GLAESER E., KALLAL H., SCHEINKMAN J., SCHLEIFER A. (1992), "Growth in Cities", *Journal of Political Economy*, vol. 100, pp. 1126-1152.

GRILICHES Z. (1995), "R&D and Productivity: Econometric Results and Measurement Issues", in: STONEMAN P. (Ed.), *Handbook of the Economics of Innovation and Technological Change*, Blackwell, Oxford.

HENDERSON V., KUNCORO A., TURNER M. (1995), "Industrial Development in Cities", *Journal of Political Economy*, vol. 103, pp. 1067-1090.

KNIGHT F. (1921), *Risk, Uncertainty, and Profit*, Houghton Muffin, Boston.

JACOBS J. (1969), *The Economy of Cities*, Vintage, New York.

JONES C., WILLIAMS J. (1997), *Measuring the Social Return to R&D*, mimeo.

MARSHALL A. (1920), *Principles of Economics*, Eight Edition, MacMillan, London.

PATEL P., PAVITT K. (1995), "Patterns of Technological Activity: Their Measurement and Interpretation", in: STONEMAN P. (Ed.), *Handbook of the Economics of Innovation and Technological Change*, Blackwell, Oxford.

PAVITT K. (1984), "Sectoral Patterns of Technical Change: Towards a Taxonomy and a Theory", *Research Policy*, vol. 13, pp. 343-373.

PORTER M. (1990), *The Competitive Advantage of Nations*, Free Press, New York.

SANTARELLI E., STERLACCHINI A. (1996), "Statistiche e banche dati sull'innovazione tecnologica a livello di impresa: la situazione italiana", *Rivista Italiana degli Economisti*, vol. 1, pp. 285-316.

SCHUMPETER J. (1934), *The Theory of Economic Development*, Harvard University Press, Cambridge.

SCHUMPETER J. (1943), *Capitalism, Socialism and Democracy*, Harper & Row, New York.

Environmental impact and innovation in industrial districts

Anna Montini[1] and Roberto Zoboli[2]

[1] University of Bologna, Strada Maggiore 45, 40125, Bologna, Italy
[2] CERIS-DSE, National Research Council, Via Bassini 15, 20133 Milan, and Catholic University, Milano, Italy [+]

1 Introduction

The impact of environmental policies on industrial activities has been studied mainly with regard to specific industrial sectors, e.g. chemical industries, and specific 'sectoral' environmental problems and policies, e.g. regulations on water pollution. Instead, the impact of environmental policies on spatially-concentrated systems of (small) industrial firms has received minor attention so far. In this work, we look at the experience of the Italian IDs to explore the possibility that industrial responses to environmental policies may assume specific features in ID-like economic 'environments'.

The starting point of our analysis is at the crossroads between the results of two lines of research (Section 2). The first one is that regarding environmental policies as a stimulus for organisational and technological innovation and, ultimately, for industrial dynamics. Based on different formulations of the 'induced innovation' hypothesis, this line of research tries to highlight that the response of industry to environmental policy and environmentally-oriented markets is the start of adaptation processes that can be either a source of net economic costs or, according to the so called 'Porter hypothesis', a source of competitive advantages. The second line of research is the 'Neo-Marshallian' perspective on IDs and LSPs, which highlights the role of positive agglomeration externalities and informal relationships in locally-concentrated systems of firms in giving rise to holistic localised competitive advantages and favourable conditions for 'collective action' – whether it be consciously pursued or not.

As a consequence, the main research hypothesis of this work is the possibility that those advantages attributed to IDs-specific socio-economic organisation (or positive externalities) may also give rise to specific innovative responses to (negative) environmental externalities and policies through IDs-specific mechanisms of leadership, diffusion, rivalry, and local cooperation or 'collective' action. Despite the relatively limited information on environmental innovation by ID firms, we shall propose some conclusions on the prevailing modes of

[+] The authors wish to thank Daniele Ferrero for providing useful information. The usual disclaimer applies.

innovative response to environmental policies and the role of IDs' 'traditional advantages' in them (Sections 5 and 6) [1].

Beforehand, however, we will explore a very basic issue, which is the quantification of environmental negative externalities arising from industrial concentration in IDs. Due to the fact that even in this area, structured information is limited, we will carry out (Sections 3 and 4) original estimates of air pollution by manufacturing in Local Labour Systems (LLSs) and IDs of Emilia-Romagna, an important Italian region in terms of ID experience and analysis. By adapting the approach of IPPS (Industrial Pollution Projection System) to Italian data (NAMEA and industrial census), we will propose a measure for the role of industrial composition and concentration for air emissions of Emilia-Romagna at the level of municipalities forming LLSs and IDs.

2 Environmental policy, industrial innovation, and the 'advantages' of IDs

The debate on economic implications of environmental policy still oscillates between two opposing views.

The first view, generally supported by firms and their organisations in the arena of environmental regulation making, is that environmental policy depresses competitiveness because compliance investments create costs either as opportunity costs compared to alternative allocations (i.e. loss of potential productivity gains) or because non-uniform legislation across countries creates competitiveness disadvantages for 'environmental leaders' and/or unfair competition by firms of emerging countries having weak regulations ('environmental dumping').

---

[1] Externalities and market failures represent basic concepts of environmental economics and policy in the neoclassical tradition (see Baumol and Oates, 1988). It is usual to refer to 'Pigouvian taxes' as an economic instrument for eliminating a negative environmental externality (a form of 'technological' external economy in consumption or production). At the same time, in industrial economics and development economics it is usual to refer to positive externalities occurring in ID-like 'environments' as 'Marshallian externalities', or 'external economies', corresponding to both 'technological' and 'pecuniary' externalities (i.e. those working through the price system) if the traditional distinction is accepted. However, in this work we will avoid labelling externalities according to these categories and we will just refer, on the one hand, to environmental negative externalities and, on the other hand, to positive ('Marshallian') externalities arising from spatial concentration of (specialized) industrial activities. The reason for a simple 'positive-Vs-negative' classification is that more sophisticated distinctions between different kinds of externalities, as well as their relationships in the theory of economic development, are still rather elusive (compare, for example, the overviews by Bohm, 1987 and Laffont, 1987). In addition, we will not explicitly refer to externalities and public goods typically arising in innovation processes, which are obviously relevant also for environmental innovation.

The second view, currently referred to as the 'Porter hypothesis' (see Porter and van der Linde, 1995; Jaffe *et al.*, 2003) and now supported by a part of the business community, suggests that firms investing in environmental modernisation (under the 'whip' of tough regulation and/or the 'carrot' of different incentives) can gain advantages for their products and technologies in both domestic and international markets because the one-way direction of environmental policies and 'green' demand by consumers across countries gives leadership advantages to 'environmental innovators'.

Despite the extensive series of studies available on costs and benefits of environmental policies, especially in the US experience, it is difficult to draw neat conclusions[2]. Both the above-mentioned positions are partly weakened by the shortcomings in defining exactly what (international) competitiveness is, and by the limited ability of cost-benefit based models to account for a typical dynamic phenomenon such as induced technological innovation.

The current debate on the economic implications of the 1997 Kyoto Protocol on climate change is an example of how the two positions (i.e. net costs Vs competitive advantages) can arise from a different blend of uncertain strategic considerations, e.g. international leadership in energy-climate technologies, and accountable economic costs and benefits. Actually, despite the diverging policy approaches taken by the USA and Europe partly shaped by diverging views about economic costs of the Protocol, all post-Kyoto strategies include additional investments in energy and climate innovation, which is already a significant part of research programmes in many countries, in the search for environmental *and* economic 'dividends' (see, among others, Edmonds *et al.*, 2000; Biondi *et al.*, 2002). The strategic dimension of the industrial response to environmental policies, as well as the dynamic nature of costs and benefits, is illustrated by several other cases, e.g. the introduction of unleaded gasoline, or the substitution of ozone-depleting CFCs. When innovation is considered, even the concept of cost of environmental policy may become elusive (see Quadrio Curzio and Zoboli, 1997).

Whatever the position in the debate, there is evidence that environmental policies, together with environmentally-sensitive consumption, can be powerful factors behind economic and technological dynamics of single firms, industrial sectors, and economic systems. Many cases support the view that the main impact of environmental policies is organisational and technological change and not the decrease in economic activity, even in sectors directly affected by policy provisions. Some policies actually caused the disappearance of specific product and industrial processes but, in most cases, the substitute products/processes

---

[2] On the US experience in evaluating industrial costs and benefits of environmental policies see Jaffe *et al.* (1995), Morgenstern *et al.* (1998), as well an extensive series of works done at Resource for the Future (www.rff.org). The well-developed state of *ex ante* and *ex post* policy evaluation in the US is stimulated, *inter alia*, by regulations imposing Regulation Impact Analysis (RIA) as a routine requirement, and further by the significant research commitment by EPA.

rapidly took over the market of phased-out products/processes and the same industrial sector was often the producer of the substitutes. This has been the case, for example, with the substitution of ozone-depleting substances by the chemical industry (see Quadrio Curzio et al., 1994; Mokyr, 1995). Innovative responses are not limited to powerful oligopolistic sectors having great R&D capabilities (and the power to influence environmental policymaking) because it also involves SMEs to a significant extent.

During the last few years, increasing research efforts have been directed towards innovation as a response to environmental policy, or to so-called 'dynamic incentive effects'. The theoretical foundations of these lines of research are rooted in the economic analysis of innovation, both of marginalist and evolutionary inspiration (see Jaffe *et al.*, 2003). Most empirical analyses, based on both models and case studies, address the mechanism of environmental policy 'induced innovation' in terms of products, processes, and industries, also giving significant attention to organisational innovation (see, among others, Hemmelskamp *et al.*, 2000; Hemmelskamp and Leone, 1998; Kemp, 1997; Zoboli *et al.*, 2000). A part of these works takes into consideration the role of policy instruments (command and control, economic instruments, voluntary agreements, etc.) in an effort to find evidence of their different capacity in creating incentive for innovation (see, among others, Hahn and Stavins, 1992; Jaffe and Stavins, 1995; Klemmer, 1999, Mazzanti and Zoboli, forthcoming)[3]. The results of these lines of research include, *inter alia*, evidence on the knowledge and competence dynamics implied by industrial responses to policy, the significant role of interaction between industry and policy makers for the timing and nature of the innovative outcome, the role of expected cost and industrial strategies in shaping the innovative response, the importance of inter-firm and inter-industry relationships in making environmental innovation possible.

Research on environment-induced innovation, however, mainly looks at specific industrial sectors or environmental issues and it does not address spatially-concentrated industrial systems or IDs to explore: (a) the environmental impacts of local industrial concentration or 'agglomeration diseconomies', and (b) the mechanism of innovative response to policy by geographically concentrated industrial clusters. In fact, even though Italian IDs have been extensively studied from most economic and social aspects, the role of environmental problems and policies for their technological, organisational, and industrial dynamics has received a very limited attention. If there are economic advantages in firms being concentrated over a 'small' geographical area, there can also be concentrated environmental impacts (or negative externalities), which

[3] Also in the case of induced innovation, a consistent research program is under way at Resource for the Future; see, for example, Parry *et al.* (2002) and Jaffe *et al.* (2000).

raise special concerns for both the population and policy makers[4]. IDs and local systems of production can have critical environmental situations because of the specific technology of the industry in which they are specialised, or because of the particularly strong local concentration of industries even in the case of the latter not being strongly polluting, or because there is a very high local concentration of highly-polluting industries.

Our key research hypothesis is the possibility that agglomeration advantages allegedly enjoyed by IDs in their economic processes, or positive 'Marshallian' externalities, can also play a role in solving growth-dependent environmental negative externalities. Extensive literature on IDs refers to the positive effects of agglomeration and Marshallian externalities (see, in particular, Brusco, 1982; Becattini, 1989). In the context of the recent developments of local externalities theory, Glaeser (2000) suggests that urban economics must focus on local non-market interactions because proximity (and hence spatial density) makes reciprocal relationships easier to start and maintain. Gaspar and Glaeser (1998) emphasise the importance of agglomeration economies both in the urban labour market and the urban ICT sector, while other authors emphasise the advantages in terms of learning and knowledge of concentrated localisation[5]. The hypothesis that Italian IDs are exploiting and can widely exploit the above-mentioned 'traditional advantages' also in responding to environmental policy requirements (and to 'environmental demand' in general) is envisaged by Iraldo (2001).

Our aim is to contribute to a further exploration of this hypothesis with a specific focus on organisational and technological innovation. It must be stressed that our analysis is still largely exploratory in nature because of the limited, albeit increasing, structured knowledge on industrial pollution at the local level (i.e. municipalities) and the limited research results available on the impact of environmental policy in IDs. We will start by an attempt, for the first time in Italy, to estimate environmental impact of certain categories of pollutants at the ID-level in a large Italian region. We will then analyse the available evidence on investments and innovation undertaken by firms in Italian IDs under the stimulus of environmental policies. We shall, finally, discuss some findings about the prevailing modes of innovative responses of IDs to environmental policies. Let us say in advance that available evidence suggests a mixed and evolving picture in which Italian IDs are still lagging behind in exploiting their own 'cooperative advantages' for environmental innovation.

---

[4] It must be noted that many environmental impacts and damages, for example on ecosystems, are non-linearly (e.g. exponentially) related to cumulative doses and spatial concentration of pollutants.

[5] See also the essays by Cainelli and Zoboli, Nuti, Cainelli and De Liso in this book for more detailed analyses.

3 Agglomeration dis-economies and the environmental impact of industry

Even though from the very beginning of industrial revolution it was clear that industrial concentration in 'Marshallian' districts was the source of significant damages to local environment, in particular acidic air depositions and ecosystem alterations (see Ponting, 1991), the environmental dimension of industrial development has been approached by environmental economists mainly from the perspective of the industrial sectors and their specific technologies rather than from the perspective of concentrated industrial location.

Interesting suggestions on the environmental significance of spatially-concentrated economic activities come from contributions in urban and regional economics. According to Verhoef, Nijkamp, Daniel (2001), two strands of literature relating to local and urban 'sustainable development' may be identified. The first strand of contributions refers to 'urban ecology' and looks at sustainable cities from the perspective of a multidimensional set of environmental, social and cultural indicators. Specific threshold values (such as carrying capacity, critical emission levels, etc.) are considered to be necessary for defining and implementing sustainability. The second class of contributions on local sustainability originates from urban economics. In this framework, Orishimo (1982) demonstrates that city size and urban environmental quality are related, and the main negative externalities are determined by urban land use, the urban transport system and the urban way of life. However, urban concentration is not only a source of environmental problems but also provides an opportunity to implement innovative solutions because it provides a wide range of possibilities for energy-efficient patterns (e.g. district heating, public transport), environmentally-benign modes of production (e.g. combined heat and power, waste water treatment), and consumption (e.g. solar energy, insulation of buildings)6.

In the framework of 'environmental Kuznets curve' literature, which addresses the non-linear relationships between economic growth and pollution (see, among many others, Arrow *et al.*, 1995; Hettige *et al.*, 1998), the work by Kaufmann *et al.* (1998) explores the effect of both the spatial density of economic activity and income on the atmospheric concentration of sulphur dioxide in some cities of 23 nations. The results indicate that there is a U-shaped relation between income and atmospheric concentration of SO_2 and an inverted U-shaped relation between the spatial intensity of economic activity and SO_2 concentration[7]. These results would

[6] Reciprocally compensating externalities of a different nature are also envisaged by Henderson (1974) who formalises Marshall's ideas of technological spillovers from one firm to another and suggests that agglomeration disamenities on households may be offset by the productivity advantages on the part of firms.

[7] SO_2 emissions decrease with increases in income between $3,000 and $12,500 while SO_2 concentrations may increase with the first increases in per capita GDP, i.e. $0-3,000.

suggest that, *ex post*, it is the spatial density of economic activity, more than income, that provides the impetus for policies and technologies reducing SO_2 emissions.

The analysis of environmental impacts of industrial concentration is complicated by the fact that emissions from manufacturing combines in an almost indistinguishable way with other local sources of pollution in determining the total environmental pressure. Considering for example air pollution, Figure 1 summarises the relevance of the macro-sectoral source of emission for the most important air pollutants. Manufacturing industries (included in "other industries") account for a significant part of total emissions but they are the main culprits only in the case of methane (CH_4). The transport sector, instead, is the major emitting sector of carbon monoxide (CO), nitrogen oxides (NO_x) and non-methane volatile organic compound (NMVOC). Agriculture is the most polluting sector with regard to the emissions of ammonia nitrogen (NH_3) and nitrous oxide (N_2O). Finally, the energy production industry is mainly responsible for the emissions of sulphur dioxide (SO_2) and carbon dioxide (CO_2).

Figure 1. Main gaseous emissions by sector in Europe (according to EEA, 1999)

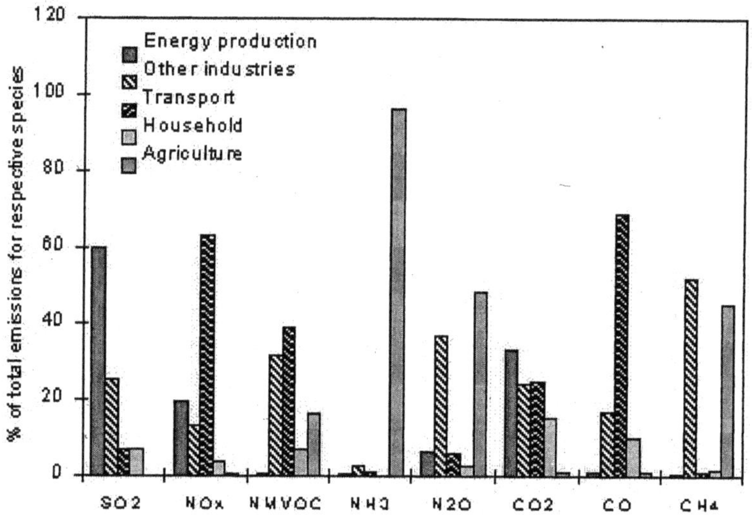

Moreover, in terms of environmental impact, each air pollutant has a different relevance and this depends on several factors, for example (i) the physical characteristics of the diffusion process, including long-distance transportation of pollutants; (ii) the removal processes which have an effect on the length of permanence in the atmosphere and the metabolism of ecosystems as measured by so called 'critical loads'[8]; (iii) the chemical transformations which can generate other dangerous substances. Therefore, at local level, damages from pollutants are strongly related to the level of emissions, local topography, population density, and other physical characteristics of the surrounding area such as the quality of soils and vegetation, and many other factors. Fig 2 summarises the most important air pollutants and their relevance to the state of the environment at local/regional and global level.

Figure 2. Airborne pollutants and their links to environmental issues (Monkhouse, 2001)

	Local-Regional			Global	
	Acidification	Tropospheric O_3	Toxic pollut.	Climate change	O3 depletion
SO_2				2	
NOx				2	
Ammonia				2	
MNVOC			1	2	
CO				2	
CO_2					
CH_4					
N_2O					
HFC, PFC					
SF_6					
Heavy metals (Pb, Hg, etc.)					
Particles					
POPs (e.g. PAH)					

1) Some specific compounds, e.g. benzene, 1.3 butadiene
2) No_x, NMVOC, CO: indirect effects via O_3 (GHG)
SO_2 and NH_3: indirect effects via aerosols, e.g. sulphates/nitrates (local cooling)

The possibility of analysing the specific role of industrial concentration for local environmental conditions is also limited by the still surprisingly scant information on air, water and toxic emissions from industrial activities at the highly disaggregated geographical level. The information which is the most lacking regards *actual* emissions. Most structured datasets on industrial pollution are built-up on indirect estimates, i.e. based on estimated technical coefficients of pollution by industry.

A significant experience in the construction of an industrial pollution database is the Industrial Pollution Projection System (IPPS) produced by the

[8] 'Critical loads' are used as reference measures in international policies for air pollution developed during the 1990s after the Convention of 1979. They represent the quantity of a pollutant that a specific natural environment can receive without suffering stable alterations, according to present state of scientific knowledge.

Infrastructure and Environment team of the Development Research Group of the World Bank (Hettige *et al.*, 1995). Starting from the fact that industrial pollution is affected by the scale of industrial activity and its sectoral composition, IPPS elaborated a wide set of estimates of sectoral pollution intensity, usually defined as pollution per unit of output or per unit of employment. As emphasised by Hettige *et al.* (1995), while physical volume of output would be the ideal unit of measurement, industries (and even establishments within a given industry) often use different units to report the volume of their production. This does not allow for comparison across industries. However, the value of output and plant-level employment do offer such common units of measurement (value of output and employment), and the ranking of industrial sectors by their pollution load is almost identical whether the value of output or employment is used as the unit of measurement. For the purpose of policymaking it is the ranking that is of relevance[9].

The IPPS estimates have been obtained from merging data from industrial activity, i.e. production and employment, from the U.S. Manufacturing Census (which contains information on 200,000 plants) and the Annual Survey of Manufactures (ASM), and data on pollution emissions per unit of industrial activity referred to a sample of 20,000 plants from U.S. Environmental Protection Agency (EPA). Pollution intensities obtained in IPPS can be used to estimate the pollution load of several industrial sectors and are presented as pollution per unit of output or per employee for various branches of industry. Data are referred to the 3-digits ISIC (International Standard Industrial Classification) industries. Based on 1987 estimates of industrial pollution in the United States, the IPPS provides pollution intensities for a variety of pollutants, including those for air, soil, water, and others[10].

Despite the fact that IPPS provides good data sets and have been used in different geographical contexts (particularly for developing countries[11]), there are several limitations in using these data outside the United States. Firstly, it could be questionable to take IPPS data to provide an estimate of emissions in other countries because it implies the assumption that every industry in all countries

---

[9] The approach based on estimated emission factors is used in different environmental areas, as in the case of industrial 'waste factors'. Observations from small sample of firms are extrapolated at the sector level combining the unit emission coefficients (per employee or per unit of production) and a scale variable (production, employment) for the whole sector (see Mertins *et al.*, 1999).

[10] The pollutants considered in IPPS are essentially water and air pollutants together with an estimation of toxics and metals with effects on air, land, and water. As water pollution data, the biochemical oxygen demand (BOD) and total suspended solids (TSS) are considered. As air pollution data, the following are considered: nitrogen dioxide (NO2), fine particulates (PM10), sulphur dioxide (SO2), carbon monoxide (CO), total suspended particulates (PT) and volatile organic compounds (VOC).

[11] For recent applications to Mexico see Aguayo *et al.* (2001), Aroche (1999), Ten Kate (1993), and for the case of Thailand see Laplante and Meisner (2001).

has the same pollution intensity as the corresponding U.S. industry. Secondly, IPPS data are available for only one year (1987) and, when using it for other years, it is implicitly assumed that coefficients of industrial pollution are stable over time.

4 Measuring air pollution by manufacturing industries in Italy

In order to estimate emissions in Italian IDs we will adopt an approach similar to that of IPPS. We will limit our analysis to air pollutants for which data and information are relatively more extensive.

An increasing amount of structured environmental datasets are available from APAT, the Italian environmental agency (see APAT, 2002), also with reference to systems of indicators (i.e. the DPSIR and the system of 'core' indicators of the European Environment Agency) and international classifications of pollutants and economic activities. At the national level, pollution data by source of emission, considering both combustion and production processes of manufacturing industry, are available for sulphur oxides, nitrogen oxides, non-methane volatile organic compounds, methane, carbon monoxide, carbon dioxide, nitrogen protoxide and ammonia.

However, at the local level, in particular at the municipality level, available data are still rather limited. Information is available on several 'high environmental risk areas' but only for the number of people living in each area, the main risk source and the presence (and the status) of specific policies. For example, in the Emilia-Romagna region two areas are classified 'with high environmental risk': the Conoidi area (76 municipalities, 1,459,951 inhabitants, 3,692 km^2, in the provinces of Bologna, Modena and Reggio Emilia), in which the main source of environmental risk are the ceramic tiles industry and the stock farm sector, and the Burana-Po di Volano area (42 municipalities, 509,635 inhabitants, 3,622 km^2, in the province of Ferrara) where the main source of environmental risk is the stock farm sector.

A source providing industrial emission data that can be useful for research purpose is the first National Accounts Matrix including Environmental Accounts (NAMEA) recently produced by ISTAT. The first NAMEA, referring to 1990 data, was published in ISTAT (2000) and ISTAT (2001)[12].

Six air pollutants[13] are considered by NAMEA and they refer to emissions from household consumption (transport, heating and other) and production activities

[12] ISTAT is currently preparing NAMEA for more recent years (around the end of the1990s). For an input-output analysis of the Italian economy including NAMEA see D'Antonio (2002).

[13] Sulphur oxides (SO_x), Nitrogen oxides (NO_x), Methane (CH_4), Carbon dioxide (CO_2), Nitrous oxide (N_2O), and Ammonia nitrogen (NH_3).

(agriculture, industry and services) according to ATECO classification. These macro sources of emission are described in Figure 3 and they are similar to those of Figure 1 referring to international data. The manufacturing industry (included in 'other industries') is the one mainly responsible for carbon dioxide emissions and accounts for about 40% of the sulphur oxide emissions too. Agriculture is mainly responsible for the ammonia nitrogen, nitrous oxide and methane pollution while transport accounts for one third in the nitrogen oxide emissions.

Figure 3. Main gaseous emissions by sector in Italy

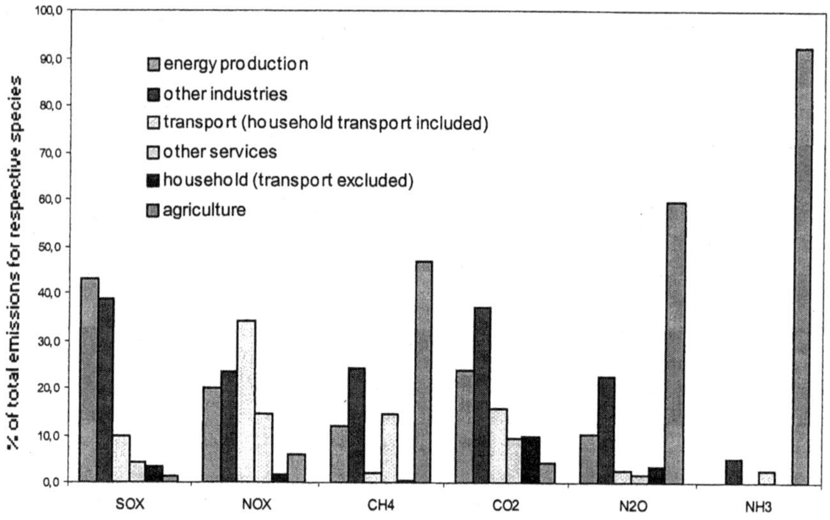

The NAMEA data allow us to estimate manufacturing industry emissions per unit of value added (or employee). By considering the ratio of emissions (six air pollutants) to employment in each industrial branch, it is possible to derive the amount of emission per employee, in a way similar to the IPPS approach. The estimate is possible at regional or sub-regional level under the implicit assumption that unit emissions of a manufacturing industry are the same across all production sites in Italy. An additional, even stronger assumption is that the emissions per employee are the same for different sizes of firm in the same manufacturing industry. We are aware that these assumptions impose limitations on the value of empirical findings, but they make it possible to explore, for the first time at the highly disaggregated geographical level, the environmental pressure due to industry specialisation and manufacturing density.

4.1 Estimates of air pollution in the 'Local Labour Systems' of Emilia-Romagna

To exploit the NAMEA data in an IPPS-like approach at the local level, we need statistical units approximating IDs and local systems of production for which a coherent set of economic data are available. In Italy, the main statistical reference for 'local systems of production' is represented by the Local Labour Systems (LLSs). LLSs are defined according to a specific procedure and are estimated on Census data. A Local Labour System is an employment gravitational area that can be an 'industrial district' if its total employment in one manufacturing industry attains a specific threshold share. The official definition of LLSs in Italy results in 784 LLSs and 199 IDs. Given the main focus of this work, we limit our analysis to just one region, i.e. Emilia-Romagna, which is particularly important for the economic analysis of IDs in Italy (26 out of 50 LLSs of region are industrial districts).

Emission data for the 23 manufacturing industries considered in ISTAT (2001) have been processed in order to obtain the air emissions per employee for each manufacturing industry in each LLS of the Emilia-Romagna region. More specifically, NAMEA's 1990 total air emissions by pollutant and by industry have been divided by total employment in the corresponding manufacturing industries based on the 1991 ISTAT census (ISTAT, 1991) thus obtaining average national technical coefficients of emissions[14]. Detailed results are presented in Tables A.2 and A.3 in the Appendix. The estimate of each emissions in each Emilia-Romagna's LLS has been derived by the following steps[15]:

1. We have computed the number of employees by NAMEA manufacturing class in each LLS by considering the employment data for all the 341 municipalities in the Emilia-Romagna region from the 1996 intermediate industrial census (ISTAT, 1996) in the 2-digits manufacturing industry (3-digits in the class 26 to derive the appropriate hybrid classes according to the NAMEA ones)[16];

---

[14] The manufacturing industries considered by NAMEA data are 20 ATECO classes at 2-digits and three 2-digits ATECO "hybrid" classes (food and beverage, code 15 and tobacco, code 16 have been aggregated; two groups have been derived inside the non metallic mineral products class, code 26: glass and products, code 26_1 and other non-metallic mineral products, code 26_2/8).

[15] As mentioned above, Italian data on emission intensities based on NAMEA are available only for 1990 whereas employment data refers to the 1996 census. This time inconsistency could be eliminated when updates of NAMEA and Census data will be available.

[16] The 341 municipalities in the Emila Romagna-region are included in 50 Local Labour Systems (LLSs); six of these LLSs, with its main centre in the region, include municipalities of neighbouring regions and two of these with the main centre outside the region include municipalities of Emilia-Romagna. To keep a regional focus, we consider all the LLSs only with regard to *regional* municipalities. It results that six municipalities included in two non-regional LLSs (according to the location of the main centres,

2. LLSs employment data have been linked to the Table A.2 data (LLS employment data by manufacturing industry/100 multiplied by air pollution emission by manufacturing industry per 100 employees) to obtain the six air pollutants' emissions in each LLS;
3. We have then normalised each LLS data obtained in the previous step by dividing the six pollutants' emissions by the total LLS territorial surface (square kilometre, sqkm) and the total LLS inhabitants[17].

Table 1 and 2 show the first ten LLSs in terms of emissions per capita and per square kilometre respectively for each of the six different pollutants. If the LLS is classified as an industrial district it is labelled by ID in brackets.

Even at a first glance, industrial districts are quite frequently on the top of the ranking of the most polluted LLSs. Considering the rank by per capita emissions, two industrial districts, Sassuolo and Castellarano, with a high specialisation in other non-metallic mineral products (ceramic tiles) are the highest emitters for sulphur oxides (SO_X), nitrogen oxides (NO_X), carbon dioxide (CO_2) and nitrous oxide (NO_2) with absolute per capita values which are double, or even more, of those ranking third. Moreover, the ceramic tiles ID of Sassuolo comes first in terms of the per area emissions of all pollutants.

Normalised data per inhabitants and territorial surface in each LLS are mapped in Figure 4 and Figure 5. As a first approximation, we consider emission per capita to be influenced by sectoral specialisation and emission per square km to be influenced by economic concentration. Some IDs can rank top in terms of emission per land area for pollutants which are not specific to their industrial specialisation due to the high economic concentration in their geographical areas.

Per capita data maps (Figure 4) show that Sassuolo (code 257) and Castellarano (code 245), the two districts of the ceramic tiles industry, have the highest emissions for all the pollutants with the exception of methane (CH_4); N_2O and NH_3 values are the highest also considering the two LLSs with the highest chemical industry presence in the region, i.e. Ferrara (code 267) and Ravenna (code 270), although they are not classified as industrial districts.

It is possible to highlight the relationship between LLSs main specialisation and the share of emissions due to this specialisation (Table A.3 in the Appendix). High specialisation (from about 30% to 60%) in other non-metallic mineral products (code 26_2/8) – i.e. the ceramic-tiles industry – results as having more than 80% in terms of share of emissions for the main gaseous air pollutants due to fossil fuel combustion i.e. sulphur oxides (SO_X), nitrogen oxides (NO_X), carbon dioxide (CO_2) and, less clearly, nitrous oxide (N_2O). Another industry in which Emilia-Romagna is highly specialised, the food industry, has the highest shares of

Cremona and Suzzara respectively) are maintained in the two corresponding LLSs but without considering the other non-regional municipalities involved. Similarly, for the six LLSs with the main centre in the region (Piacenza, Pievepelago, Porretta Terme, Ferrara, Cattolica and Morciano di Romagna), which involve non-regional municipalities, we consider only the regional ones.

[17] The results for each LLS are available from the authors upon request.

methane (CH_4) emissions in the LLSs. On the contrary, high shares of employment in machinery/mechanical industry are not associated with high shares of pollutants due to this industry, although the corresponding LLSs or IDs may or may not be highly polluted, depending on their secondary specialisation.

Square kilometre data maps (Figure 5) show that SO_X, NO_X and CO_2 have the highest values in the Central-Western LLSs while methane (CH_4) and nitrogen based gas (N_2O and NH_3) emissions have the highest values in a higher number of LLSs. A lower manufacturing density in the LLSs, such as in the 'low Ferrarese' LLSs (up-right) and in the Apennines LLSs (South-Western borderline), is related to lowest per-square-kilometre emissions.

Econometric analysis of these data is not presented in this study because it has at least two shortcomings: (i) emissions data are not *observed* data on air pollution, but *estimated* data, and (ii) the low number of LLSs considered. To regress our LLSs pollutant emissions on manufacturing employment concentration, for instance, would result in tautological findings. Each LLS emission amount is derived from the number of employees, although the employees are considered at the industry level (see Table 3 for the correlation coefficients between air pollution emissions and social indicators). With missing information about *real* pollution data in *each* area (e.g. spatially disaggregated measures of SO_2 concentration), it is difficult to estimate the relationship between pollution and income or spatial intensity of economic activity or concentration of firms belonging to the same industry[18].

[18] A comparison of our estimates based on Italian coefficients and those that would result from the Italian data by using the IPPS coefficients is presented in the Appendix (Table A1).

Table 1. First ten Local Labour Systems by per capita emissions (kg); (ID) if industrial district

Rank	*SOx*		*NOx*		*CH4*		*CO2 (a)*		*N2O*		*NH3*	
1	Sassuolo (ID)	67.13	Sassuolo (ID)	83.85	Langhirano (ID)	61.66	Sassuolo (ID)	28.54	Sassuolo (ID)	2.86	Ferrara	1.34
2	Castellarano (ID)	49.02	Castellarano (ID)	61.96	Tizzano Val Parma	33.58	Castellarano (ID)	21.13	Castellarano (ID)	1.78	Ravenna	1.30
3	Fornovo di Taro (ID)	33.19	Vignola (ID)	24.70	Santa Sofia (ID)	33.12	Vignola (ID)	7.81	Ravenna	1.62	Sassuolo (ID)	0.94
4	Fiorenzuola d'Arda (ID)	23.11	Pavullo nel Frignano	21.79	Parma (ID)	31.37	Pavullo nel Frignano	7.11	Ferrara	1.58	Correggio (ID)	0.89
5	Vignola (ID)	18.23	Bedonia	17.15	Castellarano (ID)	27.35	Fornovo di Taro (ID)	6.64	Correggio (ID)	1.33	Parma (ID)	0.65
6	Ravenna	16.71	Parma (ID)	16.54	Mirandola (ID)	27.23	Bedonia	5.17	Parma (ID)	0.99	Argenta (ID)	0.50
7	Pavullo nel Frignano	16.28	Mirandola (ID)	15.71	Vignola (ID)	21.72	Imola	4.98	Mirandola (ID)	0.89	Mirandola (ID)	0.48
8	Correggio (ID)	15.01	Fornovo di Taro (ID)	15.58	Sassuolo (ID)	20.35	Mirandola (ID)	4.95	Vignola (ID)	0.73	Lugo (ID)	0.44
9	Mirandola (ID)	13.96	Imola	15.50	Correggio (ID)	20.00	Fiorenzuola d'Arda (ID)	4.63	Imola	0.72	Busseto (ID)	0.38
10	Parma (ID)	13.66	Correggio (ID)	15.40	Lugo (ID)	19.22	Correggio (ID)	4.54	Argenta (ID)	0.68	Castellarano (ID)	0.38

(a) 1000 Kg per capita

Table 2. First ten Local Labour Systems by per square kilometre emissions (kg); (ID) if industrial district

Rank	*SOx*		*NOx*		*CH4*		*CO2 (a)*		*N2O*		*NH3*	
1	Sassuolo (ID)	41,783	Sassuolo (ID)	52,188	Sassuolo (ID)	12,666	Sassuolo (ID)	17,767	Sassuolo (ID)	1,781	Sassuolo (ID)	584
2	Cattolica	3,867	Cattolica	5,732	Parma (ID)	8,846	Cattolica	1,517	Ravenna	359	Ferrara	300
3	Parma (ID)	3,851	Parma (ID)	4,666	Modena (ID)	7,067	Vignola (ID)	1,348	Ferrara	353	Ravenna	289
4	Correggio (ID)	3,754	Vignola (ID)	4,266	Cattolica	6,481	Modena (ID)	1,241	Correggio (ID)	332	Correggio (ID)	223
5	Ravenna	3,710	Modena (ID)	4,174	Carpi (ID)	6,140	Reggio E. (ID)	1,241	Parma (ID)	279	Parma (ID)	184
6	Modena (ID)	3,474	Reggio E. (ID)	3,961	Cesena	5,625	Castellarano (ID)	1,228	Bologna	212	Bologna	140
7	Reggio E. (ID)	3,282	Correggio (ID)	3,851	Reggio E. (ID)	5,390	Parma (ID)	1,178	Cattolica	209	Lugo (ID)	87
8	Vignola (ID)	3,149	Castellarano (ID)	3,599	Correggio (ID)	5,001	Correggio (ID)	1,134	Modena (ID)	188	Modena (ID)	83
9	Castellarano (ID)	2,847	Carpi (ID)	3,374	Mirandola (ID)	4,536	Rimini	891	Reggio E. (ID)	179	Reggio E. (ID)	82
10	Carpi (ID)	577	Rimini	3,041	Rimini	4,303	Carpi (ID)	848	Mirandola (ID)	148	Mirandola (ID)	80

(a) 1000 kg per sqkm

Figure 4/a. Per capita emissions by pollutant and by LLSs in Emilia-Romagna

SOx, kg per capita

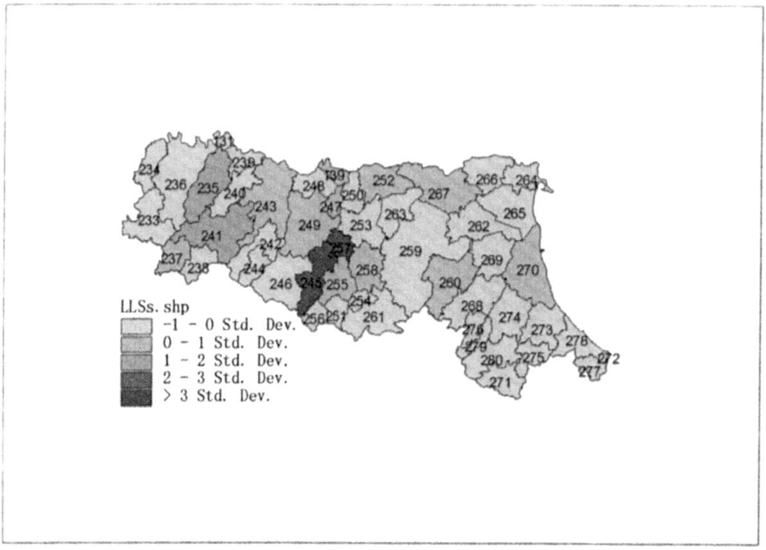

NOx, kg per capita

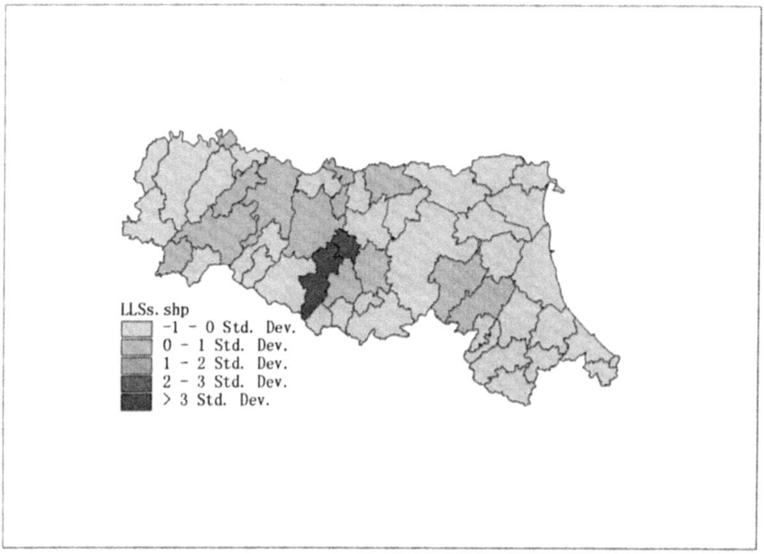

Figure 4/b. Per capita emissions by pollutant and by LLSs in Emilia-Romagna

CH_4, kg per capita

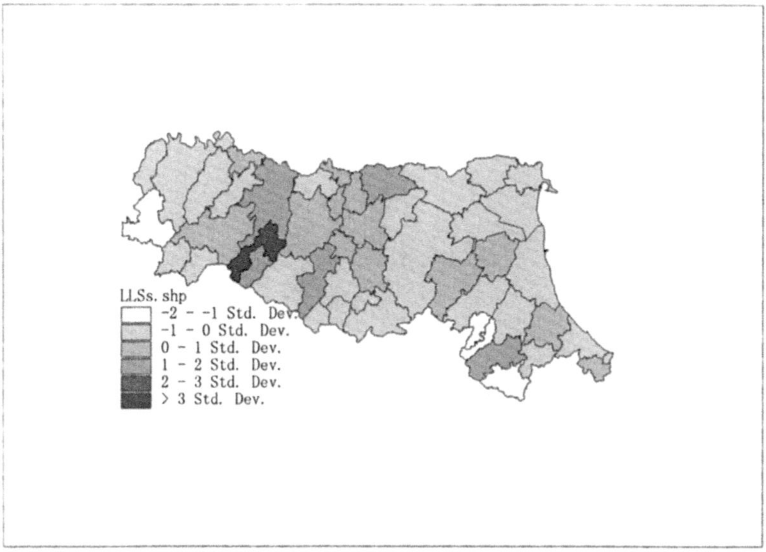

CO_2, 1000 kg per capita

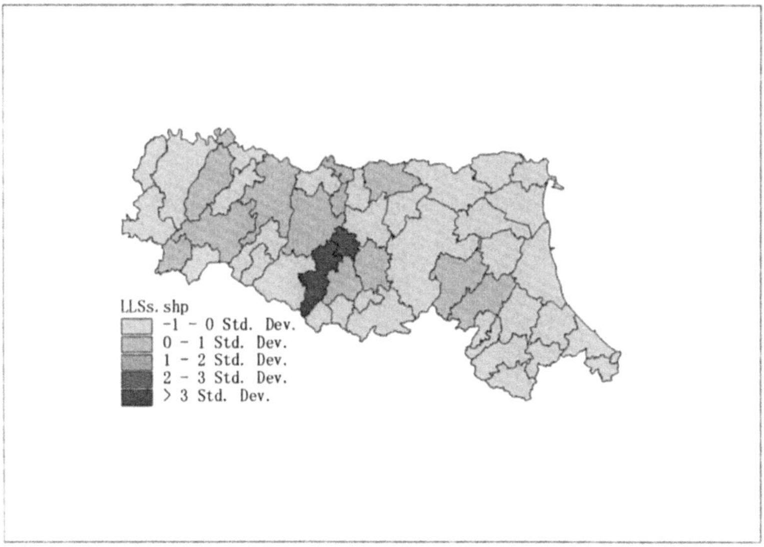

Figure 4/c. Per capita emissions by pollutant and by LLSs in Emilia-Romagna

N_2O, kg per capita

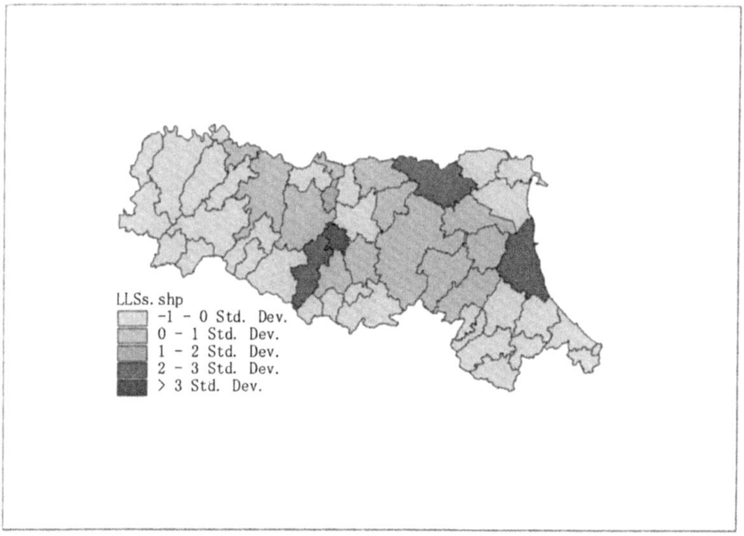

NH_3, kg per capita

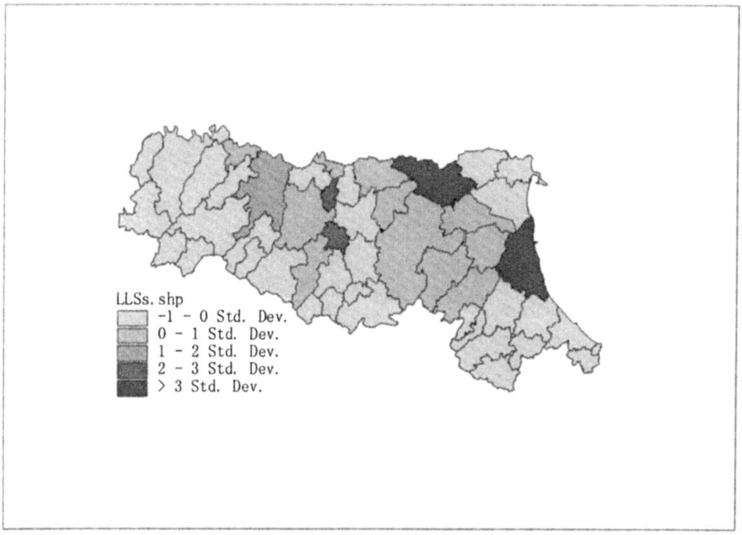

Figure 5/a. Per square kilometre emissions by pollutant and by LLSs in Emilia-Romagna

SO_X, kg per sqkm

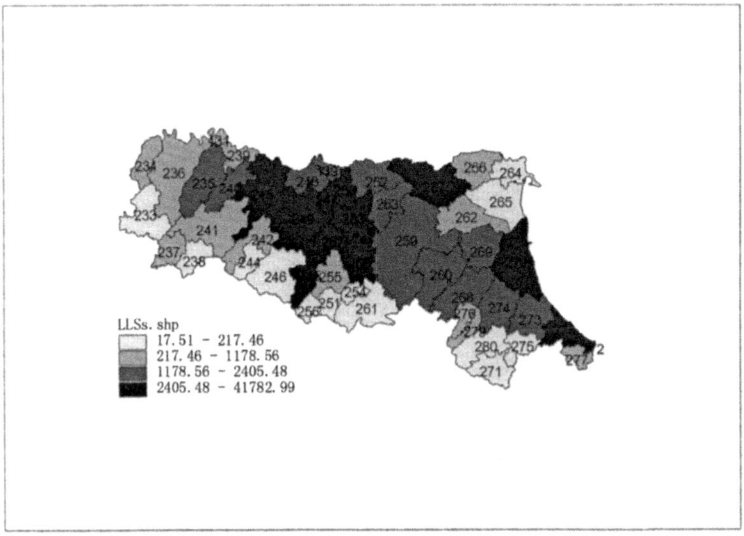

NO_X, kg per sqkm

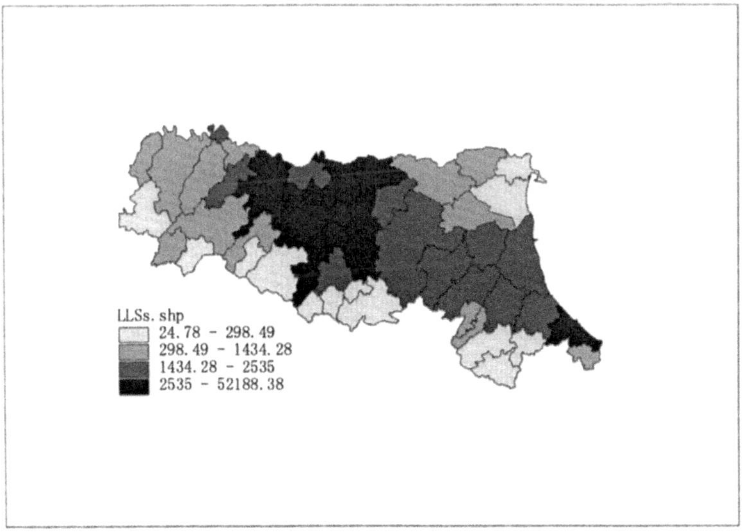

Figure 5/b. Per square kilometre emissions by pollutant and by LLSs in Emilia-Romagna

CH_4, kg per sqkm

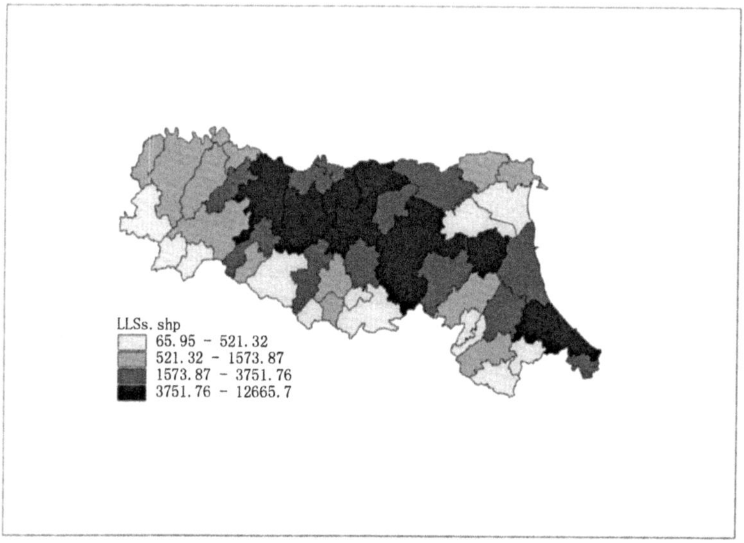

CO_2, 1000 kg per sqkm

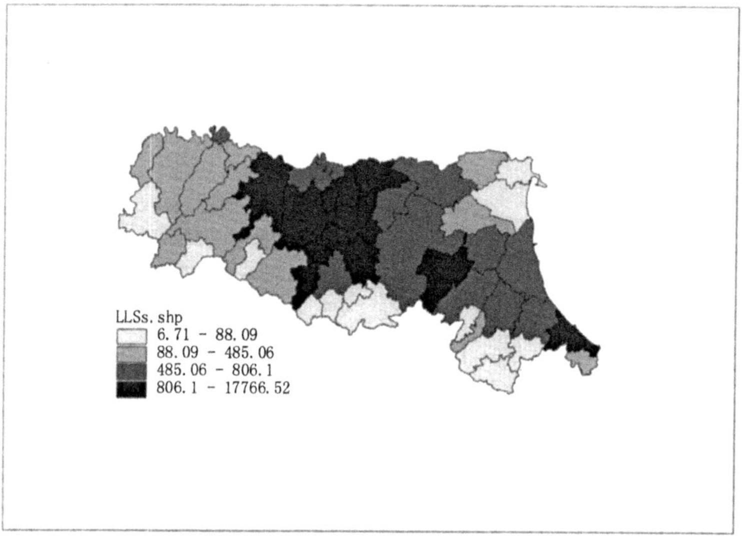

Figure 5/c. Per square kilometre emissions by pollutant and by LLSs in Emilia-Romagna

N_2O, kg per sqkm

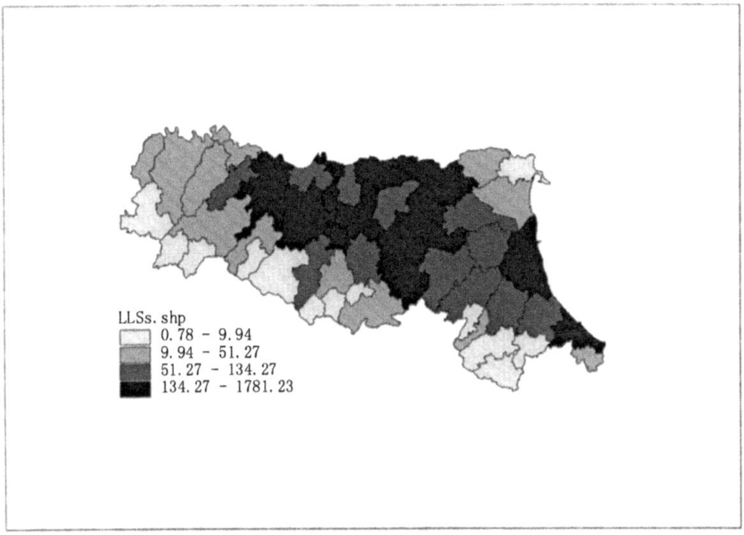

NH_3, kg per sqkm

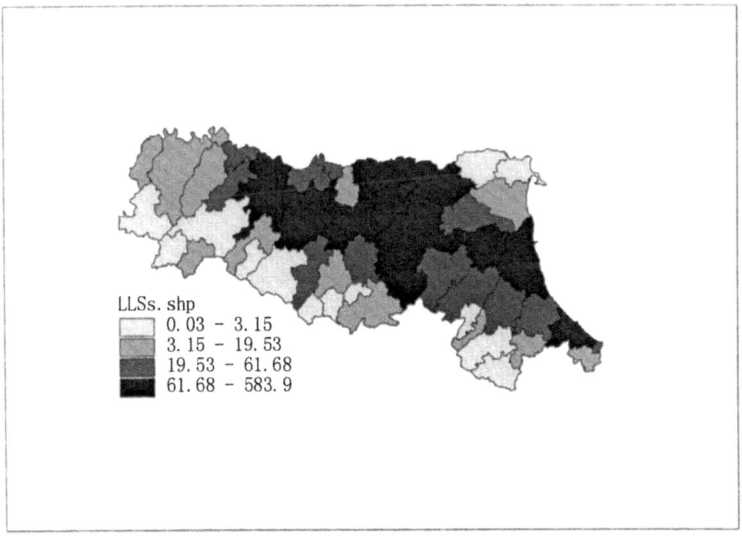

Table 3. Correlation coefficients between air pollution and social indicators in LLSs of Emilia-Romagna

	Manufacturing density	*Population density*	*Available income*
SO_X (kg per sq. km)	0.80	0.50	0.25
NO_X (kg per sq. km)	0.81	0.51	0.24
CH_4 (kg per sq. km)	0.90	0.81	0.39
CO_2 (kg per sq. km)	0.79	0.48	0.23
N_2O (kg per sq. km)	0.82	0.56	0.27
NH_3 (kg per sq. km)	0.71	0.53	0.29
SO_X (kg per capita)	0.50	0.21	0.26
NO_X (kg per capita)	0.55	0.23	0.25
CH_4 (kg per capita)	0.11	-0.03	0.34
CO_2 (kg per capita)	0.54	0.23	0.23
N_2O (kg per capita)	0.56	0.33	0.31
NH_3 (kg per capita)	0.37	0.30	0.26
Manufacturing density	1.00	0.84	0.32
Population density	0.84	1.00	0.20
Available income	0.32	0.20	1.00

Source: our processing

5 Responses to environmental policies in Italian industrial districts

The above results for Emilia-Romagna clearly suggest the importance of local industrial concentration for the local environmental pressures as represented, in our case, by air pollutant emissions. The high potential impact depends on either specific features of the sectoral production technologies or spatial concentration of industry, and it is obviously highest when there is a high local concentration of industrial activities producing high levels of specific or multiple pollutants.

Not surprisingly, historical developments of environmental policy in Italy have significantly involved manufacturing industries in IDs such as: (a) sectoral policies on a specific issue, e.g. water quality, air emissions, waste, or other environmental impacts generated by a particular industry; (b) 'agglomeration' of specific environmental policies when the dominating industry is a multiple-pollutant producer; (c) more recently, locally integrated policies addressing the state of the environment in the local system.

The history of environmental policies in IDs is very diverse and sometimes specific but, following Massarutto (1997) and Iraldo (2001), the process of the local application of environmental policies can be summarised by some major phases as defined, in particular, by the evolving interaction among industrial, social, and institutional actors[19].

The first phase, up to the early 1980s, was marked by a 'social pact' involving the full acceptance of the local environmental impact of industry due to the priority commonly being given to economic development, also leading to a benevolent attitude by local authorities and a mild application of environmental legislation. The second phase, corresponding to the 1980s, marked a change in attitude on the part of local institutions under pressure by the local population which was increasingly concerned with environmental and health impact of industry, a change in the attitude of trade unions towards the issues of labour conditions, a more extensive and institutionally strong environmental policy and an increasing number of localised environmental crises due to industrial activities. In this phase, local institutions began to assume the role of direct and indirect supplier of environmentally relevant infrastructures and services, and also started to focus on the possibly cooperative dimension of local environment policy despite the still conflictual dimension of the latter. The third phase, starting from the 1990s, can be characterised as one in which compliance with environmental legislation accelerated even in 'marginal' enterprises, local institutions increased their support and achieved results towards a locally cooperative approach to policy implementation, and some leader companies began to see environmental competitiveness as a possibility (see Iraldo, 2001).

As regards the 'third phase', the key point is exactly about how far IDs go along the above described lines and how fast the process is towards: (a) a cooperative, collective management of environmental policies at the ID level, thus exploiting the 'traditional' ID advantages, i.e. density of interactions, agglomeration economies, 'Marshallian' positive externalities; (b) the perception by companies of environmental innovation as a source of competitive advantages at the national and international level, i.e. the Porter hypothesis. These two issues are particularly important because current developments in environmental policies and environment-sensitive markets are going in the one-way direction of preventive integrated approaches – an opportunity for locally integrated systems of production. We will look at these developments from the point of view of technological and organisational innovation.

5.1 Environmental investments and policy-induced innovations

Despite the great interest in IDs in Italy, available information on the investments and innovations related to environmental policy is rather limited and the work by Iraldo (2001) is the most extensive economic analysis carried out so far. Most information comes from direct interviews and non-structured case studies

[19] The three phases suggested for IDs might referred to also in describing national environmental policy in Italy (see, in particular, Lewanski, 1997).

although some recent developments of local Agenda 21 and environmental reports by industrial associations and local authorities can supply fresh datasets and information, in particular on the state of local environment (see Provincia di Modena *et al.*, 2002 for the case of the ceramic-tiles district). Despite these limitations, some features of environment-related investments are clearly emerging and they allow us to identify the prevailing modes of innovative response and adaptation[20].

A set of information on innovations related to environmental policies in IDs comes from the preparatory work for the National Plan for Sustainable Development produced by the Ministry of Environment in 1999-2000. Some of the results in terms of organisational and technological solutions that have either been introduced or are being introduced in specific IDs are reported in Table 4 and can be summarised as follows:

- Most initiatives for introducing environmentally low-impact technologies have been taken in the IDs of furniture, leather, machinery, ceramic tiles;
- They mainly originate from command and control policies and mainly address some environmental issues for which regulation is more developed and demanding (i.e. water emissions, air emissions, and waste treatment);
- Investments and innovative solutions are mainly end-of-pipe, and generally consist of the adoption of well-known technologies for the chemical-physical cleaning of sludge, the reduction of air emissions and waste treatment before disposal; investments for the recovery of materials and energy from waste and the reuse of new/old scrap in production processes are less significant;
- In some significant cases, investments and treatment operation are locally 'integrated' or 'collective' as they are centralised in relatively high-scale plants to serve a number of companies in the ID or the local system, also as a result of initiatives taken in cooperation with local authorities in the framework of 'normal' or 'special' local plans and programmes;

---

[20] We will disregard the issue of effectiveness of environmental policies in IDs, on which structured evidence is very limited. For the Sassuolo district, probably the most well documented, Cervi (2000) shows that during the 1990s the concentration of some pollutants decreased thanks to: (i) improved abatement technologies, (ii) restricted regulations in favour of employees and the environment and (iii) very high penalties for non-compliance. New specific information on the environmental situation of Sassuolo district comes from the 'Environmental balance of the ceramic district' prepared by the regional environmental agency (ARPA) in 2002. The report, based on the time evolution of indicators referred to the DPSIR framework, shows a mixed picture of evolution of pressures, impacts, and responses during the second part of the 1990s (see Provincia di Modena *et al.* 2002). These results can be useful for updating data in analyses such as that carried out in Section 4.1, which also will have opportunities for updating from the on-going work by ISTAT for preparing NAMEA accounts for late-1990s. On the problems of environmental policy effectiveness analysis see Zoboli (2000).

- Investments in clean technologies (generally speaking: integrated production process or plants that incorporate prevention-oriented solutions to pollute less and/or save natural resources) are relatively rare and can be observed only in few IDs and companies; however, there is an increasing number of programmes and projects addressing technological initiatives that belong to the class of clean technologies; the situation is clearly one of transition;
- Similar to the average situation in Italy, environmental management systems, and in particular EMAS and ISO 14001, are not very widespread in ID companies but there is an increasing interest and a blossoming of projects under way aimed at adopting these instruments;
- There is, in general, a great number of environmental initiatives and projects under way that involve companies and local authorities or institutions; in most cases, their actual significance is mainly in creating the awareness of problems and explore the different dimensions (technical, social, economic, financial) of the solutions.

Table 4. Technologies for reducing environmental impacts in Italian IDs

Ceramics	Sassuolo: Reuse of sludge in the production cycle after treatment; recovery of ceramic scrap, used water from painting processes, ash from air emission treatment; optimisation of paint consumption
Leather	Valdarno: One plant for chromium recovery and several plants for the treatment of waste water; two plants for sludge drying and reuse are under development; a local Plan of Action for the reduction of special waste production includes several provisions for minimising waste production from leather industry Arzignano: One plant for industrial waste recovery and several plants for waste water treatment; six companies participating in a LIFE project for the development of clean technologies; the latter consist of the substitution of sulphur derivatives in the cleaning of leather, the substitution of old technologies for refining, new systems for painting, electrochemical technologies for water treatment, and recovery of chemicals; three companies participated in a LIFE project for the introduction of EMAS systems of environmental management; a protocol for the preparation of an environmental report in leather-working companies has been established by the Province and the Industry association; an initiative is under way for the definition and classification of the Best Available Technologies for the leather industry
Paper works	Città di Castello: High diffusion of technologies for waste water cleaning and sludge treatment
Marble	Apuo Versiliese: One plant for treatment and reuse of waste from marble working; a system for separate collection of working residues and a system for sludge treatment is being established in the framework of a more comprehensive plan by the Tuscany Region, including auditing, education of workers etc. Valpolicella: One plant for drying sludge from industrial water recovery; the Province proposed that companies adopt some new technologies such as

	special pressing of waste water solid content with recovery of calcium carbonate and silicates as well as the use of chemicals with low environmental impact in finishing
Machinery	Lecco: Some companies use membrane and electrochemical technologies for recycling waste water and total or partial recycling of cooling water; clean technologies are used for painting, e.g. water powder paintings, and recovery of chromium baths; two companies intend to adopt an environmental management system Valsesia: One pant for industrial waste water recovery; some companies use paints with low environmental impact, closed loop washing, water-based substances, automatic painting processes with low dispersion; the diffusion technologies for recovering heavy metals is instead very slow; some companies intend to adopt an environmental management system
Furniture	Manzano: A plant for energy recovery from wood residues is being activated by a private consortium; electrostatic and automatic painting systems are widespread among companies whereas few companies are using water-based paints and there are no examples of complete closed-loop systems of water use; a project for the environmental certification of companies is being set up. Alto Livenza: Four companies adopted overspray recovery systems, and use natural oils as substitutes for traditional solvents; some companies are experimenting with water-based paints; the Province proposed the use of some low-impact technologies to reduce VOC emission and new technologies for painting. Pesaro: The use of water based paints is widespread and there are various experiments taking place for the recovery of energy from wood residues
Textile	Biella: Systems for the recovery of cooling water from painting and conditioning are very wide spread; there area some cases of reuse of process heat. Prato: One integrated water treatment system with two centralised plants for cleaning, one plant for industrial water recycling and a special industrial water distribution system allowing for partial recovery; there are plants for the selection and separation of residues of textile natural fibres for reuse.
Glass	Murano: The use of a special oven with low energy use has been experimented; glass mixes in form of pellets containing antimony instead of arsenic and favouring less evaporation in the oven are being used; systems for recycling water are very widespread.

Source: personal communications on the preparatory works of the Plan for Sustainable Development, 2000, Ministry of the Environment.

Additional evidence is provided by the second inquiry on innovative environmental action of ID companies (‘*Ecodistretti*’) sponsored by Ente Fiera di Padova and Legambiente, and carried out by Ambiente Italia (see Cariani, 2001). The inquiry is based on questionnaires to ID-related organisations supplemented by literature and web-sites analysis. The indirect coverage (i.e. firms associated to the replying entrepreneurial organisations) is around 73,000 firms with 486,600 employees.

The main results are summarised below and Table 5:

- ‘Integrated environmental infrastructures’, i.e. those managed jointly by private and/or public companies supplying many firms in the ID, are relatively widespread for the treatment, recovery, and landfill of special waste, cleaning of waste water, drying of sewage sludge, and water supplying systems; waste is the sector with the greatest coverage by integrated infrastructures (73% of recorded replies), followed by water (25%) whereas only the remaining 3% deal with energy; most of the investments and technologies are 'end of pipe' and they are mostly (68% of replies) for treatment and cleaning, 27% are for recovery and reuse of materials and water and only 5% for the saving of resources (energy and water); the highest degree of completeness of the infrastructure (in terms of the number of companies served in the ID) is achieved by Sassuolo which is also the only one to have an integrated system for energy management;

- 'Integrated environmental services' are those infrastructures and services created by public or private initiatives that serve the companies in the ID; their diffusion in the different IDs is illustrated in Table 5; the most widespread forms of local environmental services are the supply of education and training, laboratories for analysis, and R&D; also in this case, Sassuolo seems to have the most complete system of such services;
- The adoption of clean technologies is limited but increasing; the main sectors concerned being water (especially water recycling and saving), waste, air emissions, chemical products hazard, and energy;
- The development of environmental certification in IDs is lagging behind; there are only 23 SMEs adopting ISO 14001 in IDs, corresponding to around 3% of all Italian companies using this instrument, and there are 2 ID companies adopting EMAS compared to a total of 35 sites in Italy;
- Other initiatives and projects under way include: local environmental accounting and local Agenda 21 (4 IDs); Life Cycle Analysis (4 IDs); ID-level EMAS (3 IDs); voluntary agreements (6 IDs); environmental accounting in firms (2 IDs).
- The adoption of environmental marks are being pursued by projects in 4 ID: Sassuolo (ceramic tiles), Poggibonsi Quarrada (furniture), and Arzignano and Santa Croce (leather).

The features of environmental investments and innovation emerging from the above analyses are only partly confirmed by a series of interviews we have carried out in 2001 with companies in selected IDs of Emilia-Romagna. The results for 39 companies in the ID of ceramics (Sassuolo) and 55 companies in the food industry ID of Parma are reported in Table 6. There is a significant difference between the environmental issues to be faced by the two sectors as an obvious consequence of their different technologies and products, and in particular:

In Sassuolo's ID:

- Companies must face an extensive series of issues and regulations, and those considered as most significant are air emissions, waste (in particular special waste), and noise; legislation on parks and protected land (non-metallic minerals are the main primary input) has also a role.
- None of the companies have part of their businesses in the environmental market, e.g. environmentally innovative plants and equipments or services.

Table 5. Integrated local environmental services in selected Italian IDs

ID	*Sector of specialisation*	*A*	*B*	*C*	*D*	*E*	*F*	*G*	*H*
Sassuolo	Ceramics	x	x	x	x	x	x	x	x
Poggibonsi	Furniture	x	x	x		x	x	x	
Manzano	Furniture	x	x		x	x		x	
Como	Textile	x	x	x	x			x	
Arzignano	Leather		x	x		x	x	x	
Biella	Textile		x	x		x		x	
Calangianus	Cork			x	x	x		x	
Prato	Textile	x			x	x	x		
Murano	Glass								
S. Croce sull'Arno	Leather		x			x		x	
Alto Livenza	Furniture		x			x		x	
Solofra	Leather		x			x			
Lumezzane	Metal working		x			x			

Keys: A = research on eco-products; B = research on environmental technologies; C = environmental certification; D = services for the diffusion of environmental technologies; E = laboratories for chemical and physical analyses; F = consulting service for innovation; G = education and training; H = energy auditing.
Source: Cariani (2001).

- 21 companies out of 39 make investments to respond to environmental legislation, at a level that corresponds to around 8% of total investments, a relatively high level.
- Most companies claim that environmental legislation is a source of cost disadvantage compared to other Italian areas and foreign competitors.
- Two companies are certified ISO 14000 and one adopted EMAS.

In Parma's ID:

- For companies in the food ID of Parma, compliance with environmental legislation is relevant for water emission, special waste, and health-safety regulation on processes and products.
- During the last few years, most food companies have made investments related to environmental or health policy requirements, especially the smallest companies due to the past delay in complying with evolving regulation.
- Most companies claim that environmental regulation is a source of cost disadvantages compared to companies in other regions and foreign competitors.
- Only 5 out of 55 companies carry out research on environmental problems; 5 companies are certified ISO 14000, one adopted EMAS, and one adopted LCA.
- Most companies expect that changing regulation on product quality and controls over the food production process will affect company business.

- For many companies, safety and health requirements will create the need to ask suppliers for the certification of raw materials and they expect to be forced to increase their direct control over their procurement chain.

The results suggest that, in the case of Sassuolo, despite the significant progress achieved in reducing environmental impacts – and the consequent relatively good position compared to other Italian IDs – the environment remains a continuous source of concern that calls for additional investments and the adoption of new approaches. However, it does not lead to developing businesses in environment-related sectors. At the strategic level, in the recent past the industrial association of the Italian ceramic tiles industry has tried to influence the European policy making process for the standard setting on Ecolabel for the ceramics industry, an attempt which was only partially successful and marked by resistance by European competitors.

In the case of the food district of Parma, results suggest that new or expected regulations on food health and safety are a stimulus for direct investments and organisational innovations in the relationships with suppliers (in particular, 'producer responsibility' along the production chain up to tracking all primary inputs). In this case, the attempt on the part of the Italian government to locate the European Food Agency in Parma, which may allow the Italian industry to play a significant role in the European regulation making on food safety to be developed in the next few years, has been successful only after a troubled political process.

Table 6. Main environmental issues and selected initiatives by companies in the IDs of Sassuolo (ceramic tiles) and Parma (food)

	Sassuolo (39 firms)	*Parma (55 firms)*
Area of environmental regulation of medium-to-high concern		
Air emission	36	9
Water emission	15	33
Waste classified as urban	11	10
Waste classified as special	24	44
Waste classified as hazardous	16	0
Recovery/recycling of materials	14	9
Recovery of energy	9	17
Water supply	5	11
Electro-smog	0	0
Noise	28	14
Health and safety regulation on products	16	48
Land use constraints	5	8
Certification and environmental management		
ISO 14000	2	5
EMAS	1	1
Eco label	0	0
IPPC	0	0
LCA	0	1

Source: our elaboration on direct interviews with companies (2001).

Findings from interviews and other evidence suggest that single companies have a low propensity to consider the environment as a strategic issue of competition and do not make significant investments to gain environmental advantages in their markets – if they do it can be considered the 'normal' result of a more stringent environmental regulation compared to some competitors in some foreign countries. Other answers to the questionnaire suggest that action on these issues is largely delegated to associations and other organisations, as in the case of the above mentioned (and generally unsuccessful) attempts to have a role in the European process of environmental and health regulation making, or they are expected from service structures of local institutions and 'collective' infrastructural investments in cooperation with local authorities.

Some interesting developments towards 'collective action' for the environment are taking place through local environmental 'voluntary agreements'. During the last few years, following a similar trend in many European countries, the establishment of 'voluntary agreements' (VAs) or, in general, environmental agreements, between companies and regulators have been gaining acceptance and momentum. A recent analysis of environmental agreements in Italy (see Camera, 2002), supplies evidence on the significance of environmental agreements at the local level and IDs.

Compared to 15 agreements recorded at the national level, there are 28 agreements at the local level generally involving a variable set of companies and authorities (Table 7). The forms of 'protocol of intentions', agreed programmes, VAs, framework programmes, are the prevailing ones. Most of them address air pollution and waste, and others address management and certification instruments. The agreements that can be referred to IDs are 4 out of 12 in Emilia-Romagna and 1 out of 2 in Veneto. Although these numbers are very low, they are promising for the development of such instruments. The very differentiated features of these agreements, which fall in the area of 'cooperative approaches' to environmental policy and respond to specific problem-solving aims and relationships between the parties involved, is an obstacle to perform a clear analysis of their implications for innovation[21].

---

[21] Still in search of a sound and satisfactory taxonomy, VAs and environmental agreements represent a way of complying with environmental policies and regulations while escaping the alleged (by industry) shortcomings (i.e. excessive costs and ineffectiveness) of both economic instruments and command-and-control approaches. Generally associated to "extended producer responsibility" principles (see Davis *et al.*, 1997; Palmer and Walls, 1999), they pursue economies of cooperation and self-management, possibly avoiding the issues of asymmetric information between regulators and industry. Voluntary agreements, however, are not fully comparable to other policy instruments. Their incentive structure is not standardised and it is, instead, tailored to a problem as VAs are specific reciprocal commitments among actors. VAs include not only cost sharing but also benefit sharing, be they direct economic benefits or externalities, as in the case of knowledge and capability externalities (see Aggeri and Hatchuel, 1997). However, the dominance of technical commitments in most agreements greatly limits their transparency in terms of cost-benefit implications for the different participants. A critical element for the incentive structure

Table 7. National and local environmental agreements (including VAs) in Italy in 2001

	Number	*Agreements in IDs*
National level	15	-
Regional level	27	5
Emilia-Romagna	*12*	*4*
Lombardy	*7*	*0*
Liguria	*5*	*0*
Veneto	*2*	*1*
Piemonte	*1*	*0*

Source: adapted from Camera (2002).

Environmental (voluntary) agreements in IDs actually represent the successful formalisation of more extensive efforts towards cooperation between regulating institutions and regulated companies (or among regulated companies in one industry) which local environmental polices in IDs aimed at during the above mentioned 'third phase'. The case studies (Carpi, Prato, Sassuolo) presented in Iraldo (2001) show that, during the 1990s, there was a great deal of interaction between local authorities and industry in some very important IDs with critical environmental problems. Only in few cases these processes gave rise to formal 'environmental agreements', e.g. Carpi for water quality from painting processes, but the picture for successes and failures is rather complex and evolving.

The most evident shortcomings emerging from these experiences have been: (a) the length of time needed to make local investments in environmental infrastructures despite the availability of public resources and the public authorities' commitment; (b) the prevailing pre-implementation phase in which many cooperative projects are still at present after significant efforts in terms of study and evaluation; (c) the fact that most of the experiences are not making IDs or their companies international 'environmental leaders'; (d) the fact that successful initiatives are not so numerous or widespread as compared to the role of IDs in the Italian economy (and consequently in the Italian environment). The most positive aspects are: (a) the acceleration of initiatives of local-level cooperation (both among firms and between firms and institutions); (b) the integrated approach of some recent initiatives, e.g. the 'industrial ecology' project in the area of Prato, and the district-level EMAS initiative in Sassuolo.

Despite the remarkable progress made in exploiting local economies of scale (e.g. those justifying consortia for environmental infrastructures) and the creation of a positive 'environment' for a stable interaction between institutions and firms,

associated to VAs is the participation of the public authorities and, in particular, the possibility that the latter commit themselves to supply support, services, and facilitations that allow industry to save compliance costs.

it seems that the huge potential for environmental cooperation and collective action shaped by 'traditional advantages' of IDs is still under-exploited. Furthermore, despite some remarkable innovations, it seems that the attempts by leader firms, or the ID as a whole, to become environmental leaders in search of strategic comparative advantages have been limited, thus slowing down also the possibility of imitation and diffusion processes.

5.2 Main features of innovative responses to environmental policy

Although evidence is not fully satisfactory, marked by the lack of standardised information and a limited number of case studies, it allows us to sketch some conclusions and suggestions on the prevailing modes by which IDs respond to environmental policy by investments and innovation.

- Environmental investments and innovation are more advanced in 'mature' IDs and those having more critical impacts as a consequence of sectoral technologies and products (e.g. ceramic tiles, leather);
- Most investments and innovations are end-of-pipe in response to command and control legislation and regulation; recently, a shift towards clean technologies and other forms of organisational/management innovations (certification, ecolabel, EMAS, etc.) have been taking place;
- Single ID companies have a low propensity to be 'first comers' and take an environmental leadership position by themselves; the perception of economic cost (opportunity costs, competitiveness losses) of environmental investments seems to prevail over the perception of returns from being leaders; environmental differentiation of products is generally limited as are the perceived returns from environmentally-friendly process innovation;
- The similarity to problems faced by single (SMEs) companies and the existence of economies of scale for some key environmental investment (e.g. treatment plants for sludge) leads to a considerable cooperation in making investments at the ID level to share costs; 'collective' environmental investments emerge as an efficient way to face problems and legislation requirements;
- The 'collective action' dimension is extending to a series of initiatives by intermediate industrial and local institutions (association, consortia, special projects, etc.); some of these actions are also linked to European R&D projects and/or based on government incentives to invest in cleaner technologies;
- Collective local action for the environment reveals that the concepts of agglomeration economies and local externalities may have a role also in dealing with environmental issues; however, the potential for exploiting these 'traditional advantages' of IDs seems to be greater than that which has actually been done so far in most IDs;
- Environmental initiatives and investments are often stimulated and supported by local authorities; the relationships between the latter and industry in ID is

rather complex but is now generally cooperative in nature; local authorities are, in some cases, the promoter of environmental initiatives even beyond legislation requirements, or they promote forms of legislation compliance which is not based on purely command and control (moral suasion, threats, common projects and programmes, financial resources from public budgets, 'greening' of local services to the industrial system, the promotion of voluntary agreements, etc.);

- Local level voluntary agreements (VAs) in specific sectors and problem seem to be a promising approach to give a formal dimension to 'cooperative advantages' in environmental investments and innovation; usually ID have suitable information and institutional structure to facilitate the achievement of effective VAs, and in particular to solve the problem of asymmetric information often arising in non-local sectoral VAs; however, these cooperative forms of institutional innovation seem to be less developed than suggested by other organisational and social features of IDs; the above arguments on limited 'collective action' also apply to VAs.

6 Conclusions: unexploited 'cooperative advantages'?

Environmental policies and environment-related markets can be a significant source of dynamic incentives by stimulating industrial responses based on organisational and technological innovation. The latter are also the means by which industry can prevent environmental policies from imposing net economic costs and competitiveness losses and succeed in creating competitive advantages *à la* Porter based on environmental leadership instead. In IDs and local system of production, environmental issues and responses to environmental policies have an additional dimension deriving from the spatial concentration of pollution and firms. In the Italian experience, formal and informal interrelationships among firms and active intermediate institutions created ID-specific competitive advantages able to overcome the possible weaknesses of an industrial structure based on SMEs. It can be expected that the same 'traditional advantages' can help in overcoming the possible disadvantages of SMEs in innovatively responding to environmental problems and policies.

In this work we have explored the consistency of this argument by addressing two issues. The first one is the environmental impact of concentrated local systems of industrial firms. Given the limited availability of structured sets of information at the local level, we have developed an original empirical analysis on Italian data using an approach similar to IPPS, that combines the estimated pollution coefficient by industrial sector with data on industrial employment. We have estimated six air pollutant emissions due to manufacturing activity in local systems, using the Italian NAMEA and the Italian manufacturing employment data. As a case study we have considered the 50 Local Labour Systems (341 municipalities) of Emilia-Romagna. IDs in the region are clearly highly ranking in terms of emissions of the most important air pollutants compared to non-ID areas. Although high emissions per capita and per land area in IDs can be largely attributed to the sector-specific technologies and then industrial specialisation, the manufacturing-concentration effect can be significant even for IDs with relatively a low-impact industrial specialisation.

The second issue we have addressed is how far the specificity of IDs, i.e. spatial concentration of industry leading to economies of agglomeration and positive 'Marshallian' externalities, can also shape both the ways in whch environmental policy is implemented and also the industrial responses based on innovation.

Sectoral environmental policies have already had a significant role for stimulating innovative solutions, in particular in those IDs with high specialisation and/or high impacts on the environment. In general, 'mature' IDs are more advanced in introducing innovations in response to environmental policies. Starting from end-of-pipe solutions to comply with traditional command-and-control regulations, these IDs are now moving, albeit at a slow pace and with various limitations, to more advanced approaches based on clean technologies and integrated environmental management instruments. Innovative reactions to environmental problems, however, are still characterised by a limited willingness on the part of companies to take a leading role. The possibility of developing environmental competitive advantages *à la* Porter seems to be still limited for most IDs and their firms.

The capacity to exploit the advantages from a social 'environment' with a high density of positive externalities and cooperation opportunities mostly manifested itself through the experiences of locally 'integrated' shared investments for treatments and management of waste/emissions. This path, also stimulated by economies of scale of infrastructural investments and services, is now expanding to include new instruments for environmental policy implementation, as environmental VAs and EMAS at the district level. This capacity for cooperative and 'collective action', generally guided by local authorities and increasingly endorsed by firms, suggests that economies of agglomeration, in the form of both accountable cost saving and externalities, may work also in the response to environmental policy requirements. However, the history of environmental policy experience in specific IDs suggests that the huge potential for collective action or 'cooperative advantages' is still under-exploited.

The 'cooperative advantage' of IDs can be extremely important in the perspective that decentralisation, 'Europeisation', and 'integrated approaches' will prevail in environmental policies.

The trend towards local-level environmental policies of the 'geographically integrated' kind is suggested, for example, by the diffusion of the so-called Local Agenda 21 involving an increasing number of Italian local systems[22]. The

---

[22] There are various experiences in Italy of Local Agenda 21, local environmental accounting, and similar instruments. There are 18 pilot projects on Local Agenda 21 supported by the Italian Ministry of the Environment. In the area of local environmental accounting, the CONTARE project is carried out by Regione Toscana, Emilia-Romagna,

perspectives of 'vertically integrated' policy approaches are associated to the process towards Integrated Product Policy (IPP)[23] and filière-level regulations on food safety and quality. The latter emphasise a specific cooperative dimension of environmental innovation through a 'producer responsibility chain' involving all intermediate suppliers in a production chain; the various forms of vertical integration already existing among IDs' firms (for example, groups of SMEs, sub-contracting networks, co-localisation of primary, intermediate and final phases of production) can be a favourable technical and economic framework for responding to this kind of policy. Similar considerations can apply to 'horizontally integrated' policies, as the IPPC (Integrated Pollution Prevention and Control) Directive, involving integrated permits for multiple pollutants at the plant level, an issue typically characterising various ID firms even when they are not in the area of application of IPPC Directive. Finally, the process towards integrated environmental management with regard to firms is well-advanced through the progress of EMAS at the EU and national level, and the attempts to arrive at a 'district-level EMAS' can clearly be a way for exploiting IDs' cooperative advantages.

Another area of opportunities is associated to the recent emphasis on tradable permits or 'markets for pollution' (i.e. a total amount of allowed emissions distributed among firms that can trade the corresponding permits) as the preferable policy approach to many environmental problems[24]. The permits for emissions into water, soil, and air are already a firm's asset in various IDs, e.g. Sassuolo, in which some local firm acquisitions by other local firms are alleged to be mainly motivated by the acquisition of the merged firm's emission permits, also as an instrument for creating barriers to entry. These conditions are structurally similar to those leading to the early schemes of tradable permits or emission credits in the US experience in the framework of the Clean Air Acts (see Stavins, 2003). Sophisticated schemes of tradable emission permits are now at work in the USA and some European countries, the Kyoto protocol for climate protection includes international emission trading as a core instrument, and an European emission trading scheme is being implemented in the EU in the framework of the Protocol implementation (see Biondi *et al.*, 2002). Both theory and evidence suggest that emission-trading schemes can stimulate innovation, although not necessarily more than other policy instruments. A significant potential for implementing such policy schemes in IDs is clearly associated to their economic, technological and institutional characteristics and would therefore contribute to a materialisation of their 'cooperative advantages'.

Liguria, Piemonte, and the Provinces of Trento and Bolzano. The CLEAR (LIFE) project is carried on by 18 local authorities (the municipalities of Ferrara, Bergeggi, Castelnovo nè Monti, Cavriago, Grosseto, Modena, Pavia, Ravenna, Reggio Emilia, Rovigo, Salsomaggiore, the provinces of Bologna, Ferrara, Reggio Emilia, Modena, Napoli, Torino and the Emilia Romagna Region).

[23] See European Commission (2001).

[24] For an illustration of the basic mechanism see, for example, Pearce and Turner (1990).

A very important issue not addressed in this work is that the specialisation of various Italian IDs is directly based to the conservation of material and immaterial 'assets' represented by local natural and cultural resources. For example, food-industry districts largely benefit, both historically and currently, from local environmental health and preservation (i.e. water and air quality, safety of agricultural products, typical and traditional techniques, etc.). Italian fashion industry districts largely benefit, directly and indirectly, from their cultural an historical heritage and the traditional craftsmen knowledge. The consideration of these factors would disclose an extended set of externalities and interactions among non-technically dependent sectors which would reinforce the importance of local collective action.

References

AGGERI F., HATCHUEL A., 1997, "A Dynamic Model of Environmental Policies. The Case of Innovation Oriented Voluntary Agreements", *Nota di Lavoro* n. 24.97, FEEM, Milano.

AGUAYO F., GALLANGER K.P., GONZALES A.C. (2001), "Dirt is in the Eye of the Beholder: The World Bank Air Pollution Intensities for Mexico", *Global Development and Environment Institute Working Paper* n. 01-07, Tufts University, Medford MA.

APAT (2002), *Annuario dei dati ambientali. Edizione 2002*, Agenzia per la Protezione dell'Ambiente e per i Servizi Tecnici, Roma.

AROCHE F. (1999), Reformas estructurales y composiciòn de las emisiones contaminantes industriales: Resultados para México. Serie Medio Ambiente y Desarrollo. n. 24, CEPAL, United Nations, 2000.

ARROW K, BOLIN B, COSTANZA R., DASGUPTA P., FOLKE C., HOLLING C.S., JANSSON B.O., LEVIN S., MALER K.G. PERRINGS C., PIMENTEL D. (1995), Economic Growth, Carrying Capacity and the Environment, *Science*, 268, Reprinted in *Ecological Economics*, 15, pp. 91-95.

BAUMOL W.J., OATES W.E. (1998), *The Theory of Environmental Policy*, (2nd ed.), Cambridge University Press, Cambridge.

BECATTINI G. (1989), "Sectors and/or Districts: Some Remark on the Conceptual Foundation of Industrial Economics", in GOODMAN E., BAMFORD J. (Eds.), *Small Firms and Industrial Districts in Italy*, London Routledge.

BIONDI V., ZOBOLI R., ANGHINELLI S., MAZZANTI M., PALEARI S. (2002), *Nuove prospettive del Protocollo di Kyoto: meccanismi attuativi e impatto sulla competitività*, Istituto per l'Ambiente e Ministero per l'Ambiente, Ipaservizi, Milano.

BOHM P. (1987), "External economies", in EATWELL J., MILGATE M., NEWMAN P. (Eds.), *The New Palgrave: A Dictionary of Economics*, Mac Millan, London.

BRUSCO S. (1982), The Emilian Model: Production Decentralization and Social Integration, *Cambridge Journal of Economics*, n. 2, pp. 167-184.

CAMERA R. (2002), *Gli accordi volontari come strumenti di politica ambientale. Profili teorici, istituzionali ed esperienze*, unpublished Laurea-Degree thesis, Academic Year 2000/2001, Catholic University, Milano.

CARIANI (2001), "Ecodistretti 2001: politiche ambientali innovative nei distretti industriali italiani", paper presented at the conference: "Distretti, piccole e medie imprese: ambiente, innovazione tecnologica e competitività", Fondazione Montedison, March 6th, Milano.

CERVI F. (2000), Interazioni tra inquinamento atmosferico dovuto alle polveri di piombo, boro e fluoro ed inquinamento delle acque superficiali e di falda nel comprensorio ceramico di Sassuolo (Modena), at www.dsa.unipr.it.

D'ANTONIO M. (2002), Produzione, consumo e inquinamento atmosferico: un'analisi del caso italiano, *Economia Pubblica*, XXXII(1).

DAVIS G.A., WILT C.A., DILLON P.S., FISHBEIN B.K. (1997), "Extended Product Responsibility: A New Principle for Product Oriented Pollution Prevention", University of Tennessee, Center for Clean Products and Clean Technologies, in cooperation with the US EPA, Office of Solid Waste, June.

EDMONDS J., ROOP J.M., SCOTT M.J. (2000), *Technology and the Economics of Climate Change Policy*, PEW Center on Global Climate Change, Washington D.C.

EUROPEAN COMMISSION, 2001, *Green Paper on Integrated Product Policy*, COM(2001)68 final, Brussels, 07.02.2001.

EUROPEAN ENVIRONMENT AGENCY, EEA (1999), *L'ambiente nell'Unione Europea alle soglie del 2000*, at: reports.eea.eu.int.

GASPAR J, GLAESER E.L. (1998), "Information Technology and the Future of Cities", *Journal of Urban Economics*, 43 pp. 136-156.

GLAESER E.L. (2000), "The Future of Urban Research: Nonmarket Interactions", *Brookings-Wharton Papers on Urban Affairs 2000*, pp.101-138.

HAHN R.W., STAVINS RN., (1992), Economic Incentives for Environmental Protection: Integrating Theory and Practice, *American Economic Review*, vol. 82.

HEMMELSKAMP J., LEONE F. (Eds.) (1998), *The Impact of EU-Regulation on Innovation of European Industry*, IPTS-JRC, Seville.

HEMMELSKAMP J., RENNINGS K., LEONE F. (Eds.) (2000), *Innovation-Oriented Environmental Regulation*, ZEW Economic Studies n. 10, Physica-Verlag, Heidelberg.

HENDERSON V. (1974), "The Size and Types of Cities", *American Economic Review,* 64, pp. 640-656.

HETTIGE H., MANI M., WHEELER D. (1998), "Industrial Pollution in Economic Development (Kuznets Revisited)", *Working Paper Series,* n. 1876, Development Research Group, The World Bank, Washington, D.C.

HETTIGE H., MARTIN P., SINGH M., WHEELER D. (1995), "The Industrial Pollution Projection System", *Working Paper Series, n. 1431*, Development Research Group, The World Bank, Washington, D.C.

IRALDO F., (2001), *Ambiente, impresa e distretti industriali*, Franco Angeli, Milano.

ISTAT (1991), *VII° Censimento dell'industria e dei servizi*, Roma.

ISTAT (1996), *Censimento intermedio dell'industria e dei Servizi*, Roma.

ISTAT (2000), *Contabilità ambientale: dagli schemi alle realizzazioni*, Roma.

ISTAT (2001), *Statistiche ambientali 2000*, Roma.

JAFFE A.B., STAVINS R.N. (1995) "Dynamic Incentives of Environmental Regulations: The Effects of Alternative Policy Instruments on Technology Diffusion", *Journal of Environmental Economics and Management*, 29:S43-S63, 1995.

JAFFE A.B., NEWELL R.G., STAVINS R.N. (2003), Technological Change and the Environment, in: Mäler K-G. and Vincent J.R. (eds.), *Handbook of Environmental Economics. Volume 1*, Elsevier, Amsterdam.

JAFFE A.B., NEWELL R.G., STAVINS R.N. (2000), "Technological Change and the Environment", *Discussion Papers*, 00-47, Resources for the Future, Washington D.C.

JAFFE A.B., PETERSON S.R., PORTNEY P.R., STAVINS R.N. (1995), "Environmental Regulation and the Competitiveness of US Manufacturing: What Does Evidence Tell Us?", *Journal of Economic Literature*, 33, March.

KAUFMANN K., DAVIDSDOTTIR B., GARNAHAM S. PAULY P. (1998), The Determinants of Atmospheric SO2 Concentrations: Reconsidering the Environmental Kuznets Curve, *Ecological Economics*, 25, p. 209-220.

KEMP R. (1997), *Environmental Policy and Technical Change*, Edward Elgar, Cheltenham.

KLEMMER P. (Ed.) (1999), *Innovation and the Environment*, Analytica-Verlag, Berlin.

LAFFONT J.J. (1987), "Externalities", in EATWELL J., MILGATE M., NEWMAN P. (Eds.), *The New Palgrave: A Dictionary of Economics*, Mac Millan, London.

LAPLANTE B., MEISNER C. (2001), "Estimating Conventional Industrial Water Pollution in Thailand", *Working Paper Series*, Development Research Group, The World Bank, Washington, D.C.

LEWANSKI A. (1997), *Governare l'ambiente*, Il Mulino, Bologna.

MASSARUTTO A. (1997), "Le problematiche ambientali nei distretti industriali e agro-industriali", in atti del convegno "Problemi e prospettive dei distretti industriali in Friuli Venezia-Giulia", Udine, 11 aprile.

MAZZANTI M., ZOBOLI R. (forthcoming), "Environmental Policy Instruments and Induced Innovation. The EU Directive on End-of-Life Vehicles", *Ecological Economics*.

MERTINS L. *et al.* (1999), "Development and application of waste factors: An overview", EEA, *Technical Report* n. 37, Copenhagen.

MOKYR J. (1995), Environmental Crises and Technological Change, in: Quadrio Curzio A. and Zoboli R. (eds.), *Science, Economics, and Technology for the Environment*, Cariplo Foundation for Scientific Research, Milan.

MONKHOUSE P. (2001), Environmental Aspects of Energy Related Activities, paper presented at the Marie-Curie Conference, Steyr/Upper Austria, 16-19 May.

MORGENSTERN R.D., PIZER W.A., SHIH J-S. (1998), "The Cost of Environmental Protection", *Discussion Paper* 98-36, RFF, Washington D.C.

ORISHIMO I. (1982), *Urbanisation and Environmental Quality*, Kluwer, Dordrecht.

PALMER K., WALLS M. (1999), "Extended Product Responsibility: An Economic Assessment of Alternative Policies", *Discussion Paper* 99-12, Resource for the Future, Washington D.C.

PARRY W.H., PIZER W.A., FISCHER C. (2002), "How Large are the Welfare Gains from Technological Innovation Induced by Environmental Policy?", *Discussion Papers* 02-57, Resources for the Future, Washington D.C.

PEARCE D.W., TURNER K.R. (1990), *Resource and Environmental Economics*, Harvester Wheatsheaf, New York.

PONTING C. (1991), *A Green History of the World*, Penguin, New York.

PORTER M.E., VAN DER LINDE C. (1995), "Towards a New Conception of the Environment-Competitiveness Relationships", *Journal of Economic Perspectives*, 9: 97-118.

PROVINCIA DI MODENA, PROVINCIA DI REGGIO EMILIA *et al.* (2002), *Bilancio ambientale del distretto ceramico. Province di Modena e Reggio Emilia*, ARPA, Bologna.

QUADRIO CURZIO, A., FORTIS, M., ZOBOLI, R. (Eds.) (1994), *Innovation, Resources and Economic Growth*, Springer-Verlag, Berlin.

QUADRIO CURZIO, A., ZOBOLI, R. (1997), "The Costs of Sustainability", in: Proceedings of the International Conference 'The 50th Anniversary of the United Nations and the Italian Contribution to the 'Earth Charter', Accademia Nazionale delle Scienze, Roma, May 1996.

QUADRIO CURZIO, A., ZOBOLI, R., (a cura di) (1995), *Ambiente e dinamica globale. Scienza, tecnologia ed economia a confronto*, Il Mulino, Bologna.

STAVINS R.N. (2003), Experiences with Market-based Environmental Policy Instruments, in: Mäler K-G. and Vincent J.R. (eds.), *Handbook of Environmental Economics. Volume 1*, Elsevier, Amsterdam.

TEN KATE A. (1993), "Industrial Development and Environment in Mexico", *Working Paper Series*, n. 1125, Development Research Group, The World Bank, Washington, D.C.

VERHOEF E., NIJKAMP P., DANIEL J. (2001), Urban Sustainability, Agglomeration Forces and the Technological Deus ex Machina, *Tinbergen Institute Discussion Paper*, TI2001-117/3

ZOBOLI R., BARBIROLI G., LEONCINI R., MAZZANTI M., MONTRESOR S. (2000), *Regulation and Innovation in the Area of End-of-Life Vehicles*, EUR 19598 EN, IPTS-JRC, Seville, at: www.jrc.es.

ZOBOLI R. (2000), "Valutazione delle politiche ambientali e sistemi di indicatori", in: LUPÒ AVAGLIANO M.V. (a cura di), *Efficienza della pubblica amministrazione. Misure e parametri*, Franco Angeli, Milano.

Appendix

Table A.1. Comparison between Italian NAMEA-based estimation and US IPPS-based estimation for air pollution in the Sassuolo district

		Using US IPPS data		Using IT NAMEA data		
		Kg per sqkm	Kg per-capita		Kg per sqkm	Kg per-capita
ToxAir	Toxics: air	5,667.2	12.3		n.a.	n.a.
ToxLand	Toxics: land	8,114.0	17.5		n.a.	n.a.
ToxWat	Toxics: water	892.2	1.9		n.a.	n.a.
ToxTot	Toxics: total	14,673.4	31.7		n.a.	n.a.
MetAir	Metals: air	49.2	0.1		n.a.	n.a.
MetLand	Metals: land	872.9	1.9		n.a.	n.a.
MetWat	Metals: water	9.9	0.0		n.a.	n.a.
MetTot	Metals: total	932.1	2.0		n.a.	n.a.
BOD	Water: biochemical oxygen demand	4,974.5	10.8		n.a.	n.a.
TSS	Water: total suspended solids	21,504.8	46.5		n.a.	n.a.
NO_2	Air: Nitrogen dioxide	43,978.0	95.1	NO_X	35,744.8	77.27
PM10	Air: Fine Particulates	52,313.8	113.1		n.a.	n.a.
SO_2	Air: Sulphur dioxide	74,385.9	160.8	SO_X	28,324.2	61.23
CO	Air: Carbon monoxide	19,182.3	41.5	CO_2	12,122.2	26.20
PT	Air: Total Suspended Particulates	50,143.8	108.4		n.a.	n.a.
VOC	Air: Volatile organic compounds	7,952.3	17.2		n.a.	n.a.
CH_4	Air: Methane	n.a.	n.a.		12,279.1	26.54
N_2O	Air: Nitrous oxide	n.a.	n.a.		1,141.6	2.47
NH_3	Air: Ammonia nitrogen	n.a.	n.a.		325.2	0.70

n.a.= not available

Source: elaboration on Hettige *et al.* (1995), ISTAT (2001) and ISTAT (1996) data

Note: The estimation of pollutants' emissions in Italy by industry obtained through the US IPPS data results in higher values than those obtained with national NAMEA data (it should be noted that the reference year for pollutant' emissions is 1987 for US data and 1990 for Italian data). Although comparison between the two estimates is possible only, partially, for nitrogen oxides, sulphur oxides and carbon oxides, it seems that IPPS data reveal that US industrial firms produce higher pollution than the Italian ones: sulphur oxides emissions estimated through the US data are more than doubled with respect to those derived from the Italian data, while the nitrogen oxides and the carbon oxides are greater by 23% and 58% respectively.

Table A.2 Estimation of air pollution emissions by manufacturing industry in Italy (kg per 100 employees)

Ateco classification	*SOX*	*NOX*	*CH4*	*CO2(a)*	*N2O*	*NH3*
15&16 Food, beverages and tobacco	5,853.7	9,417.5	59,079.1	2,370.1	217.3	2.3
17 Textiles	3,818.3	3,582.4	936.3	1,006.0	122.6	0.6
18 Wearing apparel, except footwear	987.7	2,080.4	36.0	279.9	48.5	0.5
19 Leather products and footwear	2,067.4	2,786.2	61.0	501.8	72.9	0.8
20 Wood products, except furniture	4,329.7	4,581.5	109.9	822.7	113.9	1.4
21 Paper and paper products	15,314.0	12,012.6	257,123.0	5,562.2	410.3	1.0
22 Printing and publishing	1,290.1	2,195.0	68.4	495.4	64.9	0.5
23 Coke production, petroleum refineries, nuclear comb. treatment	1,015,129.2	190,058.2	22,010.2	129,658.6	10,340.7	0.7
24 Chemical products	61,766.1	21,602.6	1,550.3	10,226.2	10,927.6	9,800.3
25 Rubber and plastic products	3,547.8	3,978.8	97.1	873.8	108.8	0.9
26_1 Glass and glass products	13,640.4	51,076.2	401.3	7,351.0	350.7	1.8
26_2/8 Other non-metal mineral products	38,517.5	50,566.7	655.5	17,321.9	1,133.9	19.9
27 Iron and steel	23,989.5	18,118.2	3,852.7	10,159.8	220.0	16.8
28 Metal products, except machinery	974.4	3,001.0	75.8	748.5	90.5	0.7
29 Machinery, except electrical	621.7	2,242.4	62.3	389.9	51.6	0.6
30 Machinery electric	362.6	2,750.6	89.2	626.2	67.7	0.8
31 Other machinery	1,080.1	2,338.3	52.2	400.6	58.5	0.4
32 TV and communication apparels	920.4	1,770.9	48.0	331.4	47.2	0.4
33 Professional and scientific equipment	277.8	1,660.8	42.3	251.6	35.4	0.4
34 Means of transport (motor vehicles)	1,658.5	4,270.3	235.3	1,331.0	119.0	0.8
35 Other means of transport	1,552.8	4,379.5	275.5	2,491.9	445.0	0.6
36 Furniture; other manufacturing industries	1,332.8	3,174.2	72.3	666.9	74.6	0.8
37 Recover and recycling preparation	1,614.7	8,866.1	147.0	1,095.8	81.6	3.6

(a) 1000Kg/100 employees

Source: our elaboration on ISTAT (2001) and ISTAT (1996).

Table A.3. Share of emissions by pollutant due to the highest share of industry' employees in each LLS

LLS code	*Name*	*ID*	*Special. industry*	*Empl. share*	SO_X	NO_X	CH_4	CO_2	N_2O	NH_3
131	Cremona	ID	26_2/8	36.06	83.0	81.7	1.7	86.4	65.2	5.6
139	Suzzara	ID	29	36.37	6.3	17.7	0.4	10.8	8.6	0.2
233	Bobbio		29	26.50	3.9	10.0	0.1	6.4	7.3	0.3
234	Castel San Giovanni	ID	15&16	18.85	14.4	13.4	79.5	13.6	17.3	1.0
235	Fiorenzuola d'arda	ID	29	24.78	0.6	4.6	0.1	2.0	2.7	0.2
236	Piacenza		29	24.27	2.4	6.8	0.2	3.8	4.1	0.1
237	Begonia		26_2/8	31.41	83.9	73.9	2.3	84.0	80.3	89.2
238	Borgo Val di Taro		15&16	28.21	15.4	20.8	48.9	16.4	10.3	0.2
239	Busseto	ID	15&16	26.02	22.1	31.4	96.1	26.9	9.9	0.2
240	Fidenza		26_1	17.35	25.2	59.9	0.5	38.7	14.4	0.2
241	Fornovo di Taro	ID	15&16	27.50	5.0	17.1	87.6	10.1	12.2	4.6
242	Langhirano	ID	15&16	66.53	79.0	83.3	94.9	82.2	61.6	2.8
243	Parma	ID	15&16	28.56	18.7	24.9	82.3	24.8	9.6	0.2
244	Tizzano Val Parma		15&16	82.75	85.4	87.2	100.0	86.0	86.4	83.7
245	Castellarano	ID	26_2/8	62.21	89.3	92.8	2.7	93.2	72.4	6.0
246	Castelnovo ne' Monti		15&16	25.83	15.1	18.4	94.9	14.5	16.6	1.3
247	Correggio	ID	29	17.57	1.9	6.8	0.1	4.0	1.8	0.0
248	Guastalla	ID	29	31.01	4.3	11.0	0.4	6.4	7.7	0.3
249	Reggio E.	ID	29	21.88	2.0	6.0	0.1	3.3	3.0	0.1
250	Carpi	ID	17	33.29	40.7	29.2	4.2	32.6	31.1	0.8
251	Fanano		15&16	29.09	25.5	28.1	99.1	24.2	27.8	20.8
252	Mirandola	ID	33	16.90	0.7	3.6	0.1	1.7	1.4	0.0
253	Modena	ID	29	30.07	3.9	11.6	0.2	6.8	5.9	0.1
254	Contese		15&16	25.65	38.9	45.5	99.2	45.2	22.2	0.5
255	Pavullo nel Frignano		26_2/8	35.18	84.9	83.2	1.7	87.4	71.5	9.0
256	Pievepelago		32	28.09	6.1	8.1	0.1	5.4	8.8	5.7
257	Sassuolo	ID	26_2/8	50.21	86.4	90.8	4.9	91.4	59.7	3.2
258	Vignola	ID	28	21.25	2.1	4.8	0.1	3.8	4.9	0.2
259	Bologna		29	24.08	3.0	11.4	0.2	6.3	2.8	0.0
260	Imola		29	24.94	2.1	6.0	0.2	3.2	2.9	0.1
261	Porretta Terme		29	37.85	7.2	18.1	0.5	10.7	11.0	0.3
262	Argenta	ID	28	17.15	1.6	7.9	0.2	5.6	1.9	0.0
263	Cento	ID	28	20.69	3.5	9.2	0.3	7.0	5.1	0.1
264	Codigoro		18	36.52	10.8	16.2	0.1	9.3	15.4	16.7
265	Comacchio		29	22.87	2.6	6.8	0.1	4.1	4.0	0.1
266	Copparo		29	47.67	14.0	28.8	0.5	20.5	21.0	1.0

Cont. Table A.3.

LLS code	*Name*	*ID*	*Special. industry*	*Empl. share*	SO_X	NO_X	CH_4	CO_2	N_2O	NH_3
267	Ferrara		24	18.18	78.0	46.0	2.9	59.8	94.6	99.9
268	Faenza	ID	28	17.81	1.8	4.7	0.1	3.6	2.8	0.0
269	Lugo	ID	15&16	22.24	20.0	29.2	86.2	25.0	9.0	0.1
270	Ravenna		15&16	20.75	4.8	15.6	85.0	9.6	1.9	0.0
271	Bagno di Romagna		18	30.31	8.6	11.4	0.3	5.8	11.1	9.2
272	Cattolica		17	20.94	22.2	14.0	3.2	14.9	13.2	0.2
273	Cesena		15&16	24.47	25.1	32.8	86.2	30.2	17.6	0.4
274	Forli'	ID	36	18.19	3.7	9.4	0.1	6.2	3.9	0.1
275	Mercato Saraceno	ID	19	21.17	9.1	11.3	0.3	7.3	5.8	0.1
276	Modigliana		20	63.44	79.5	70.9	4.4	65.8	70.3	74.3
277	Morciano di Romagna	ID	36	18.42	3.6	6.4	0.1	4.5	4.8	0.2
278	Rimini		29	19.20	2.2	6.4	0.1	3.8	3.1	0.1
279	Rocca San Casciano	ID	28	35.47	4.9	11.3	0.8	9.0	9.4	0.2
280	Santa Sofia	ID	15&16	41.62	66.6	64.7	99.8	65.8	62.2	60.7

Source: our processing on ISTAT (2001) and ISTAT (1996) data

Part IV. Internationalisation

Local engines of global trade: the case of Italian industrial districts

Stefano Menghinello

Department of Economic Statistics, ISTAT, Via Tuscolana 1782, 00173, Rome, Italy

1 Introduction

In the past decade, the economic policy debate in EU countries has increasingly focused on the role of local industrial systems as a key factor both in improving national competitiveness in international markets, and in stimulating economic growth in less developed regions (Cowling and Sugden, 1999).

Although the results of this debate are still to be considered controversial (Amin, 1992), there is an almost universal agreement that some particular kinds of local industrial systems, including industrial districts as the most remarkable examples, seem to give a specific competitive edge, based on both tangible and intangible factors, to the enterprises located therein. In particular, the intangible factors, closely related to specific skills and competencies embedded in the local labour force and industrial culture, provide flexibility and a significant ability to meet new customer needs, These are all crucial elements for competing at an international level successfully.

This chapter provides empirical evidence on the role played by industrial districts as regards Italian competitiveness in international markets. In particular, it analyses a quantitative measure of industrial districts contribution to Italian exports, that has recently been elaborated by the researchers of ISTAT on the basis of a new methodological approach. The empirical analysis also highlights a more complex role for industrial districts in promoting output variety and raising national standards of product quality, at least for exported goods, thus redefining new trends for Italian competitiveness in international markets.

The analysis of Italy is important in more general terms because it provides original and strong empirical evidence to support theoretical studies stressing the crucial role played by industrial districts and industrial district-like local industrial systems as specific engines of economic growth and international competitiveness. In effect, the lack of extensive applied studies on local economies, reflecting a world wide information gap in terms of national statistics at local level, can be considered one of the major limits of this particular field of economic research.

The chapter will proceed as follows: section 2 focuses on industrial districts as a specific source of competitive advantages, Section 3 stresses the main statistical and methodological issues underlying the empirical analysis, Section 4 shows empirical evidence on the real contribution of industrial districts to Italian exports, also highlighting some relevant case studies in terms of exported goods' quality and variety, Section 5 provides some final comments.

2 Industrial districts as a source of competitive advantage

The role played by industrial districts as a source of competitive advantage can be summarised in a simple and self-evident statement: the remarkable economic growth and strong international competitiveness shown by some specific kinds of local industrial systems cannot be explained only in terms of the behaviour of single enterprises located within these particular areas. Although the mainstream neo-classical framework has for a long time ignored the role of industrial districts as specific engines to stimulate competitiveness and economic growth, a large body of economic literature has already been developed following Marshall's pioneering works. In effect, most of the current work in this field of research draws, directly or indirectly, on Marshall's thoughts about the determinants of industrial concentration. These factors are: a local labour market of specialised workers, greater availability and higher efficiency in the production of intermediate and instrumental goods and, finally, a pool of specific competencies along with a high aptitude for innovation. Recent debate in literature is focusing attention both on the nature of the positive link between competitiveness and local dimension, and on the possibility to extend the field of interest of the analysis to local industrial models other than industrial districts.

In particular, economic literature is increasingly turning to the possibility that the main source of local based competitive advantage is due not only to tangible factors[1], but rather to intangible assets, related to a set of knowledge, competencies and skills embedded within each industrial district. In effect, whilst today's international competition arena is characterised by an increasingly international mobility of traditional inputs (capital, labour and standardised technology), regional or local specific inputs, in terms of not standardised knowledge, competencies and skills, cannot be easily transferred, and thus traded in international markets. Furthermore, these regional or local based intangible assets are the main resources for learning and innovating. Competencies (Prahalad and Hamel, 1990), in their different forms (Dosi and Malerba, 1996), have increasingly become a crucial feature underlying enterprise competitiveness, and thus the enterprise processes or strategies for obtaining these specific inputs are becoming more important.

The creation of new knowledge for an enterprise is a rather complex task (Nonaka, 1994) which is characterised by a peculiar mix of two different kinds of intangible assets. The first is represented by the so-called codified knowledge, realised on the basis of standardised scientific protocols and then easily tradable on international markets. The second refers to the uncodified knowledge

[1] In effect, they refer mainly to productivity gains obtained, in terms of technical efficiency and innovatory path improvement, through the role of traditional external economies.

(sometimes called in literature *tacit* knowledge). This kind of knowledge is embedded within single enterprises or specific geographical areas, and thus cannot be easily separated from the particular industrial places where it has been created and improved. Within this framework, the industrial district has proven to be one of the best kinds of local industrial models to stimulate the creation of new knowledge and a fast lane for innovation, starting from a common and shared pool of competencies and skills (Becattini and Rullani, 1993).

Recent works by Amin and Wilkinson (1999), Belussi and Gottardi (2000), including a fascinating applied and theoretical study realised by Lipparini and Lorenzoni (1996), represent an important step forward in the analysis of the relationship between local learning processes and industrial performance. Given the variety of particular issues and specific connections highlighted by these interesting studies, it seems clear that the creation of new knowledge within an industrial district or a *district like* cluster of SMEs relies not only on traditional externalities but rather on a complex network of social and economic relationships, based on a common pool of values and institutions. In particular, De Propris (2001) focuses on the connections between cluster governance. systemic flexibility and production fragmentation, highlighting how industrial districts, compared to other local based industrial models, guarantee flexibility to a greater extent in terms of both output quantity and variety.

The investigation into industrial districts as specific source of comparative advantage means widening the field of interest of the analysis to firms' behaviour in international markets. Although economic literature does not explicitly refer to exports as a key factor in explaining enterprise competitiveness, some basic remarks can be made on the relationship between industrial districts, exports and international competitiveness.

The concept of international competitiveness can not be limited to the enterprise's general attitude for reaching new countries or improving market shares with already established partner countries. It refers above all to a firm's ability to meet new customer needs world wide with a wide and high quality range of products. Given that the general issue of stressing the superior capability of local industrial systems to meet new customer needs by means of a flexible production structure, Becattini (1998) defined an industrial district-specific source of competitive advantage by introducing the concept of industrial district cluster of products (IDCP). An IDCP represents a sort of enlarged production *filière*, which includes products on the basis of either technological connections or customers' preferences. For instance, in the case of industrial districts specialised in furniture production, the following products are considered within the same IDCP: glass and glass products, ceramic tiles and flags, cutlery, tools and general hardware, machine tools, other special purpose machinery, domestic appliances, lighting equipment and electric lamps and, of course, furniture[2]. In

[2] As regards the empirical definition of industrial district cluster of products, an attempt was made by Becattini and Menghinello (1998), starting from the 3 digits level of the Italian merchandise trade classification (236 merchandise groups) available for the exports of each of the 103 Italian provinces. These authors defined, on the basis of *a priori*

effect, an IDCP exploits a industrial district's specific capability (i.e. competitive advantage) to satisfy customers' changing preferences with a wide range of products. These products are realised within the industrial district on the basis of a proper mix of technological competencies and intangible assets, which may cross sector boundaries.

Recent results from both applied and theoretical literature have enlarged the field of interest of this research to a wider range of local industrial systems. In effect, until few years ago the economic literature on local development was limited to a small range of industrial models, basically industrial districts or *district like* local industrial systems, mainly localised in advanced countries and specialised in traditional and low technology sectors. A limited but still significant number of applied studies have shown that local industrial systems of SMEs can be found even in less developed countries such as Pakistan, Mexico, Brazil, India (Di Tommaso and Rabellotti, 2000). Other applied studies, carried out in advanced countries (Scott, 1992 and Saxenian, 1994), have shown that local industrial systems of SMEs characterised not only traditional sectors but high technology sectors as well. The empirical evidence of *district like* local industrial systems of SMEs across different sectors seems to support a pioneering theoretical work of Becattini (1979), suggesting that the role of the industrial district in economic development must not be limited by a sector based approach. Another significant achievement in this field of analysis, based on both theoretical and empirical approaches, has been the widening of the taxonomy of different local industrial systems, generally defined by Markusen (1996) as firm clusters. However, as far as analytical goals or policy measures are concerned, it is always important to distinguish between different kinds of local industrial systems, for their particular economic features have different effects on employment, economic growth and competitiveness in international markets.

3 Methodological background

The great lack of a large set of relevant economic variables at a local level is the main reason why applied studies, and especially model-oriented empirical studies, are very limited within this specific field of research[3]. Furthermore, the

---

information, some clusters of products related to typical 'made in Italy' industrial sectors. A single product is classified within a given cluster not only on the basis of technical reasons (intermediate or instrumental goods which characterise a given industrial *filière)* but also according to similar customer needs (personal fashion needs, household needs etc).

[3] As regards previous empirical studies on the role of industrial districts in the Italian economy, two different approaches can be considered. The first approach seeks to measure the 'industrial district effect' on the enterprise performance. This approach, based on an enterprise panel survey carried out by the Italian Mediocredito Centrale, includes the analysis, at a microeconomic level, of relevant issues such as technical efficiency (Fabiani *et al.*, 2000) and foreign trade performance (Bagella *et al.*, 2000). An alternative

statistical definition of the best territorial unit to identify local industrial systems in general, and industrial districts in particular, is still an open and quite controversial issue in most countries.

The empirical analysis adopted in this work basically follows an exploratory approach, exploiting fully the availability of a new economic variable at local level: exports. Exports allow us to investigate the relevance of local dimension as a specific factor for competitiveness to support national performance in international markets. However, the use of a single economic variable, outside a well defined analytical framework, may lead to both striking and potentially misleading results. As a consequence, the specifications mentioned below must be carefully read in order to analyse the empirical results correctly.

The territorial unit used in the analysis to identify local industrial systems, and in particular industrial districts, refers to Italian local labour markets. Local Labour Markets (LLM) are officially defined by ISTAT as 784 functional areas on the basis of the Census of population, reference year 1991 (ISTAT, 1997).

Industrial districts are defined, again for the year 1991, as a subset of LLM using a two-phase procedure based on concentration indexes. First, LLM are selected on the basis of a local concentration index of manufacturing activities, measured in terms of persons employed at industrial plant level. Second, the selected subset of LLM is further split with respect to a SMEs concentration ratio. This methodology is applied using standard EU classification criteria to classify enterprise dimension, and it identifies 199 local labour systems that specialise in manufacturing and characterised by the significant role of SMEs (ISTAT, 1997). Although this approach has some difficulty in identifying all the Italian industrial districts successfully, it is still the best way to identify industrial districts location by the means of a rigorous and consistent statistical procedure.

All other kinds of local industrial systems are residually defined as non industrial districts LLM. This definition is not very satisfactory as, indeed, in effect it may refer to non manufacturing local systems or to industrial areas dominated by single large firms. However, the combination of export data (i.e. only manufacture products exporting LLM are considered) and a more

---

approach, based on a more organic prospective which considers the industrial district as a single player in today's competition, highlights, on the basis of ISTAT territorial statistics, the particular role of industrial districts in the Italian economy in terms of industrial structure and export performance. As far as the first issue is concerned, the relative empirical work by Sforzi (1990), using ISTAT local labour systems data, must be considered a benchmark for the analysis of the Italian industrial structure on a territorial base. On the basis of traditional ISTAT export statistics by Italian provinces, Viesti (1993) and Conti (1995) have tried to highlight relevant connections between the Italian international specialisation model and the local concentration of export activities within *district like* provinces. Finally, a significant but not completely satisfactory attempt to estimate the real contribution of industrial districts to the Italian exports has been made by Becattini and Menghinello (1998), on the basis of the territorial distribution of Italian exports by provinces.

geographical disaggregation as opposed to a strictly industrial approach makes this assumption quite reasonable.

Export data at LLM level are currently available, for methodological reasons, only for the year 1996 and the sector breakdown refers, at most, to the 100 merchandise groups belonging to the European classification of products related to enterprise economic activity (CPA-NACE Rev.1). The methodological approach, underlying these interesting results, relies on the combination, at a single enterprise level, of Census and foreign trade registers, in order to merge export data with territorial information at single plant level. Critical problems, such as multi-plant manufacturing enterprises or wholesalers, are dealt with using proper estimation techniques. The empirical evidence is based on a provisional issue of Census data on local enterprise units and is limited to the year 1996. In particular, all the statistical results used in this chapter have already been published by ISTAT in official publications or in the 1999-2000 editions of the ISTAT Report on the State of the Nation.

The empirical analysis is performed on the basis of a set of simple statistical indicators: market shares, localisation quotients, export performance and product quality indicators.

Market share is defined as the ratio between a part and the total of a given export flow. Market share may refer to a specific product, partner country or territorial area or a combination of these variables. In particular, within the framework of this research, it provides interesting information on the contribution of industrial districts to the Italian exports as a whole and across merchandise groups. Market share ratios are also used to measure the contribution of a single industrial district to world exports[4] with respect to a specific group of products.

Location quotient (LQ) is used both to measure the territorial concentration of export activities at local labour market level and to identify the local based export specialisation. Furthermore, within the analytical framework of international economics, it corresponds to the revealed competitive advantages measured at local level, LQ expressed as:

$$LQ_{k,j} = \frac{\dfrac{Exp_{k,j}}{Exp_{k,IT}}}{\dfrac{Exp_{j}}{Exp_{IT}}}$$

[4] This latter ratio is obtained, for a given group of products, simply by multiplying the industrial district share of national exports by the Italian share of world exports. The possibility of matching territorial and international shares is made possible since this information refers to the same merchandise classification scheme (the breakdown of Italian share of world exports by CPA merchandise groups is estimated on the basis of the harmonised system nomenclature). Of course, the meaning of industrial districts national and world share refers to exports outside Italian national boarders and thus do not include trade with other Italian regions.

In this formula, $Exp_{k,j}$ refers to export value in national currency and $k = 1, \ldots, n$ and $j = 1, \ldots, m$ identify, respectively, merchandise sectors and local labour markets, while *IT* stands for national values. The location quotient is calculated by dividing the local share of national exports with respect to a selected group of products by the local share of national exports in general terms, Index values greater than one identify a local concentration (export specialisation) with respect to a given group of products.

A very simple export performance (EP) indicator is also defined on the basis of the export value per person employed ratio, calculated at a local level and then normalised with respect to the national average ratio:

$$EP_j = \frac{\dfrac{Exp_j}{Emp_j}}{\dfrac{Exp_{IT}}{Emp_{IT}}}$$

In this formula, Emp_j refers to the number of person employed while $j = 1 \ldots, m$ identifies, as above, different local labour markets. This indicator is affected by different phenomena including labour productivity, propensity to export, labour to capital ratio, product composition and quality, etc. Therefore. it can only be used in a very broad sense as an export performance indicator[5]. According to this indicator, four different classes of values have been considered: *strongly competitive areas* (for index values above 120), *competitive areas* (for index values between 80 and 120), *weakly competitive areas* (for index values between 40 and 80) and *not competitive areas* (for index values less then 40).

In order to measure the quality standard of exported products at the local level, an index has been defined on the basis of export unit values, Italian provinces have been used as a reasonable proxy for some well known industrial districts, while sector breakdown refers to 236 merchandise groups. Quite a long range of time - from 1985 to 1999 – enables us to develop a dynamic analysis.

This index measures the local quality standard of a given exported product (i.e. its export unit value calculated at the province level) with respect to the national average (the same export unit value calculated at the national level) and it is defined as follows:

$$RQ_{k,j} = \frac{\dfrac{Exp\ Val_{k,j}}{Exp\ Vol_{k,j}}}{\dfrac{Exp\ Val_{k,IT}}{Exp\ Vol_{k,IT}}}$$

---

[5] In effect, it includes a wide range of variables related to technical efficiency, product quality and industry structure which all contribute, in different ways, to determine the overall level of competitiveness of the local area.

In this formula, *Exp* $Val_{k,j}$ refers to monetary export values in national currency at current prices and Exp Vol identifies export volumes in terms of standard weight units (kg) or, in some relevant cases, other weight units (pair of shoes etc). The codes k =1...,n and j= 1...m identify, respectively, merchandise sectors and provinces, while IT denotes the average national values. Index values greater than one identify standards of quality of local exported goods higher than the national average.

The empirical analysis of the next section focuses on some relevant issues related to the role of local industrial systems, and in particular of industrial districts, for Italian exports and competitiveness in the international markets. In particular, section 4.1 illustrates the main features of local industrial systems' export performance using the EP indicator; section 4.2 focuses, on the basis of markets shares, on the contribution of industrial districts to Italian exports of manufactured goods, as regards total and across merchandise groups; while section 4.3 analyses two relevant issues, related to variety and quality of industrial district exported goods, by the means of the LQ and RQ indexes respectively.

4 Some empirical evidence[6]

4.1 Export performance of local industrial systems

Some basic remarks on the role of local industrial systems for Italy's competitiveness in the international markets can be made from the map of the export performance (EP) of Italy' local labour markets[7] (LLM), as shown in Figure 1. This map illustrates local labour markets distinguishing between strongly competitive, competitive, weakly competitive and not competitive ones[8].

The map clearly shows that the highest levels of export performance appear to be concentrated in specific LLM across the national territory. Although these LLM are mainly located within the northern and central regions, a significant variability characterises these regions. Furthermore, a limited number of LLM with relevant export performance are located in the southern regions of Italy.

---

[6] All data used in this paper refer to official statistics or research studies already published by ISTAT. In particular, the distribution of Italian exports by local labour systems used in this study refers to provisional data already published in the 1999 and 2000 editions of the ISTAT Report on the state of the Country (*Rapporto annuale sulla situazione del Paese*). This data is based on a provisional issue of the 1996 Census data on industrial and market services firms and local units.

[7] On the use of local labour markets (LLM) as a proxy of local industrial systems see the methodological discussion in section three.

[8] Different kinds of LLM in terms of export performance are identified, on the basis of the EP index, according to value classes described in section three.

Many of the LLM characterised by high levels of export performance can be classified as ID[9], while in the others large enterprise or MNEs owned industrial plants, and emerging areas of SMEs can be identified. In particular, ID account for more than 44 per cent of strongly competitive LLM, and more than 50% of competitive LLM. The industrial districts share of weakly competitive LLM is about 46% and it sharply drops to about 18% as far as not competitive LLM are concerned. Within the list of the 199 LLM classified as industrial districts, more then 50% can be classified as competitive or strongly competitive, and the share reaches 80% when weakly competitive LLM are considered.

The empirical evidence suggests that Italy's capacity to sell goods in foreign markets is driven by the strong export performance of some of its local industrial systems, and proves that industrial districts can be considered one of the most powerful local engines of Italian export performance in international markets.

Figure 1. The export performance of local labour markets (1996)

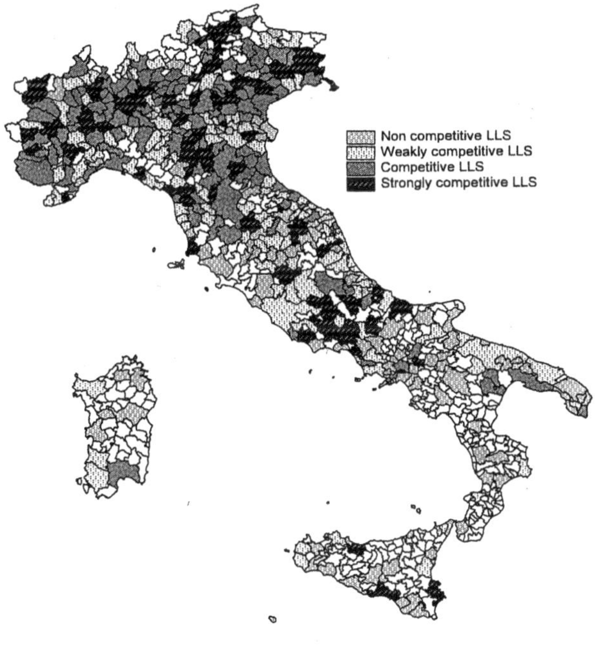

[9] Information on the statistical criteria used to identify industrial districts as a subset of LLM are included in section three. The definition includes Marshallian and industrial district like local labour systems.

4.2 Contribution of industrial districts to Italian exports

In 1996, the contribution of industrial districts[10] to the Italian exports of manufactured goods well exceeds 40%[11]. Table 1 illustrates the contribution of industrial districts to national exports across merchandise groups. Industrial districts account for more then 70% of Italian exports for sports goods, leather, ceramic tiles and flags, jewellery and related articles, musical instruments, agricultural and forestry machinery, knitted and crocheted articles. Once more, for some sectors industrial districts are responsible for more then 50% of exports: these include footwear, textile weaving products, furniture, reservoirs and containers of metal, central heating radiators, leather clothes, weapons and ammunitions, cutlery, tools and general hardware, basic iron, steel and iron-alloys, meat and meat products, tubes, lighting equipment and electric lamps, stationery and cardboard, machine tools, accumulators, primary cells and primary batteries, other special purpose machinery.

This wide range of products proves that the contribution of industrial districts to the national exports of manufacture goods is not limited to a narrow set of non technology- driven traditional sectors but it also includes advanced sectors with a strong core of competencies oriented towards mechanical and electromechanical technologies.

It is also interesting to single out the role of some individual industrial districts with respect to the Italian and world exports for a given group of products. Table 2 illustrates that some well known Italian industrial districts hold remarkable shares not only in terms of national exports for a given group of products, but which is also significant as far as their shares of world exports are concerned.

For instance, the industrial district of Sassuolo, located in the Emilia Romagna region which specialises in the production of ceramic tiles and flags, accounts for 51.3% of national exports and 28.1% of world exports for the sector. Prato, the well known industrial district in Tuscany specialising in woven textile products, accounts for 24.6% of national exports and 3.5% of world exports. Vicenza. famous for he production of jewellery an related articles, is responsible for 17.1% of national exports and 2.5% of world exports for the jewellery sector. Finally, Treviglio, an industrial district located in Lombardy which specialises in the production of agricultural and forestry machinery, constitutes 15.7% of national exports and 1.8% of world exports.

[10] Information on the statistical criteria used to identify industrial districts as a subset of LLM are included in paragraph three. The definition includes Marshallian and industrial district like local labour systems.

[11] This result, based on the release of provisional Italian Census data for the year 1996 is affected both by the statistical criteria chosen to identify industrial districts and by the methodological approach used to estimate exports at the local level.

Table 1. The contribution of industrial districts to the Italian exports of manufactured goods by economic activity groups (1996)

Merchandise groups (a)	*Market share in % of national exports*
Sports goods	93.5
Leather	84.2
Ceramic tiles and flags	82.0
Jewellery and related articles	74.6
Musical instruments	72.0
Agricultural and forestry machinery	71.5
Knitted and crocheted articles	71.1
Footwear	69.0
Textile weaving products	66.9
Furniture	65.8
Textile fibres	65.5
Tank. reservoirs and containers of metal. central heating radiators etc	65.4
Veneer sheets, plywood, laminated and chipboard	64.8
Knitted and crocheted products	64.3
Leather clothes	61.8
Weapons and ammunitions	59.8
Cutlery, tools and general hardware	58.6
Basic iron, steel and iron-alloys	57.0
Meat and meat products	56.2
Tubes	54.9
Clothing and accessories (not including leather clothes)	54.4
Non-refractory ceramic goods other than for construction purposes	53.1
Other fabricated metal products	53.0
Lighting equipment and electric lamps	52.7
Stationery and cardboard	52.0
Machine tools	51.5
Accumulators, primary cells and primary batteries	51.5
Manufacture of articles of fur	51.1
Other special purpose machinery	50.5
Cement, lime and plaster	49.7
Optical instruments and photographic equipment	48.0
Ornamental and building stones	46.0
Electric motors, generators and transformers	46.0
Wooden containers	45.8
Games and toys	45.7
Plastic products	44.2
Total manufactured goods	43.3

(a) Merchandise groups are identify on the basis of the 3 digits level of the CPA (European classification of products associated to economic activities)

This empirical evidence suggests that individual districts by themselves, rather than regions, can be considered to be major players in the international markets. In addition, it must be considered that this analysis refers only to trade flows coming from territorial area within Italy, thus it does not take into account more complex phenomena related to ID outward FDI.

Table 2. Contribution of some industrial districts to Italian and world exports for typical 'made in Italy' products (1996)

Industrial district	Merchandise groups	Market share in % of national exports	Market share in % of world exports (a)
Sassuolo (Emilia Romagna)	Ceramic tiles and flags	51.3	28.1
Sant'Ambrogio di Valpolicella (Veneto)	Ornamental and building stones	25.6	8.9
Arzignano (Veneto)	Leather	26.8	5.1
Santa Croce sull'Arno (Tuscany)	Leather	21.5	4.1
Montegranaro, Porto Sant'Elpidio e Monte San Giusto (Marche)	Footwear	8.5	1.5
Pietrasanta (Tuscany)	Ornamental and building stones	8.3	2.9
Solfora (Campania)	Leather	16.3	3.1
Prato (Tuscany)	Textile weaving products	24.6	3.5
Arezzo (Tuscany)	Jewellery and related articles	34.3	3.6
Treviglio (Lombardy)	Agricultural and forestry machinery	15.7	1.8
Udine (Friuli Venezia Giulia)	Furniture	12.1	1.8
Vicenza (Veneto)	Jewellery and related articles	17.1	2.5
Como (Lombardy)	Textile weaving products	11.7	1.7
Lecco (Lombardy)	Cutlery, tools and general hardware	5.8	2.2
Montebelluna (Veneto)	Footwear	9.5	1.6
Modena (Emilia Romagna)	Agricultural and forestry machinery	10.1	1.2

(a) The calculation of this share combines national with international statistical sources. In particular, international source refers to COMTRADE, the UN database on foreign trade statistics.

4.3 Output quality and variety of industrial districts exports

The analysis of the role played by industrial districts in their contribution to Italy's competitiveness in international markets can be widened from a strictly quantitative approach to include a more complex set of issues, including output quality and variety related to industrial districts export flows. In effect, given the remarkable contribution of industrial districts to national exports for some typical 'made in Italy' sectors, higher quality standard and strong output variety can be considered key factors for improving and maintaining the competitive edges they have in international markets.

The analysis of industrial districts output variety in terms of exports is carried out using the LQ index at LLM level. This index detects significant levels of local concentration of export activities across different groups of products, thus highlighting a possible range of different export specialisation models, reflecting in effect the comparative advantages of industrial districts in international markets[12]. Table 3 illustrates rather complex export specialisation models which go far beyond the boundaries of a single sector specialisation model.

Table 3. LQ index of export activities by group of products for some industrial districts (1996)

Merchandise groups	Bassano del Grappa	Udine	Sassuolo	Pesaro	Biella	Treviso	Carpi
Textile weaving products					S/S		
Knitted and crocheted articles	S				S	S/S	S/S
Leather clothes	S					S	
Leather	S				S		
Paints, varnishes and similar coating		S	S/S	S			
Glass and glass products				S			
Ceramic tiles and flags			S/S				
Cutlery, tools and general hardware		S		S		S	
Machine tools		S	S	S/S			S/S
Other special purpose machinery		S	S		S		
Domestic appliances	S					S/S	S
Lighting equipment and electric lamps	S			S		S/S	
Furniture	S	S/S		S/S		S	
Jewellery and related articles	S/S						
Sports goods	S	S			S		

S/S identifies strongly specialised areas (LQ>1.5) while S denotes specialised areas (LQ in between 1 and 1.5)

For instance, Bassano del Grappa, an industrial district located in the Veneto region, presents a wide range of export specialisations, related to different traditional products and to domestic appliances as well. Carpi, an industrial district located in Emilia Romagna, presents a narrow set of specialisation which anyhow includes, besides knitted and crocheted articles, machine tools and domestic appliances.

It is interesting to note that, in some specific cases, output variety processes seem to be affected by the local industrial culture. For instance, the industrial district of Pesaro, located in the Marche region, which traditionally specialises in the production of furniture, presents a wide range of exported products. These products include paints, varnishes and similar coatings, glass and glass products,

[12] In effect, industrial districts specialisation indexes or comparative advantages are calculated with respect to Italian rather then world export structure.

cutlery, tools and general hardware, machine tools and lighting equipment and electric lamps, which define a quite complex IDCP[13] related to the local industrial culture. In effect, the industrial culture of the industrial district of Pesaro is mainly aimed at satisfying an increasingly sophisticated international demand in terms of household goods, its related appliances and processing instruments.

Within the framework of the role played by industrial districts towards Italian competitiveness in international markets, another interesting issue to be considered is related to the quality standard of exported products. In effect, the ability to maintain a relatively high standard of quality is crucial for winning today's competitive game in international markets. The quality of exported products is tested on the basis of merchandise unit values, measured at a local level and thereafter normalised with regard to the average national level. In order to consider a more detailed classification of merchandise and to perform a dynamic analysis, the data used refer to the territorial breakdown of Italian exports into administrative provinces[14].

The analysis is limited to some specific Italian provinces for which the *district like* proxy can be considered reliable[15]. It also takes into account only particular groups of products that can be considered homogeneous enough to calculate and not significantly bias export unit values. We can assume that although the national average of product quality has improved with the passing of time, due to nation-wide changes in technologies and standards of production, it may be considered a dynamic benchmark to test individual ID performance in terms of quality improvement.

In effect, the single industrial district must keep its quality standards constantly above the national average in order to maintain its specific competitive advantage. Figure 2 shows the dynamic trends of product quality, with respect to the national average, for some ID and related specialisation products. Given a certain variability across different products and in time series, it is clearly proved that every ID has improved or, at least, maintained its quality standards above the national average (equal to 1). Similar conclusions apply to the footwear ID as shown in Fig.3. Here, we considered ID that specialise in the exports of leather and non-leather footwear. All ID seem to have systematically raised or maintained their product quality standard above the national average.

---

[13] See Section 2 for some theoretical remarks on this topic.

[14] The territorial analysis of foreign trade statistics at province level (about 100 provinces) has officially been carried out by ISTAT since 1985. A very detailed sector classification of exports at province level, about 230 merchandise groups, is available until 1999. The current dissemination of foreign trade territorial statistics refers to about 100 merchandise groups, which corresponds to the third digit level of the CPA- NACE Rev.1 classification

[15] The identification of district like provinces is carried out on a quantitative basis (Becattini and Menghinello, 1998). In particular, an ID over total coverage ratio is calculated in terms of persons employed and with respect to individual province and sector (introducing estimation hypotheses when a single ID LLM overlaps different provinces). Only provinces performing 80% or more coverage ratio are considered district like provinces with respect to an individual sector.

Figure 2. Product quality index at the local level by merchandise groups in the textile industry for some *district like* Italian provinces (1985-1999)

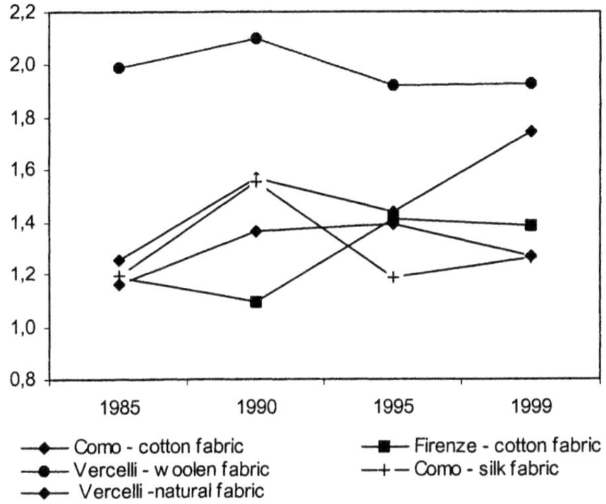

Figure 3. Product quality index at the local level by merchandise groups in the footwear industrial for some *district like* Italian provinces (1985-1999)

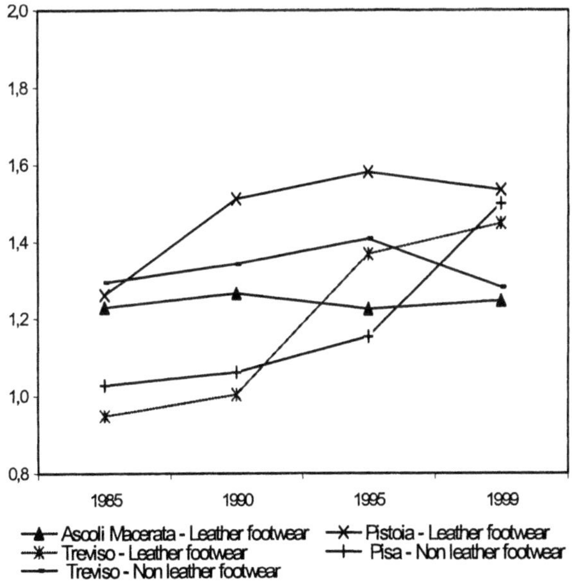

5 Conclusions

This study explores the role of industrial districts as specific engines of Italy's export performance and international competitiveness.

It is proved, on a quantitative basis, that industrial districts rank among the best export performing local industrial systems, and they account for, on aggregate, a significant share of national exports in manufactured goods.

The breakdown by groups of products shows that industrial districts play a significant role not only for a narrow set of non technology-driven traditional sectors but that they also attain significant shares in relatively advanced productions, such as mechanical and electro-mechanical instruments and apparels. Furthermore, individual industrial districts sometimes attain, for a given group of products, significant shares of world exports, thus acting as major players in international markets.

Some further empirical evidence shows that, in some cases. Industrial districts are characterised by a wide ranging output variety related to exported products, which seems to be affected by specific local competencies rather than being technology or capital-driven. We have also shown that product quality standards of exported goods are for most industrial districts systematically higher than the national average. We think that this finding can be generalised at least to the industrial districts that we classified as competitive or strongly competitive with respect to export performance.

These results highlight a more complex role for industrial districts as powerful engines in supporting and improving Italy's competitiveness in international markets. The empirical evidence can be extended to support theoretical studies on the crucial relevance of industrial districts and other kinds of local industrial systems for economic development and international competitiveness.

References

AMIN A. (1992) "Big Firms Versus the Regions in the Single European Market", in: DUNFORD M.,KAFKALAS G.(Eds.) *Cities and Regions in The Global-Local Interplay and Spatial Development Strategies*, Belhaven Press, London.

AMIN A., WILKINSON F. (Eds.) (1999) "Special Issue on Learning. Proximity and Industrial Performance", *Cambridge Journal Of Economics*.

BAGELLA M., BECCHETTI L., SACCHI S. (2000) "The Positive Link Between Geographical Agglomeration and Export Intensity: The Engine of Italian Endogenous Growth?" in: BAGELLA M., BECCHETTI L. (Eds.), *The Competitive Advantage of Industrial Districts. Theoretical And Empirical Analysis*, Heidelberg , Physica-Verlag.

BECATTINI G. (1979) "Dal "settore" industriale al "distretto" industriale. Alcune considerazioni sull'unità di indagine dell'economia industriale", *Rivista Di Economia e Politica Industriale*, n. 1.

BELUSSI F., GOTTARDI G. (Eds.) (2000) *Evolutionary Patterns of Local Industrial Systems. Toward a Cognitive Approach to the Industrial District*, Aldershot, Ashgate.

CONTI G. (1995) *I sistemi esportativi italiani: Un'analisi per province 1985-1993*, in: "Rapporto sul commercio estero 1994", ICE, Roma.

COWLING K., SUGDEN R.(1999) The Wealth of Localities. Regions and Nations: Developing Multinational Economies, New Political Economy. 4. 3. 361 - 378.

DE PROPRIS L. (2001) "Systemic Flexibility, Production Fragmentation and Cluster Governance", *European Planning Studies*, vol. 9, n. 6, pp. 739-753.

DI TOMMASO M.R., RABELLOTTI R. (a cura di) (2000), *Efficienza collettiva e sistemi di imprese. Oltre l'esperienza italiana,* Bologna, Il Mulino.

DOSI G., MALERBA F. (Eds.) (1996) *Organization and Strategies in the Evolution of the Enterprise*, London. Macmillan.

FABIANI S., PELLEGRINI G., ROMAGNANO E., SIGNORINI L.F. (2000) "Efficiency and Localisation: The Case of Italian Districts", in: BAGELLA M., BECCHETTI L. (Eds.), *The Competitive Advantage of Industrial Districts. Theoretical and Empirical Analysis*, Heidelberg , Physica-Verlag.

ISTAT (1997) *I sistemi locali del lavoro 1991*, in: "Argomenti", N.10, Roma.

ISTAT (1999) *Rapporto annuale. La situazione del paese nel 1998*, Roma, Istituto Poligrafico dello Stato.

ISTAT (2000) *Rapporto annuale. La situazione del paese nel 1999*, Roma, Istituto Poligrafico dello Stato.

LIPPARINI A., LORENZONI G. (1996) "Le organizzazioni ad alta intensita' relazionale. Riflessioni sui processi di learning by interacting nelle aree ad alta concentrazione di imprese", *L'industria*, n.4.

MARKUSEN A. (1996) "Sticky Places in Slippery Space. A Typology of Industrial Districts", *Economic Geography*, vol. 72, n. 3, pp. 293-313.

MENGHINELLO S. (1998), "Il Made in Italy distrettuale", *Sviluppo Locale*, vol. 1, n. 9.

NONAKA I. (1994) "Come un'organizzazione crea conoscenza", *Economia e Management*, 3.

PRAHALAD C., HAMEL G. (1990), "The Core Competence of Corporations", *Harvard Business Review,* vol. 68, n. 3.

RULLANI E. (1993), "Sistema locale e mercato globale", *Economia e Politica Industriale*, n. 80.

SAXENIAN A. (1994) *Regional Advantage, Culture and Competition in Silicon Valley and Route 128*, Cambridge, Harvard University Press.

SCOTT A.J. (1992) "The Role of Large Producers in Industrial Districts: A Case Study of High Technology Systems Houses in Southern California", *Regional Studies*, vol. 26, n. 3.

SFORZI F. (1990) "The Quantitative Importance of Marshallian Industrial Districts in the Italian Economy", in: PYKE F., BECATTINI G., SENGENBERGER W. (Eds.). *Industrial Districts and Inter-Firm Co-Operation In Italy,* Geneva, International Labor Office.

VIESTI G. (1993) *La geografia delle esportazioni italiane: un'analisi sui dati 1985-1991,* Bari.

Small multinational groups in the Italian industrial districts: interpretations and empirical evidence from the mechanical engineering industry

Sergio Mariotti[1] and Marco Mutinelli[2]

[1] Polytechnic of Milan, Via Giuseppe Colombo 40, 20133, Milan, Italy
[2] Faculty of Engineering, University of Brescia, Via Brianze 38, 25123, Brescia, Italy

1 Introduction

The formation within industrial districts of 'multinational groups' composed of local enterprises and units with a parent company in Italy and affiliates in other countries is an issue closely linked to internationalisation of the small and medium-sized enterprises (SMEs), that constitute the fundamental ecological units of the country's industrial districts.

This chapter examines several aspects of the question starting from a series of stylised facts showing that the key features of Italy's internationalisation pattern – exports versus foreign direct investments (FDIs) – make it atypical when compared to the major industrialised countries. It is then asked whether the industrial districts – a somewhat peculiar facet of the Italian economy – have helped to determine this historical divergence.

The subject is initially investigated in the light of some distinct features of the Italian districts as discussed in the abundant literature, especially since the 1980s. Some interpretations providing a promising standpoint for understanding the current situation are formulated. Lastly, empirical evidence relating to districts specialising in branches of the mechanical engineering industry is presented. This sectoral picture supports the theoretical assumptions put forward and asks for more investigation extended to the whole of the Italian industrial districts.

2 The atypical nature of Italy's internationalisation

Italy is the world's eighth exporter of merchandise and the sixth exporter of manufactured goods. In the list of outward investors, however, it has moved up a place or two in recent years, but still ranks only thirteenth and is trailing behind nations such as Spain, Belgium, Denmark, Sweden and Switzerland that generate substantial foreign direct investiments (FDI) flows even though they are much smaller in economic terms (UNCTAD, 2001). The multinational reach of Italian enterprises is still rather limited when compared with both their more dynamic

European competitors and the strong foreign investment positions acquired by nations such as Singapore, Taiwan, Hong Kong and Korea, whose industrialisation has been the work of a single generation. In a scenario increasingly characterised by open and internationally integrated markets, Italy's sluggish advance from an exporter to an exporter-investor country reflects the inability of its industries to seize the many chances of gaining a stable foothold on both the large, but difficult oligopolistic markets of the Triad, and those parts of Eastern Europe, the Mediterranean, Asia and Latin America where the competitive cut and thrust of strong players is being increasingly welcomed through liberalisation of the economy, lowering of customs barriers and encouragement of the inflow of foreign investments, including the offer of attractive financial and tax advantages to companies prepared to set up manufacturing facilities.

Italy's atypical position in this respect has raised voices of concern on many sides. Globalisation is rapidly reshuffling economic relationships between countries. One cannot turn a blind eye, therefore, to the fact that in the absence of an appropriate multinational approach the complementary nature of FDIs and exports constitutes a very real threat to Italy's performance also as an exporter and her competitive standing as a whole.

Two aspects of the structure of Italy's manufacturing framework underlie the current contrast between her strong ability to export and the low degree of internationalization through FDI of Italian manufacturing firms (Mariotti, 1999; Onida, 1999): (i) firstly, her oligopolistic core is frail and feeble: changes in the structure of most of the country's major industrial groups, both public and private, have led to the shedding of foreign equity investments, including those only recently acquired; advanced technology sectors have been redimensioned and the sound strategies needed for their relaunching are lacking; the results of some recent privatisations have revealed the insufficient ability of managers to steer the growth of large-sized enterprises; banking and insurance firms are engaged in ownership and managerial changes to secure their inclusion in a network of international alliances, but are still far from the levels of experience, culture and organisation which allow financial brokers and institutions in other countries to offer an effective strategic support to their customers competing abroad; (ii) secondly, the dark side of Italy's long-standing excellence as an exporter of consumer and other light and intermediate technology goods, machinery and mechanical engineering components is her all too evident inferiority in the very sectors that are typically the leading players in the multinational oligopolistic arena (high-tech, mass production, public utilities, advanced services, etc.). An inherited sectoral composition founded on the building up of cumulative competitive advantages, in other words, is now proving a barrier to internationalisation whose impact may well be felt by the country as a whole.

As matters now stand, it would seem that Italy's prospects of becoming more substantially multinational must be primarily entrusted to her SMEs and the districts in which they are so thick on the ground. Her strength has so far been largely due to the sturdiness and competitiveness of the 'family capitalism' that forms these districts. It is on their ability to grasp new opportunities for growth abroad, in fact, that the remaining hopes of closing the internationalisation gap separating Italy from the other industrialised countries in no small measure depend.

In this connection, emphasis must first be placed on the strong export propensity developed by Italian industrial districts in the recent past. The ISTAT data show that the districts account for 43.3% of the nation's exports. Many traditional 'made in Italy' and specialised mechanical engineering sectors, on the hand, achieve even higher percentages: 67% for textiles, 69% for footwear, 82% for tiles, 66% for furniture, 75% for jewellery, 71% for farm equipment, 59% for knives and 51% for machine tools and their components (ISTAT, 1999).

The optimistic conclusion induced by these levels of performance is that the competitive factors underlying past successes can be carried over from the mercantile side of internationalisation to more demanding forms of involvement abroad, be they 'light' in nature, such as agreements and partnerships, or of 'deeper' extent, such as FDIs in joint ventures and the establishment of directly controlled subsidiaries. Reasons for such optimism can also be drawn from empirical evidence illustrating the emergence in the industrialised countries of a substantial class of 'baby multinationals' formed of parent companies with a workforce of fewer than 500 employees, but with a spread of operations abroad requiring financial, technological, manufacturing and managerial resources sometimes even disproportional to the assets they control in their own country (Mutinelli, 1997; Buckley *et al.*, 1988; Fujita, 1995a and 1995b; Cominotti *et al.*, 1999; UNCTAD, 1993). Some of the driving factors of globalisation, too, underlie the increasing spread of internationalisation: better worldwide communication and transport facilities, expansion of the financial markets, a fuller and more far-reaching supply of business services and externalities, the spread of user-friendly and low-cost information technologies free from the previous space and time constraints associated with long-distance negotiations, and ongoing liberalisation in many countries.

A glance at the Italian scenario, however, will quickly show that this 'multinational transfer of the country's competitive edge' is certainly not something that can be taken for granted. Quite the opposite, indeed, since propensities and mechanisms impeding an international outlook on the part of the Italian districts are evidently at work. Onida (1999) remarks that a district's prospects of international growth are often strongly resisted by the firms directly concerned, since their thinking is basically commercial (produce at home and sell abroad) and averse to the idea of using organisational, managerial and financial resources to promote the transfer of their manufacturing capabilities to a different and distant environment and overcome the difficulties posed by a foreign culture and acquisition of an understanding of foreign institutions and their organisation. Furthermore, at a time when the highly labour-intensive stages of manufacturing are being increasingly shifted from the industrialised to the emerging nations, retaining control of the production stream, especially its nobler phases (innovation, design, differentiation, finishing, etc.), in an integrated logistic and quality scenario where suppliers are no longer just around the corner demands organisational and managerial innovations that pose an extremely arduous and exacting challenge for a district composed of family SMEs which have often grown up from nothing in the course of a very few years.

An intrinsic and genetically rooted weakness thus looms over any prospect of entrusting the district system with the task of blazing the trail for the internationalisation of Italian industry. What, however, are the reasons for this weakness and what counter-conclusions can be drawn from the state of the art illustrated in the economic literature on the districts? The answers will be sought in the next section.

3 Internationalisation of production by the districts

Despite their recent dynamism, small enterprises in all countries are faced with many difficulties and limitations when they seek to operate in a truly multinational style, i.e. as insiders on distant markets and in unfamiliar environments (UNCTAD, 1993). High information costs, the risks for sunk costs coupled with the start-up investments and a shortage of both financial resources and human resources capable of handling complex situations and managing transnational organisation are all factors that impede the multinational growth of SMEs and result in their failure to acquire a cosmopolitan culture and fully understand the workings of international markets.

SMEs have coped with these difficulties in peculiar ways: step-by-step and evolutive growth abroad (Johanson and Vahlne, 1977); decision making aimed at capturing contingent opportunities, rather than thorough exploration of the alternatives available; initial and subsequent investments abroad sub-optimal in both quantity and quality (Barkema, 1995; UNCTAD, 1993); search for partnerships and other forms of cooperation with indigenous firms endowed with local knowledge so as to hasten learning and the building up of experience on foreign markets (Mutinelli and Piscitello, 1998; Sachwald, 1998). These strategies, however, are the fruit of *bounded rationality*. They thus result in initiatives that are more likely to prove unsuccessful and increase the mortality and failure rate of the enterprises established abroad (Mariotti and Piscitello, 1999). All international studies, in fact, have shown that the size of parent company is a positive determinant in the formation of affiliates stably rooted in the foreign countries to which their production and commercial investments are directed.

Contributors to the literature about the districts, however (suffice it to mention Pyke *et al.*, 1990; Brusco, 1986 and 1999), have proposed a different view. An SME is no longer regarded as 'solitary', nor is its potential establishment and growth restricted by its size. In the Italian districts, a somewhat esoteric set of factors would seem to have blended over the course of time to ensure the long-term competitiveness of their local SMEs: economies of specialisation and networking; sharing of knowledge and technologies; skilled, flexible and mobile human resources; a widespread entrepreneurial vocation and a spirit of emulation; familiarity with strong domestic competition; and an interweaving of the economy with society encouraged by a local culture with an age-old traditional background.

As Onida (1999) has rightly observed, close examination of such manufacturing microcosms reveals many of the competitive advantages described by Porter (1990).

There can be no doubt that these factors have helped a district to make headway and consolidate its achievements as an exporter. Even so, they mark the limits within which it can act on the international stage. Two points must be made in this connection. In the first place, this view of SMEs brings out the *localised nature* of their competitive advantages. Economies of proximity are geographically local by definition.[1] The integrated and complementary system of customers, suppliers, services and knowledge to which they give rise is *unique* and would be very hard to reproduce abroad. The dependence of SMEs on such idiosyncratic externalities compels them to be local, as occurs in other economies with a highly networked structure.[2]

In the second place, this 'localism' encourages districts to grow on their own foundations and prefer the deepening and incremental improving of their specific skills and structures, rather than changes and discontinuities. Their progress is thus marked by behavioural inertia and path-dependence. The ensuing *lock-in effect* prevents them from reaching out beyond their domestic environment to take advantage of the opportunities offered elsewhere and open their doors to new and radically different experiences.

The end result, therefore, is that the strength of a district may prove its weakness and a barrier to change and multinational expansion at a time when firms wishing to compete internationally are being compelled to be present as insider on markets and able to manufacture their products in countries other than their own.

It must none the less be conceded that districts enjoy certain structural prerogatives that help them to at least partly overcome these obstacles to their international growth, since this is presumably promoted when their context comprises *leader enterprises* and is the original home of firms with an *international standing*.

In some districts, in fact, leader enterprises have provided the driving force in the local industry through their generation of innovation, broadening and opening of the new markets, spillovers of human capital, adoption of efficient international production set-ups (including delocalisation measures) and the implementation of market-seeking multinational strategies (Lazerson and Lorenzoni, 1999). Enter-

---

[1] One feels that metaphors such as 'virtual districts', where the relations on the territory leave room for the representation of an abstract space hinged on network architectures variously supported by the new information technologies, should be avoided.

[2] Japan's success as a competitor, for example, has been greatly assisted by certain features of her industrial structure and the behaviour patterns of her people. Japanese firms, however, have found that this scenario cannot be reproduced abroad. See Abo *et al.* (1991).

prises of this kind can clearly serve as a bridgehead for the international launching of a district provided their formation does not result in excessive horizontal and vertical concentration, since this would stifle Porter's 'domestic rivalry' and convert local SMEs into exclusive sub-suppliers as opposed to independent, specialised manufacturers capable of going it alone on the world's markets.

Of equal importance in a district's multinational vocation is the influence during its evolution of genetic factors that promoted the establishment of an international culture *ab initio* and induced its SMEs to look for possibilities of setting up stable relationships with specific markets and foreign industrial structures. This role may have been played by *foreign-controlled* industrial and commercial firms whose presence – directly, in customer-supplier relations and through spillovers – increases the chances of contacts with other countries, instills a cosmopolitan outlook, reduces the lock-in danger and encourages a district's enterprises to explore the possibilities of growth abroad.

If the picture thus painted is correct, one may expect to find that the empirical evidence from the districts discloses the formation of multinational groups significantly differing in function of their local structure and historical background, whereas the multinational propensity of their SMEs is less than the national average.

4 Empirical findings

The correctness of the hypotheses advanced in section 3 was assessed by collecting empirical data for twenty-one industrial districts engaged in mechanical engineering (see Figure 2).[3]

The ISTAT Intermediate Industry and Services Census data showing the number of enterprises in each district and their workforces at 31 December 1996 were processed. Individual international manufacturing initiatives were identified, namely: *i*) district enterprises with foreign manufacturing affiliates, *ii*) district manufacturing enterprises with foreign multinational shareholders.[4] This information related to the position on 1 January 1998 and was obtained from the Reprint database elaborated by the Polytechnic of Milan as part of research on the internationalisation of Italian industry promoted by the CNEL (National Council of Economics and Labour).[5]

---

[3] The following sectors and districts were considered: general purpose machinery (Ateco 29.1 & 29.2: Borgomanero, Lumezzane and Reggio Emilia); agricultural machinery (Ateco 29.3: Cuneo, Cento and Modena-Reggio Emilia); machine-tools and special purpose machinery (Ateco 29.4 & 29.5: Biella, Santa Croce, Arzignano, Vigevano, Brianza, Rimini-Pesaro, Sassuolo, Val Seriana, Brescia-Palazzolo, Legnano, Gallarate, Parma); arms and ammunition (Ateco 29.6: Valtrompia).

[4] Account was taken of majority, parity and minority interests.

[5] The method used to construct this database is described by Cominotti *et al.* (1999).

The main findings are illustrated in Figure 1 and 2. They show that these districts and their enterprises are very little inclined towards internationalisation of production. By comparison with their 14.2% incidence on overall employment in the Italian mechanical engineering industry, the weight of the districts falls to 4.7% of the national total in terms of the number of employees in the foreign manufacturing affiliates (Figure 1). They are also relatively closed with respect to inward internationalisation, as shown by their 8.9% incidence in relation to the number of employees in industrial enterprises with foreign shareholders.

Figure 1. Incidence of the districts on domestic employee and on internationalisation of manufacturing in the Italian mechanical engineering industry.

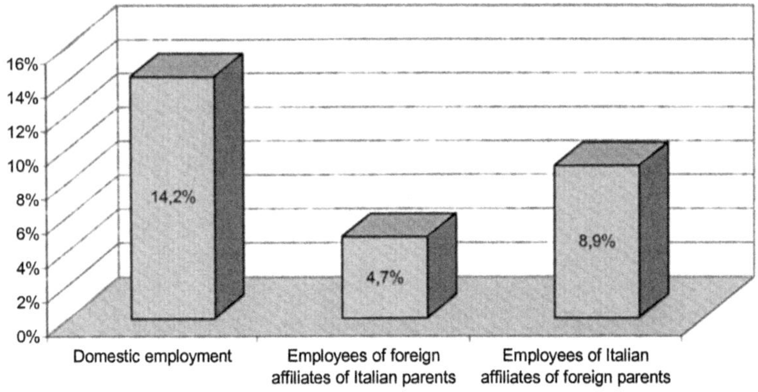

Source: ISTAT and Reprint database (CNEL–R&P–Politechnic of Milan)

Closer examination showed that nine districts displayed no evidence of outward internationalisation (no enterprises with foreign manufacturing affiliates), while another four had only one such enterprise and their foreign manufacturing units were small. Only one district had founded four multinational firms, whereas the remaining seven had two or three.

Figure 2. Outward and inward internationalisation in 21 Italian industrial districts in the mechanical engineering industry

Degree of outward internationalisation	*Degree of inward internationalisation*		
	Low	Medium	High
Low	Tapes and valves (Borgomanero) Tapes (Lumezzane) Agricultural machinery (Cuneo) Agricultural machinery (Cento) Textile machinery (Biella) Textile machinery (Brescia–Palazzolo) Tanning machinery (S. Croce) Tanning machinery (Arzignano) Footwear machinery (Vigevano) Woodworking machinery (Brianza) Woodworking machinery (Rimini–Pesaro)	Diesel engines (Reggio Emilia) Textile machinery (Val Seriana)	–
Medium	Ceramics machinery (Sassuolo) Arms and ammunition (Valtrompia)	Agricultural machinery (Modena–Reggio Emilia) Transmission parts (Bologna) Textile machinery (Legnano) Machinery for the plastics industry (Gallarate)	Machinery for the food industry (Parma)
High	–	Packaging machinery (Bologna)	–

Notes: low degree of internationalisation = 0/1 affiliates (inward: Italian affiliates of foreign parents; outward: foreign affiliates of Italian parents); medium degree of internationalisation = 2/3 affiliates; high degree of internationalisation: 4 or more affiliates.

Sources: ISTAT 1996 Intermediate Industry and Services Census data for the domestic employment figures, and the Reprint database (CNEL–R&P– Politechnic of Milan) for the workforces of the Italian affiliates of foreign parents and the foreign affiliates of Italian parents (as at 1.1.1998).

The district with the heaviest internationalisation, namely the packaging machinery district in the Bologna area, is the home of some large Italian concerns, such as Sasib, G.D. and IMA, all of which also manufacture abroad. It is interesting to note that most sectors with 'medium' and 'high' degree of outward internationalisation (two or more firms with production units abroad) are also those with a rooted presence of firms subject to foreign control. Six out of these eight districts, in fact, have hosted a foreign firm for twenty years or more, whereas only Sassuolo (ceramics machinery) has no enterprise with a foreign holding. By contrast, seven of the thirteen districts with 'low' outward internationalisation have no inward FDI, and in another four the foreign presence is relatively recent (1991 in one case; second half of the 1990s in the other three). An econometric exercise could be carried out for a closer empirical assessment of the correctness of the conclusions drawn in Section 3. The following equation was estimated:

$$\textit{Outward}\ \mathrm{FDI}_i = a_0 + a_1 \cdot \textit{Herfindhal}_i + a_2 \cdot (\textit{Inward}\ \mathrm{FDI}_i \cdot \textit{Age}_i) + \varepsilon_i$$

where:

Outward FDI_i = degree of outward internationalisation of the i-th district. (calculated as the ratio between the number of employees in the district enterprises' foreign invested firms and the district's total domestic workforce);

$Herfindhal_i$ = value of the industrial concentration in the i-th district (measured by Herfindhal's index, regarded as a proxy for the presence of leader enterprises within a district);[6]

$Inward\ FDI_i$ = degree of inward internationalisation of the i-th district (calculated as the ratio between the number of employees in the district enterprises with foreign shareholders and the district's total domestic workforce);

Age_i = number of years since the first inward FDI in district industrial enterprises.[7]

---

[6] This index was calculated from the Intermediate Industry and Services Census data split into 13 'number of employees' classes. For each district, the following formula was used:

$$\text{Herfindhal} = \Sigma_j\, n_j \cdot \left(\frac{A_j / n_j}{\Sigma_j A_j}\right)^2$$

where n_j= number of district enterprises in dimensional class j, and A_j = total number of employees in such enterprises (j = 1, … , 13)

[7] In order to take into account the saturation effect, Age was set equal to 20 for all FDIs made before 1980.

The first hypothesis arising from the discussion in section 3 is that there is a positive and statistically significant correlation between a district's degree of outward internationalisation and its industrial concentration. In the second place, the propensity of district enterprises to look for ways of internationalisation is fostered by the local presence of firms held by foreign direct investors. Since this effect, whether directly due to customer-supplier relations or through spillovers, depends on both the size of this foreign presence and the time of its rooting in the local economic fabric it can be represented by the product of the two independent variables $Inward\ FDI_i$ and Age_i.

The results of the econometric model as estimated with an OLS linear regression support the hypotheses put forward (Table 1). The coefficient of both variables, in fact, is positive and significantly different from zero and the R^2 coefficient is close to 56%.

Table 1. The econometric estimates

Variable	*Coefficient*	*t–test*	*Significativity*
Constant	–0.149	–1.336	
Herfindhal	1.308	2.073	$p < 0.05$
Inward FDI · Age	0.264	3.294	$p < 0.01$
R^2	0.559		
F	11.426		

5 Conclusions

Profound changes in the way firms compete with each other are making it necessary for them to be ever more directly present on their outlet markets and optimise the international location of their manufacturing operations. It would thus appear that the traditional strong points of the Italian districts – a somewhat peculiar facet of the country's industrial structure – are at the same time weaknesses and a barrier to the multinational expansion of their industrial enterprises. Competitive advantages generated by the economies of proximity typical of districts *cannot be readily transplanted* from their original environment and this encourages growth paths marked by behavioural inertia. In other words, lock-in mechanisms prevent an enterprise from reaching out beyond its domestic environment to take advantage of the opportunities offered by contexts other than those in which it was born and bred.

The investigation of twenty-one districts in the mechanical engineering sector has provided empirical evidence of their reduced multinational growth capacity.

Their propensity for internationalisation, indeed, is even lower than the already limited propensity of the Italian mechanical engineering industry as a whole.

The detailed analysis also showed how the formation of multinational groups differs from one district to another.

A simple econometric exercise provided empirical support for the view that some structural features can partly overcome the lock-in effect within districts. International growth, in fact, is relatively encouraged in districts with a higher degree of concentration – itself a sign of the presence of *leader enterprises* – and/or the original establishment of *foreign multinationals*, owing to their driving force as bridgeheads for the international launching of other district enterprises.

These findings are an encouragement to extend this type of investigation to all Italian districts and the present authors have such an objective in mind. If further results of the same kind are forthcoming, useful indications will be provided for the elaboration of an industrial policy capable of boosting the international reach of Italian industry, which to no small extent depends on the competitiveness and growth of the country's SMEs organised in districts.

References

ABO T., ITAGAKI H., KAMIYAMA K., KAWAMURA T., KUMON H. (1991), *Japanese–Style Production Systems in the U.S.,* Toyo Keizai, Tokyo.

BARKEMA H.G., BELL J.H.J., PENNINGS J. (1996), "Foreign Entry, Cultural Barriers and Learning", *Strategic Management Journal,* 17, pp. 151–166.

BRUSCO S. (1986), "Small Firms and Industrial District", in: KEEBLE D., WEVER E. (Eds.), *New Firms and Regional Development in Europe*, Croom Helm, London.

BRUSCO S. (1999), "The Rules of the Game in Industrial Districts", in: GRANDORI A. (Ed.), *Interfirm Networks: Organization and Industrial Competitiveness*, Routledge, London.

BUCKLEY P. J., NEWBOULD, G.D., THURWELL J. C. (1988), *Foreign Direct Investment by Smaller UK Firms*, MacMillan, London.

COMINOTTI R., MARIOTTI S., MUTINELLI M. (Eds.) (1999), *Multinational Italy 1998*, CNEL, Roma.

FUJITA M. (1995*a*) "Small and Medium–Sized Transnational Corporations: Trends and Patterns of Foreign Direct Investment", *Small Business Economics*, vol. 7, n. 1.

FUJITA M. (1995*b*) "Small and Medium–Sized Transnational Corporations: Salient Features", *Small Business Economics*, vol. 7, n. 4.

ISTAT (1999), *Rapporto annuale. La situazione del paese nel 1998,* Roma.

JOHANSON J., VAHLNE J.E. (1977), "The Mechanism of Internationalisation", *International Marketing Review,* 7, pp.11–24.

LAZERSON M. H., LORENZONI G. (1999), "The Firms that Feed Industial Districts: A Return to the Italian Source", *Industrial and Corporate Change*, vol. 8, n. 2.

MARIOTTI S. (1999), *The Internationalisation of Italian Industry*, in: COMINOTTI R., MARIOTTI S., MUTINELLI M. (Eds.) (1999), *Multinational Italy 1998*, CNEL, Roma.

MARIOTTI S., PISCITELLO L. (1999), "Is Divestment a Failure or Part of a Restructuring Strategy? The Case of Italian Transnational Corporations", *Transnational Corporations*, vol. 8, n. 3.

MUTINELLI M. (1997), *Le piccole multinazionali italiane*, Seat, Torino.

MUTINELLI M., PISCITELLO L. (1998), "The Influence of Firm's Size and International Experience on the Entry Mode Choice. Evidence from the Internationalisation of Italian Industry", *Small Business Economics*, vol. 11, pp. 43–56.

ONIDA F. (1999), "Quali prospettive per il modello di specializzazione dell' Italia?", *Economia italiana*, n. 3, pp. 573-629.

PYKE F., BECATTINI G., SENGENBERGER W. (Eds.) (1990), *Industrial Districts and Inter-firm Cooperation in Italy*, International Labour Office, Geneva.

PORTER M. (1990), *The Competitive Advantage of Nations*, McMillan, London.

SACHWALD F. (1998), "Cooperative Agreements and the Theory of the Firm: Focusing on Barriers to Change", *Journal of Economic Behavior and Organization*, vol. 35, n. 2.

UNCTAD (1993), *Small and Medium–sized Transnational Corporations. Role, Impact and Policy Implications*, United Nations, New York.

UNCTAD (2001), *World Investment Report 2001*, United Nations, New York.

Printing and Binding: Strauss GmbH, Mörlenbach

GPSR Compliance

The European Union's (EU) General Product Safety Regulation (GPSR) is a set of rules that requires consumer products to be safe and our obligations to ensure this.

If you have any concerns about our products, you can contact us on ProductSafety@springernature.com

In case Publisher is established outside the EU, the EU authorized representative is:

Springer Nature Customer Service Center GmbH
Europaplatz 3
69115 Heidelberg, Germany

Batch number: 10725973

Printed by Printforce, the Netherlands